Brilliant Lives

Brilliant Lives
THE CLERK MAXWELLS AND THE SCOTTISH ENLIGHTENMENT

John W. Arthur

*This book is for my children,
for my children's children,
and for the generations thereafter.*

First published in Great Britain in 2016 by
John Donald, an imprint of Birlinn Ltd

West Newington House
10 Newington Road
Edinburgh
EH9 1QS

www.birlinn.co.uk

ISBN: 978 1 906566 97 5

Copyright © John W. Arthur 2016

The right of John W. Arthur to be identified as the author
of this work has been asserted by him in accordance
with the Copyright, Designs and Patents Act, 1988

All rights reserved. No part of this publication may
be reproduced, stored, or transmitted in any form, or
by any means, electronic, mechanical or photocopying,
recording or otherwise, without the express written
permission of the publisher.

The publishers gratefully acknowledge the support of
the James Clerk Maxwell Foundation
towards the publication of this book.

British Library Cataloguing-in-Publication Data
A catalogue record for this book is available on
request from the British Library

Typeset by Mark Blackadder

Printed and bound in Britain by Bell and Bain Ltd, Glasgow

Contents

Preface	vii
Acknowledgements	xi
Abbreviations	xii
Genealogical Tables	xv

1	Introduction	1
2	James Clerk Maxwell's Life and Contribution to Science	10
3	The Early Clerks of Penicuik	60
4	The Baron	65
5	William Clerk and Agnes Maxwell	81
6	The Maxwells	90
7	Agnes Maxwell and James Le Blanc	102
8	The First Clerk Maxwells	110
9	The Successors of George and Dorothea	152
10	Sir James Clerk, 3rd Baronet of Penicuik	174
11	From Weir to Irving	183
12	The Enlightenment of Edinburgh	193
13	Dr Thomas Weir and Anderson's Patent Pills	209
14	The Cays and the Hodshons	219
15	James Clerk Maxwell's Scottish Homes: 14 India Street and Glenlair	255
	Epilogue	269

Appendix 1: Maxwell's Innovative Method of Drawing True Ovals	272
Appendix 2: Awards and Commemorations for James Clerk Maxwell	273
Appendix 3: James Clerk Maxwell's Legacy to Science and Mankind	275
Appendix 4: The Birth Dates of George and Dorothea's Children	281
Appendix 5: Some Notable Scots of the Age of Enlightenment	283

Glossary	285
Notes	288
Bibliography	325
Index	346

Preface

The story told in this book is not so much about the scientist James Clerk Maxwell himself but, as the subtitle suggests, it is that of his family and of their network of friends and connections. It goes sideways in time crossing from one family or friend to another and, in the case of his most important family connections, follows them through the generations from their origins. Any discussion of science arises only in trying to explain two things: what it was that inspired James Clerk Maxwell to be a physicist, and why his key discoveries were so significant. The timespan is more or less from the Reformation to the end of the nineteenth century. The location is primarily the Scottish Lowlands, with specific focuses in Edinburgh, the Upper Clyde Valley, and Dumfries and Galloway. While the families mainly stem from landed gentry, they did not live in idle docility; they endeavoured not only to survive but to go beyond and *improve*. The result of the collective efforts of many such families was astonishing. In fact, for some of their leading lights, mere improvement was not enough; nothing less than the breaking of the mould would suffice for them.

James Clerk Maxwell was born in 1831 at 14 India Street, Edinburgh. Nevertheless, through his family he also had connections to numbers 18 and 38 in that street, and to 11, 27 and 31 Heriot Row, 6 Great Stuart Street, 25 Ainslie Place and 43 Moray Place. Each one is in Edinburgh's Second New Town and within a five-minute walk of 14 India Street. While we do know the names of the people who lived at these addresses, who exactly were they? What part, if any, did they play in Maxwell's life? How did they come to live in such fine Edinburgh New Town homes? What were their life stories? What did they do for a living to be able to afford such splendid houses? What other properties did they have? What sort of lives did they lead? And whatever happened to them? It is with questions like these that our story begins.

For all that has been written about the life of James Clerk Maxwell, much of it, particularly in the case of his early years, is to be found in the original biography by Lewis Campbell and William Garnett (1882), which is supplemented with copies and excerpts of many original letters and papers. Even so, some of the details Lewis Campbell recalled from interviews, diaries and memory are open to question, and so for a more accurate picture corroboratory sources are desirable. While many of Maxwell's scientific papers were also published in this first edition of the biography, this part was superseded by the more complete work of W. D. Niven (WDN1 and WDN2, 1890). Much more recently, Harman (1990, 1995, 2002)[1] has also given wider access to much of Maxwell's scientific correspondence. But, as far as Maxwell's biography is concerned, one must go looking further afield.

It has consequently been a key aim in writing this book to avoid simply repeating information from various secondary sources and to look further afield at original records such as manuscripts, letters and official documents preserved in archives held by the National Library of Scotland, the National Archives of Scotland, National Records of Scotland, Dumfries and Galloway Archives, Scotland's People (births, deaths and marriages), RCAHMS (buildings and monuments) and others. With such records generally being limited and often fragmentary, we next look to those near-contemporary published works that give factual accounts of the information being looked for. It often happens that there is more than one such source, and the differences in detail may give a guide as to whether they are independent, or whether one has simply been copied or recounted from another. Plagiarism does not seem to have been considered much of a sin during the times that are of interest to us, which is understandable given that the availability of books was comparatively limited. Whenever such sources have been used, an attempt has been made to recover what seems to be the original version, or at least to sift out the credible facts. As far as possible, links to sources on freely available current websites such as the Internet Archive (www.archive.org) and Project Gutenberg (www.gutenberg.org) are given. However, when the cited version is not available, the closest available online version is offered, for example, a later edition.

There are situations in which it is not possible to get all the information that we would like, leaving us with questions that we can only try to answer by analysing whatever information we do have. How it was that some prominent Roman Catholic families managed to pass on their lands from one generation to another in times when the law forbade it is just one such question. It appears that they managed to protect their position by hiding the truth, dissembling in public while privately following God in their own way. When such answers are proposed, it falls to the reader to decide what is reasonable.

As to gathering information about our main characters, while some family

PREFACE

trees are available they are often only partial; only the most meagre glimpses of their lives are to be obtained here and there. For example, there are few details on Janet Irving, James Clerk Maxwell's paternal grandmother, and this is one of the gaps that this book attempts to address. On the other hand, Sir John Clerk, 2nd Baronet of Penicuik (James Clerk Maxwell's great-great-grandfather along the male line) blessed us with a memoir from which we learn a great deal not only about his family but about many other connected people and events.

Looking back at these sources, it is an uncomfortable fact how little direct information there is about the women who figure in the story. Generally speaking, such information comes only tangentially and must be gleaned from reports mentioning the father, husband, brother or son. There are a few exceptions, for example the cases of Jemima Blackburn, Elizabeth Grant and Mary Dacre but, in the main, the dearth of direct information is simply a reflection of the way things were. Unfortunately, one of the consequences is that it has been impossible to prevent stories about men from predominating in this work.

In the case of properties, official records kept over the centuries are held in the Register of Sasines kept at Register House, Edinburgh. The firm of Millar and Bryce kindly provided searches for 14 India Street within these records. A further search, on 31 Heriot Row, helped to confirm the answer to the crucial questions, whose house was it originally, and did Janet Irving ever live there? The interpretation of the critical record of the Sasine of 1851, on the occasion of Isabella Wedderburn selling the house, was kindly confirmed by Mr Kenneth Robertson of Balfour and Manson, LLP.

As to Edinburgh addresses, and who lived at them, we have a wealth of information in the postal directories of Edinburgh and surrounding counties from the late eighteenth century through to the early twentieth century. Most of these have been scanned by the National Library of Scotland and are readily available online at their website www.nls.uk. Since the collection comprises a variety of directory sources, such as those by Williamson, Aitchison, Gray and the Post Office, they are simply referred to collectively as the Edinburgh Post Office Directories, or 'EPOD'. The detail of how these directories have been used and interpreted is to be found in the notes.[2]

Other sources of information have been books such as peerages and biographical dictionaries; family trees from various publications and sources, many of them online; websites such as www.gravestonephotos.com and www.archive. stjohns-edinburgh.org.uk ; maps and places from Google Maps, Bing Maps (Ordnance Survey), the collection of Scottish maps at www.nls.uk and the extremely useful combination of RCAHMS Canmore websites

ix

(www.rcahms. gov.uk/ and http://canmoremapping.rcahms.gov.uk/), where the details and locations of almost every historic building and monument in Scotland may be found. Many internet searches have also thrown up odd bits of information in old editions of gazettes, periodicals and newspapers.

In some relatively unimportant areas we have found a great wealth of detail, while conversely we have a dearth of information in some quite crucial areas. This, however, is only to be expected. Occasionally a single scrap of information has proved to be providential in opening up a significant new lead. The main weakness of miscellaneous snippets is frequently the lack of corroboration. This need not concern the reader too greatly, however, for all such information has been treated with caution and we have endeavoured to be clear about the quality of critical evidence. When the evidence supporting an unfolding story is somewhat thin or open to question in some way, we should regard any conclusions drawn as being a hypothesis to be put to the test. Their value is that they are a potential advance in what we know about Maxwell, his forebears and their family connections. Hopefully, the questions raised will be of value in their own right as a basis for further studies to be taken up by others.

On the whole, however, it has been possible to put together a serious and broad-ranging sketch of our subject matter, James Clerk Maxwell's family and friends, their connections and their homes in Edinburgh's New Town and beyond.

Acknowledgements

I would like to express my sincere thanks to the Trustees of the James Clerk Maxwell Foundation for their encouragement of this endeavour, for access to materials at 14 India Street, for their permission to use photographs taken there, and for obtaining the records of 31 Heriot Row. For their individual contributions I would particularly like to thank Dr Dick Dougal, Professor Chris Eilbeck, Mr David Forfar, Prof Peter Grant, Professor Malcolm Longair, Professor David Milne, Professor Roland Paxton, Dr James Rautio and Professor John Reid. I am also grateful to the following people and organisations who have generously provided information and assistance of various kinds:

>   Aberdeen Art Gallery
>   Birmingham Museums Trust
>   Honor Clerk and Sir Robert Clerk of Penicuik
>   Dr William Duncan, of the Royal Society of Edinburgh
>   Captain Duncan Ferguson, present owner of Glenlair and trustee of the Maxwell at Glenlair Trust
>   Dr Michael Geselowitz
>   Mrs Catherine Gibb and Mr Graham Roberts, of Dumfries and Galloway Libraries, Information and Archives
>   Mrs Vicki Hammond, of the Royal Society of Edinburgh
>   Mrs Frances Macrae, archivist of the Corstorphine Trust.
>   Mr Basil Mahon
>   Registers of Scotland
>   Dr W. Ross Stone

The extracts from Dyce & Cay (1888) on pages 252 and 253 are reproduced courtesy of Birmingham Museums Trust.

Lastly, my warm thanks and appreciation go to my wife, Norma, for her help in producing the book, and not least for allowing me the many long hours spent sequestered away in researching and writing it.

Abbreviations

General:

BA: British Association for the Advancement of Science
HEICS: Honourable East India Company Service
HMSO: Her Majesty's Stationery Office.
IEEE: Institute of Electrical and Electronic Engineers (USA).
NLS: National Library of Scotland
RSE: Royal Society of Edinburgh

The following abbreviations are used for sources frequently referred to in the text:

BJC: *Memoirs of the Life of Sir John Clerk, Bt* (Clerk & Gray, 1892)
C&G: *The Life of James Clerk Maxwell* (Campbell & Garnett, 1882)
CANMORE: The online catalogue to Scotland's archaeology, buildings, industrial and maritime heritage. Available at http://canmore.org.uk/
References are given in the notes in the form CANMORE ID [site reference number]
DGA: Dumfries and Galloway Archives. Available at http://archives.dumgal.gov.uk/DServe/dserve.exe?dsqApp=Archive&dsqDb=Catalog&dsqCmd=Search.tcl
References are given in the notes in the form DGA: [accession number].[item number], [title], [date]
DGFHS: Dumfries and Galloway Family History Society. Available at www.dgfhs.org.uk
DNB00: Dictionary of National Biography, 1885–1900, 63 vols. London: Smith Elder & Co. Available at https://en.wikipedia.org/wiki/Dictionary_of_National_Biography

DNB12:	*Dictionary of National Biography*, Supplement, 1912. London: Smith Elder & Co. Available at https://en.wikipedia.org/wiki/Dictionary_of_National_Biography
DSA:	*Dictionary of Scottish Architects*, Available at www.scottisharchitects.org.uk/architect_list.php
EPOD:	Edinburgh Post Office Directories (irrespective of publisher). Available at www.nls.uk/family-history/directories/post-office/index.cfm?place=Edinburgh
NAK:	National Archives, Kew. Available at www.nationalarchives.gov.uk/ References are given in the notes in the form NAK: [accession number], [title], [date]
NAS:	National Archives of Scotland. Available at www.nas.gov.uk/onlineCatalogue/ References are given in the notes in the form NAS: [accession number].[item number], [title], [date]
ODNB:	*Oxford Dictionary of National Biography*. Available at www.oxforddnb.com/public/index.html (subscription required)
OS:	Ordnance Survey Maps
RCAHMS:	Royal Commission on the Ancient and Historic Monuments of Scotland. Images, mapping and records available on the Canmore website www.rcahms.gov.uk/
RPS:	Records of the Scottish Parliament to 1707. Available at the University of St Andrews website www.rps.ac.uk/ References are given in the notes in the form RPS: [reference number]
RS:	Registers of Scotland (property records). References are given in the notes in the form RS: [reference numbers], [record type], [date]
SC:	Scottish Censuses, available on the Scotland's People website, www.scotlandspeople.gov.uk/ Registration is required and charges apply
SCAN:	Scottish Archive Network online catalogue. Available at http://195.153.34.9/catalogue/welcome.aspx
SOPR:	Scottish Old Parish Records (Births, Deaths, Marriages). Available on the Scotland's People website www.scotlandspeople.gov.uk/

	Registration is required and charges apply. References are given in the notes in the form
	SOPR: [event], [reference number], [parish/county]
SRO:	Scottish Record Office
SRS:	Scottish Record Society. An index, and online copies of the 'Old Series' publications, are available at www.scottishrecordsociety.org.uk/home/publications
SSR:	Scottish Statutory Records (Births, Deaths, Marriages since 1855). Available on the Scotland's People website www.scotlandspeople.gov.uk/
	Registration is required and charges apply. References are given in the notes in the form
	SSR: [event], [reference number], [parish/county]
SWT:	Scottish Wills and Testaments, available on the Scotland's People website, www.scotlandspeople.gov.uk/
	Registration is required and charges apply.
WDN1 and WDN2:	*The Scientific Papers of James Clerk Maxwell* (Niven, 1890). Volume 1 is available at https://archive.org/details/scientificpapers01maxw while volume 2 is at https://archive.org/details/ scientificpapers02maxwuoft

Genealogical Tables

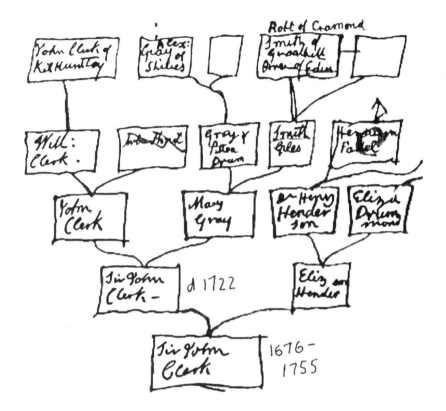

Family Tree 1 Mary Gray, wife of John Clerk of Penicuik
In this excerpt from John Clerk Maxwell's draft family tree, the name Giles Mowbray should fill the blank box on the top right. She was the attendant of Mary Queen of Scots by whom the 'Penicuik Jewels' came into the Clerk family. She married Robert Smith of Cramond, and their daughter, Eligida or Giles Smith, was said by Wilson (1891, pp. 150–151) to be 'Mary Gillies'. This may have been a misreading of 'Smith Giles', as it is written in the central box here. (From DGA: RGD56/13, nineteenth century, by courtesy of Dumfries and Galloway Libraries Information and Archives.)

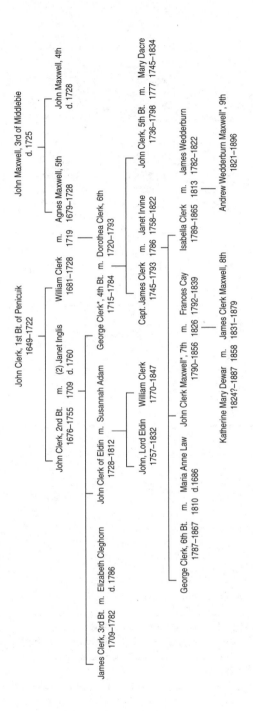

*George and Dorothea Clerk took the surname Clerk Maxwell upon Dorothea's succession to Middlebie in 1738. Likewise, John Clerk Maxwell had been John Clerk until 1803, when he succeeded his grandmother and, on succeeding James in 1879, Andrew Wedderburn also took the Maxwell name. The exception was James, who was baptised as Clerk Maxwell.

Family Tree 2 Succession of the Clerks of Penicuik and the Maxwells of Middlebie

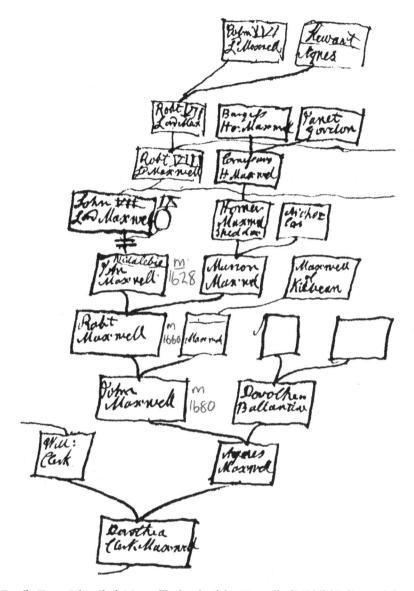

Family Tree 3 John Clerk Maxwell's sketch of the Maxwell of Middlebie line and the connection to the Lords Maxwell and the Maxwells of Speddoch

John Clerk Maxwell had problems depicting the line of the Lords Maxwell as it was actually John 4th Lord Maxwell who married Agnes Stewart. The 5th, 6th and 7th Lords were all called Robert, but the 7th died an infant, leaving only the two shown in the sketch. Having got to John the 8th Lord, he made amendments as a result of missing out the 9th. The Middlebie line more or less fits the known facts, and the Speddoch line is at least consistent with the few fragments we have. (From DGA: RGD56/13, nineteenth century, by courtesy of Dumfries and Galloway Libraries Information and Archives.)

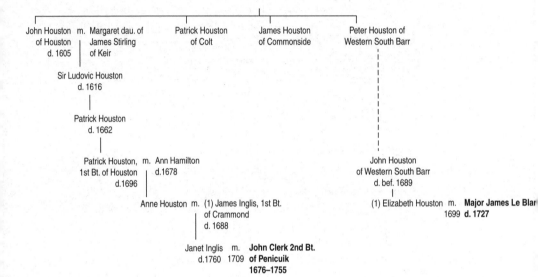

Family Tree 4a The connection between Baron Sir John Clerk and James Le Blanc
Anne Houston's brother, Sir Patrick Houston, went to Georgia in 1735 and became a major landowner near Savannah. His son John was mayor and twice governor of the state (Bulloch, 1919, pp. 27–28). Elizabeth Houston had a brother Patrick and a younger sister Agnes who laid claim to the Major's heritable property after Agnes Maxwell's death (NAS: GD18/2602, 13/6/1735).

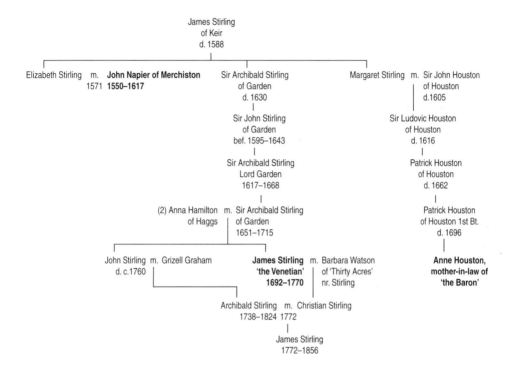

Family Tree 4b The connection between the Clerks and the Scottish mathematicians John Napier and James Stirling

Sir John Houston of Houston (d. 1605) was the eldest brother of Peter Houston of Western Southbarr, ancestor of John Houston or Huston of Western Southbarr, who died about 1688 (Family Tree 4a). Curiously, Archibald Stirling was well known to the Irvings of Newton.

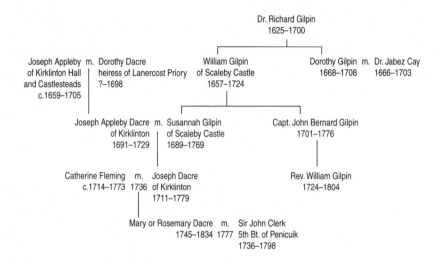

Family Tree 5 The Appleby Dacres and their connections to the Cays and Gilpins
Dr Richard Gilpin was a well-known Puritan leader. His son William Gilpin of Scaleby Castle was a friend of the Baron. William's grandson, Rev. William Gilpin, was ordained in 1746 by the Bishop of Carlisle, father of Catherine Fleming. Rev. Gilpin's 'picturesque' concept influenced John Clerk of Eldin's drawings. Dr Jabez Cay's brother John was the grandfather of John Cay (1727–1782) who in 1756 married Frances Hodshon of Lintz and started thereby the Edinburgh Cay family.

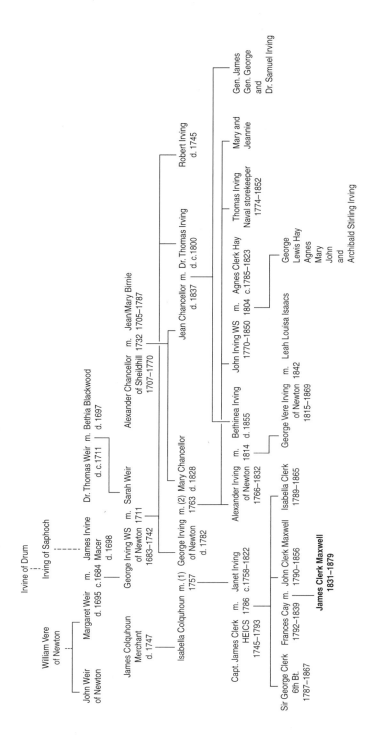

Family Tree 6 The Irvings of Newton and the Chancellors of Shieldhill

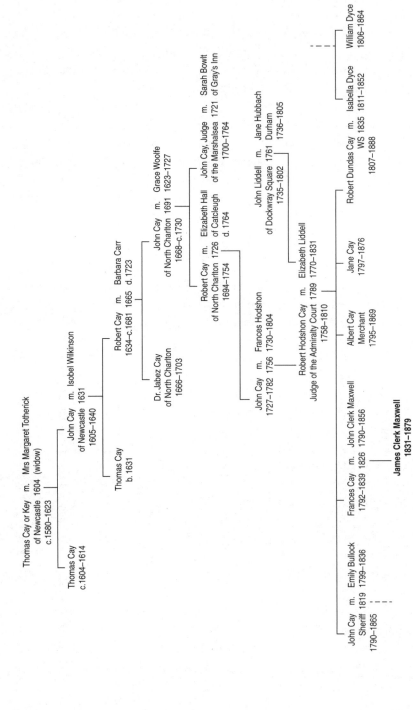

Family Tree 7 The Cays of North Charlton

The line from Thomas Key of Newcastle down to the generation of John Cay, Sheriff of Linlithgow, the last Cay to possess North Charlton (Bateson, 1895, pp. 297–300). Jabez Cay was the first owner of a moiety of North Charlton, and after his death his brother John obtained the whole estate by buying out the second moiety.

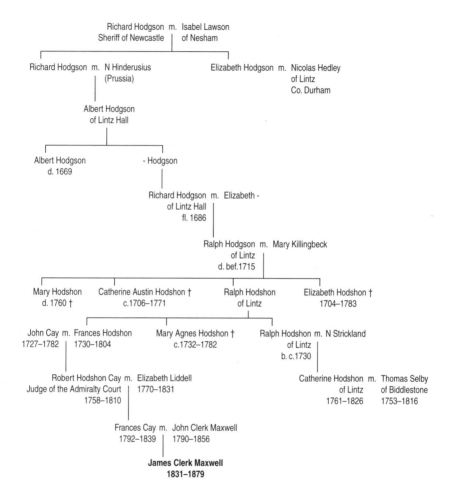

Family Tree 8 The Hodshons of Lintz
From Richard Hodgson, Mayor of Newcastle, down to the last of the line, Ralph, father of Frances Hodshon, who married John Cay. Those who were nuns (Bowden & Kelly, 2013) are marked †.

CHAPTER 1
Introduction

If James Clerk Maxwell was the reason for writing this book, it is not so much about the man himself as it is about whence he came. It deals with the origins of his predecessors and of the families, some rich and some brilliant, with whom they were connected. It also gives us a picture of what life might have been like through the generations of any well-to-do Scottish dynastic family. From the seventeenth to the nineteenth century, it reveals how they lived, or perhaps more accurately, how they managed to survive. If birth was the first great struggle, living beyond infancy was the next. For those who could have children, bringing eighteen of them into the world (at an average rate of about one per year!) was not particularly unusual, nor was it unusual to see half of them dead before they reached adulthood. For those who did survive, life was still uncertain, for they would not escape the visitations of terrible illnesses such as consumption, smallpox and typhoid, for which there was no cure, even for the rich. A simple injury such as a fractured limb could be a death sentence. They would live or die according to their relative strength or weakness, and their luck. Sometimes fate was cruel indeed, for it is not unusual to see burial records noting the death of a parent and several children all within the space of a few weeks.

Property was the traditional basis of wealth. The eldest son of a comfortably-off family could expect to be comfortable indeed. He would be well educated, have fine clothes, eat good food, and entertain himself with art and music. It was expected that he should marry and his wife would be chosen from another respectable family in the prospect of bringing him some property. In due time, he would become the new laird and begin planning a future for his own children. Surprisingly, in Scotland it seemed to be the custom for such men to have a profession, for example, as an advocate. If the laird had many brothers, we could also expect to find amongst them a solicitor, a doctor, a soldier, a sailor and a minister of religion. His sisters would be gentlewomen, some of whom would be married off onto gentlemen similar to himself, while the others would remain part of his household.

Although such people might have been regarded as being very well-off, this was a relative term. Even with good management, money was frequently in short supply, and marrying off sisters and daughters was an expensive business. Likewise, while a laird generally passed on his property to his eldest son alone, he would hope to create the prospects of property for his second son by finding him a convenient marriage. Sometimes, however, he would also contrive to give him a small parcel of land. As to the others, they usually had to take their chances in the world with little more than a good education, good connections and the good name of their family to help them. There was always the chance of falling heir to some property or other, but otherwise they simply joined the ranks of professional gentlemen one tier below their eldest brother.

For the population at large, in addition to the perpetual scourges of poverty and disease were two others, religion and war. The former was entangled with politics and the latter was often the outcome. To belong to a non-conformist religion was to set oneself against the state, for which the consequences could be severe. Non-conformists were harshly repressed, and in the face of such injustice their reactions gave rise to an undercurrent of rebellion, which now and then would come to the surface. But rebellion also presented the opportunity of settling old scores between sides that were mainly defined by history and religious differences.

Our story unfolds during the religious and political turmoil that spanned most of the seventeenth century, thence through the Union of the Scottish and English Parliaments and two Jacobite uprisings, after which came the relative tranquillity that engendered an age of enlightenment. Scotland had long been overshadowed by its larger neighbour England, which was much richer and, by ruling the seas, dominated wealth-producing foreign trade. Much of Scotland, namely the Highlands to the north and the islands to the west, was even in the seventeenth century largely untamed and ruled along old feudal lines by independent, fractious chieftains whom the Lords of the Realm struggled to keep at peace. Only after the last Jacobite rising of 1745 did things really begin to change. For those dwelling in the south and on the east coast, apart from the fact that they were generally poorer, life was much more like that of their English cousins. Nevertheless, there seems to have been a certain drive amongst the Scots to work hard in the face of their perpetual struggles and to do better, and by one small achievement after another, they began to improve. There were of course failures along the way, but they did in the end succeed, for within a century the social landscape was transformed from one of relative poverty and instability to one of greatness and splendour. They had created an era that has been justly called the Scottish Enlightenment, in which Scots seemed to come to the forefront in everything. Indeed, they succeeded not only at home, but in London, Europe, and such far-flung places (as they truly were at the time) as Russia, India and America.

Given that the population of Scotland was small, and that the number of those who were well enough off to be able to play some part in this enlightenment was smaller still, it is amazing to see how many 'giants' were thrust onto the stage in the arts, sciences and humanities to play some vital part that will never be forgotten. Nor did the Scots neglect business, exploration, medicine, engineering and architecture, where they also left lasting legacies. In architecture, in particular, this can be seen in the form of stately houses and public buildings all around Great Britain, but no greater concentration of them will be seen than in Edinburgh's New Town, which was at the very heart of that period of enlightenment.

James Clerk Maxwell was just one of the great men thrown up by the Scottish Enlightenment. Much has already been written about his life and works (see bibliography) and so here we give a mere sketch of them to set him in his proper place in our story which spans the lives and deeds of the Clerks, the Maxwells and Clerk Maxwells, the Cays, the Wedderburns and the Mackenzies, who were all part of his family network. It also touches on his two close school friends, Peter Guthrie Tait and Lewis Campbell, with the connection between all of them being Edinburgh's New Town, or more precisely, Edinburgh's Second New Town, which began about 1800, just a third of a century after the start of James Craig's original New Town of 1767 (Youngson, 1966). To provide some orientation for the reader unfamiliar with Edinburgh, Figure 1.1 shows the layout of the original Old Town of Edinburgh, together with these two consecutive new town developments to the north, and an earlier one to the south.

The connection with the Second New Town in particular comes about because at some time or another, between roughly 1800 and 1865, that is where most of these families lived; some for nearly all of that time, others, like Maxwell himself,[1] for just a few years. Irrespective of how long, they were all indelibly products of Edinburgh's New Towns and all that went with them. By tracing where they lived we bring new light to their family history and to the Clerk Maxwell connection with 14 India Street, which was Maxwell's birthplace, and 31 Heriot Row, where he lived during term-time from 1841 until 1850. The locations of these key addresses, and many more that we will encounter, are indicated in Figure 1.2.

In order to understand something of the reason why Edinburgh proved to be such a fertile breeding ground for men like Maxwell, the circumstances behind it need some explanation. The growing atmosphere of peace and

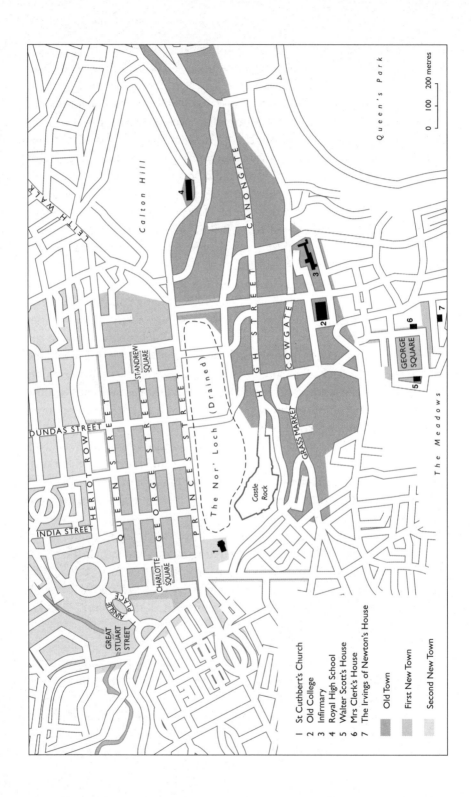

prosperity that began in the middle of the eighteenth century seemed to open doors for men of enterprise and genius to come to the fore, particularly in Edinburgh. Despite the Scottish Parliament itself having been dissolved in 1707, Edinburgh was still the Scottish capital and the legal, financial and commercial centre of the nation. It was home to the cream of the nation: landed gentry; judges, lawyers and politicians; bankers and businessmen; doctors, scientists, engineers and philosophers; artists, architects, writers and poets, many of whom were both willing and able to reach for the top. There began an outpouring of ideas, initiatives and progress.

Although these circumstances created the motherlode of this progress, what is hard to explain is how, at a time when the population of the capital was still less than 100,000, there were enough such men to produce the end result. Moreover, we can only speculate that it was the breaking free of the intellectual and financial fetters of the past that provided the catalyst that sparked the yield of so many remarkable people. Remarkable also, in that they reached the highest rank, not just at home, but on the world stage. A few of the better known of them are David Hume (philosopher), Adam Smith (economist), James Hutton (geologist), James Boswell (diarist and biographer of Johnston), Robert Adam (architect), Walter Scott (novelist) and Henry Raeburn (portrait painter), but there were many others who are better known by their contributions, for example, William Smellie, the founding editor of the *Encyclopaedia Britannica*.

In the early Enlightenment, the youth of the upper classes not only travelled in Europe but were frequently educated in Europe; for example,

Figure 1.1 (opposite) Old and New – the three towns
The Old Town spills down both sides of the Royal Mile, a long and ancient street that extends down the crest of the ridge that slopes steadily downwards from the Castle Rock, at its western apex, to Holyrood Palace. In bygone times the valley to the north, now Princes Street Gardens, had been flooded to form a defensive moat called the Nor' Loch (Daiches, 1986). To the north of this valley George Street lies along the line of a flatter ridge, beyond which the ground runs ever downhill to the distant sea, the Firth of Forth. To the south of the Old Town, however, was the Burgh Loch, which had been progressively drained to make meadows. Around 1760, the burgesses of the Old Town looked to these sites on the north and south for expansion. Although the New Town refers to the planned development to the north that commenced in 1767, the small development around George Square at the edge of the Meadows started in the previous year. Within just thirty years, the mid-grey area (first New Town) was almost filled and a second New Town was called for. Building began in this area about 1800, but another thirty years saw that largely built up too. Today, the New Town refers to the entire development characterised by buildings of quality executed in a harmonised Georgian style and set amongst wide streets and numerous gardens.

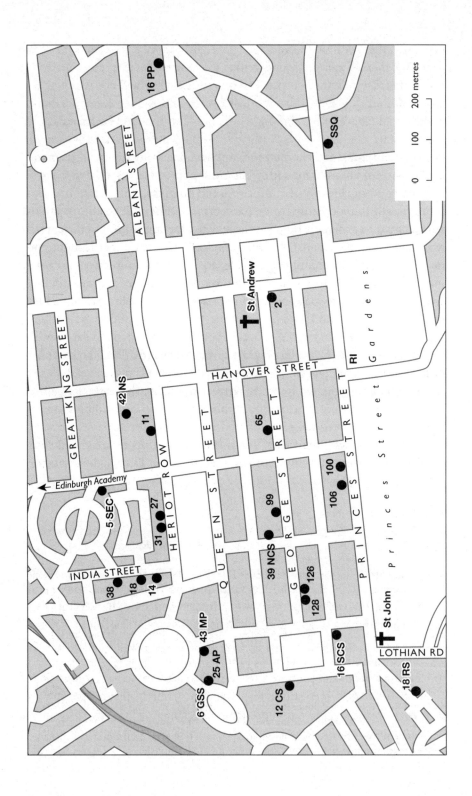

INTRODUCTION

James Boswell studied law at Utrecht before engaging on his 'grand tour', all of which he recorded for us in his journals. When they came into their fortunes, these young men already had aspirations, and compared with what they had seen on their travels, Scotland had a lot of catching up to do. In order to achieve their desire to progress, they could rely on a network of family and friends who were ready and willing to join in on the act. Marriages between families, and even within families, were a way of consolidating wealth, in particular heritable property. Marriage contracts afforded the means of securing these things in much the same way as merger contracts of commercial companies do today. In this story we will hardly encounter a person with a middle name that is not the surname of a connected family, and the Clerk in James Clerk Maxwell is no exception.[2]

The well-to-do families of Edinburgh counted numerous lawyers amongst them who had the means and the power to bring such things about, and when it was necessary they had the wherewithal to obtain the force of law and the occasional act of parliament. So it was that over several generations, the lands of a Maxwell in Dumfriesshire and of a Clerk in Midlothian were brought together into one family. Dorothea Clerk was heiress to the Maxwell estate of Middlebie.[3] In 1735 she married George Clerk, her first cousin and second son of the 2nd Baronet of Penicuik, with the eventual settlement bringing to him the Clerk lands of Dumcrieff, near Moffat, as well as the right of Middlebie. George Clerk of Dumcrieff, or George Clerk Maxwell of Middlebie, as then he alternatively styled himself, got into financial difficulties as a result of being too 'deeply interested in promoting the commerce and industries of his native land' (Wilson, 1891) and in order to pay off his creditors he had to sell off both Dumcrieff and the part of the Middlebie estate that included Middlebie itself. In due course, however, he inherited the Penicuik baronetcy from his brother to become Sir George Clerk, 4th Baronet of Penicuik; the baronetcy was his for just one scant year before passing to his eldest son John in 1784.

James Clerk Maxwell's father (Plate 1) was baptised as plain John Clerk. He was the grandson of this Sir George Clerk, or Clerk Maxwell, but it was

Figure 1.2 (opposite) The residences of the Clerk-Maxwells, Wedderburns and the Cays in the New Town
Also shown are the houses where Lewis Campbell, Alexander Graham Bell (16 SCS) and Sir Walter Scott (39 NCS) lived. Robert Louis Stevenson was at 17 Heriot Row, and Henry Mackenzie at number 6.
Key: AP: Ainslie Place; CS: Charlotte Square; GSS: Great Stuart Street; MP: Moray Place; NCS: North Castle Street; NS: Northumberland Street; PP: Picardy Place; RI: Royal Institution; RS: Rutland Street; SCS: South Charlotte Street; SEC: SE Circus Place; SSQ: Shakespeare Square.

John's older brother, named George after his grandfather, who became the 6th Baronet.[4] The entail of the Middlebie estate (Chapter 6), however, required that it be held separately from the baronetcy of Penicuik (and vice versa for Penicuik) and for the holder to take the surname Maxwell. And so it was arranged that the older brother George should inherit the Penicuik estate together with the baronetcy, while the younger brother, John, should be the one to change his name and inherit the Middlebie estate. In 1808, George Clerk came into his majority as 6th Baronet of Penicuik and similarly by 1811 his brother had officially become John Clerk Maxwell of Middlebie.

George and John took part in the Age of Enlightenment in their own different ways. George had a busy and distinguished political career, rising to Cabinet rank in several different government posts and becoming master of the Royal Mint, yet in spite of this he had the time and interest to become president of the Royal Zoological Society. John settled for being an advocate, but was also a enthusiast of science and technology, the rapid advances of which held a particular fascination for him, which in due course he sought to pass on to his son, James. If he held any ambition at all for a high-flying career, he did not show it, rather his aspirations were closer to home: his family and his small estate. In 1826 he had married the 'girl along the street', Frances Cay, who was from a family background not dissimilar to his own. Her mother was a talented amateur portrait artist, her father had property and was also a judge, one brother was a sheriff, another a merchant, while a third was yet another lawyer.

It was through the influence of his wife that John Clerk Maxwell was motivated to shake off his inertia and *do* something. He therefore forsook the life of a city lawyer and concentrated on a part of his estate called Nether Corsock, in Kirkcudbrightshire,[5] where he decided to build a decent house and become a farmer. The rest of the Middlebie estate brought him a reasonable income, and so, ever cautious and 'judicious', he must have realised that there was no great risk involved. Nevertheless, he did not lack the inclination to apply himself and so the project proved a success; his real tendencies were to the practical, and the land suited better than the law.

Until fairly modern times, the lives of the gentler classes were generally divided between spending the summer months at their country seats and overwintering in their town houses. The courts, schools and universities all ran roughly from November to June, with some breaks in between. Once the Clerk Maxwells were ensconced at Nether Corsock, they too would return to Edinburgh, a journey of two days, for the worst of the winter season when nothing could be done on the land. But when in the early months of 1831 Frances must have known she was pregnant, the clear decision was that she should remain in Edinburgh for the birth of the child. And so it was into a genteel but modest family that James was born on 13 June 1831,[6] not on his father's estate but in the Clerk Maxwells' town house at 14 India Street[7] in Edinburgh's Second New Town. This was to be James' principal home for the next two years, until it was at last considered that he was thriving, and that the house that John Clerk Maxwell had been building at Nether Corsock was fit and ready to be taken over as their permanent family home.

So began the era of James Clerk Maxwell. We shall follow the connections back to his roots, a mere hint of which we have mentioned above in the coming together of the name Clerk Maxwell. And from the roots we shall follow back the branches, to the families connected with the Clerks and the Maxwells, families that brimmed with enterprise and took their own places in the Scottish Enlightenment alongside the best of them. They include:

Sir John Clerk, 2nd Baronet of Penicuik (1676–1755) antiquarian and writer;
Allan Ramsay (1686–1758) poet;
James Stirling (1692–1770) mathematician and lead mine manager;
Adam Smith (1723–1790) economist;
Sir George Clerk Maxwell, 4th Baronet of Penicuik (1715–1784) improver;
James Hutton (1726–1797) founder of modern geology;
Robert Adam (1728–1792) neo-classical architect;
John Clerk of Eldin (1728–1812) artist and naval tactician;
Sir Walter Scott (1771–1832) romantic novelist and historian;
William Dyce (1806–1864) artist and scholar;
James D. Forbes (1809–1868) glaciologist;
William Thomson (1812–1897), Lord Kelvin, physicist and engineer;
Jemima Blackburn (1823–1909), née Wedderburn, water colourist;
Peter Guthrie Tait (1831–1901) mathematician and natural philosopher;
James Clerk Maxwell (1831–1879) physicist;
William Dyce Cay (1838–1925) civil engineer.

It will be noted that four of the names above have already been cited as leading lights of the Scottish Enlightenment. The name of James Clerk Maxwell, however, has never been ranked beside the likes of Hume and Smith. Considering what he achieved, this raises awkward questions about the nature of fame, and about whom the population at large chooses to celebrate or ignore. We will therefore give him his day by beginning with his story.

CHAPTER 2
James Clerk Maxwell's Life and Contribution to Science

You will sometimes hear it said that James Clerk Maxwell is one of the greatest physicists that ever lived. For many people this will be a surprise, for in comparison with Isaac Newton, Albert Einstein and Stephen Hawking, he is practically unknown amongst the general public. This is most keenly demonstrated by an anecdote related by R. V. Jones (1973, p. 67),[1] who was some years ago the Professor of Natural Philosophy at Aberdeen, where a century before Maxwell himself had been the Professor.[2] Professor Jones had been told the story of an advertisement that had been placed some years before in the *Aberdeen Press and Journal*, by a solicitor seeking information on the whereabouts of a certain James Clerk Maxwell. The solicitor dealt with the funds of the local music hall, which happened to pay out a small annual dividend to its subscribers, one of whom had been Maxwell. Unfortunately, Maxwell's share of the distribution was now being returned as undeliverable. While it was bad enough that the solicitor had been unaware that James Clerk Maxwell was 'the most famous man to walk the streets of Aberdeen', his letters to Maxwell at the university were being returned marked 'NOT KNOWN'!

Behold the Child

Most of what we know about James Clerk Maxwell is to be found in his original biography[3] by Lewis Campbell, a lifelong friend, and William Garnett, who worked alongside Maxwell at the Cavendish Laboratory, first as his demonstrator and then as a lecturer. Although he was born at 14 India Street in Edinburgh (Colour Plate 1) and wanted for nothing there, his parents' lives centred on John Clerk Maxwell's 'family seat' at Nether Corsock, a piece of some rough farmland set in a remote area nearly twenty miles west of the county town of Dumfries (Figure 2.1). Nether Corsock was part of the estate that John had inherited from his paternal grandmother, Lady Dorothea Clerk or Clerk Maxwell of Middlebie. The rest of his estate consisted of a few farms scattered about nearby, and some properties in and near the town of Dumfries but, by the time he inherited it, the farms and properties at Middlebie were long gone, sold off in a public roup to pay off his grandfather's debts.

After their marriage, John Clerk Maxwell and his wife Frances resolved to become country gentlefolk, alternating between living at Nether Corsock from spring to autumn and overwintering in Edinburgh beside their friends and family. For the well-to-do this was in any case the general pattern of life; in the Clerk Maxwells' case the only difference was that John's business focus was no longer in Edinburgh. The journey back and forth, nearly 100 miles over rough roads and tracks, took two entire days by horse and carriage. Such a journey was not to be undertaken on a whim, but for a definite purpose. Life at Nether Corsock, however, was not ready-made for them. They had first to turn rough pasture into a viable farm, and they had no more than a gardener's cottage to live in. A proper house had to be designed and built, a project that John Clerk Maxwell would have entirely relished, for as a boy he had often dreamed of such things.

Amidst all this, in the summer of 1829 Frances fell pregnant. By the time their child Elizabeth was born, they were back in Edinburgh for the winter. Sadly, the child died shortly after birth, and the bereft parents, having married a little late in life, must have wondered if they would ever be lucky enough to have another. Happily, within the space of a year Frances was pregnant again, but since the building of a new two-storey house at Nether Corsock[4] to John's own design was still then only just about to be started, they decided to stay on in Edinburgh. It seems natural that a man in John Clerk Maxwell's position would have insisted on his wife staying put in Edinburgh where she could be looked after by nearby family, and the best of doctors. Nothing would be spared, no risks taken. On the other hand, he would have had business to carry on with at Nether Corsock; farmhands and workmen to supervise; and rents to gather in. In the ways of these times, a man decided for himself what had to be done; Frances had to stay and he would just go and get on with things. In any case, what practical use could he be to Frances until it was near the time for the child to be born?

In due course Frances was safely delivered of a son whom they named James after his paternal grandfather, as was then the custom. James Clerk had died many years before while his son John Clerk was just a child of seven; the reason why John came to have a different surname from his father will become clearer later on. By the time James was two years old and thriving, his parents felt that they could return to Nether Corsock and make it their permanent base. The town house at 14 India Street was rented out and they

11

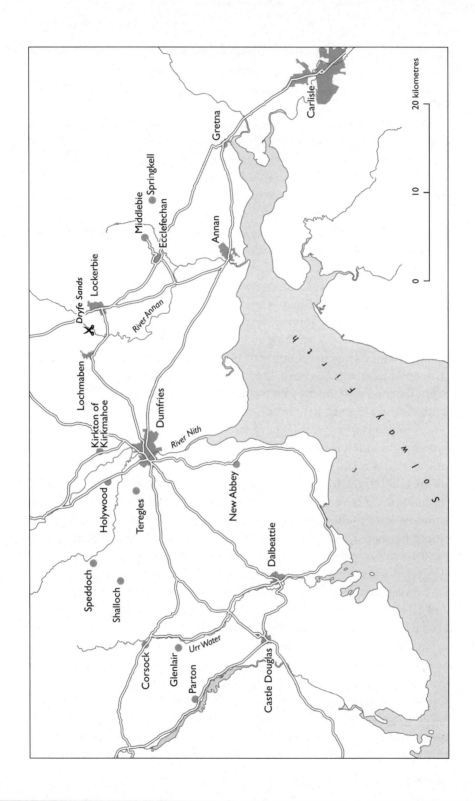

were never to live there again. The earliest demonstration we have of young James definitely at his new country home is a letter from his parents to Frances' sister Jane back in Edinburgh (C&G, p. 27):

> To Miss CAY, 6 Great Stuart Street, Edinburgh.
> Corsock, 25th April 1834.
>
> Master James is in great go [. . .] He is a very happy man, and has improved much since the weather got moderate; he has great work with doors, locks, keys, etc., and 'Show me how it doos' is never out of his mouth. He also investigates the hidden course of streams and bell-wires, the way the water gets from the pond through the wall and [. . .] down a drain into [the] Water Orr [. . .] As to the bells, they will not rust; he stands sentry in the kitchen, and Mag runs thro' the house ringing them all by turns, or he rings, and sends Bessy to see and shout to let him know, and he drags papa all over to show him the holes where the wires go through.

Fortunately, we have an image of James in infancy, with his mother, in their portrait by William Dyce (Colour Plate 2), the story of which we will arrive at later. James' cousin, Jemima Wedderburn, later Mrs Blackburn, was to tell Lewis Campbell (C&G, p. 28) that throughout his childhood his constant question was, 'What's the go o' that? What does it do?' and he would drill on in this vein until he got an answer that satisfied him. There was not always such an answer to be had, for he could not be satisfied by a mere form of words. His nurse recalled to Campbell (C&G, pp. 30–31), having been questioned by James in this manner about the colour of some pebbles he had collected on a walk together, that she answered, 'That (sand) stone is red; this (whin) stone is blue.' He retorted, 'But how d'ye know it's blue?'

From Campbell's own recollection (C&G, p. 28) we hear that James' earliest memory was of lying on the grass to the front of the house at Nether Corsock 'looking at the sun, and *wondering*'. These are Campbell's italics, for he particularly wanted to impress upon the reader this aspect of James' character. There would be hardly anything, it seems, that he did not wonder about, and amid all the other opportunities for indulging in some form of play or excitement, he would come to rest, and pause, and simply *wonder*.

Figure 2.1 (opposite) Key locations in and around the Maxwell Homelands in and around Dumfries
For Nether Corsock, see Glenlair, which is in almost the same location.

As well as having a seemingly endless appetite for enquiry, the young boy also had a keen ear for music and a prodigious, highly active memory (C&G, p. 32):

> His knowledge of Scripture, from his earliest boyhood, was extraordinarily extensive and minute; and he could give chapter and verse for almost any quotation from the Psalms [...] These things [...] occupied his imagination, and sank deeper than anybody knew.

In most other respects, James was a normal, happy boy. Although born of the landed gentry, his sensible parents did not keep him apart. Nor, as an only child, did they treat him as being overly precious. If there was any local parish school that he could have gone to, however, he did not attend; he was taught by his mother. But there was no haughty separation between masters and servants; in all other things he mixed in with the local boys, for the most part sons of his father's tenants and workers, taking part in their play and speaking their broad Gallowegian brogue. He had no fear of getting his hands dirty, for he had learned that 'country dirt' was 'clean dirt' (C&G, p. 34). He rambled in field and wood, climbed trees, took a washtub as a craft to be sailed in the duck-pond, hunted frogs and insects, and by the age of ten he could ride a pony. He was by then in every way a thorough-going country boy.

The single unhappy event that overtook his childhood was the illness and eventual death of his mother. By the time of his eighth birthday it must have been evident that all was not well with her, and in due course she was diagnosed with stomach cancer from which, all efforts to save her having failed, she eventually died at Nether Corsock in December 1839. James must have been all too aware of her suffering, for on being told that his mother was now dead he had declared 'Oh, I'm so glad! Now she'll have no more pain' (C&G, p. 32). Nevertheless, the loss of his mother at such a tender age took its toll and naturally enough disorientated him: 'his activities were apt for the time to take odd shapes [...] bright and full of innocence as they were [...] produced an effect of eccentricity on superficial observers' (C&G, pp. 45–46). A different side of James' character was to come to the fore in the days that followed, when his father brought in a young, inexperienced tutor to carry on with his education. The two did not hit it off, but whereas it seems that James had hitherto been a fairly rational and obedient boy, now he rebelled. The tutor had somewhat surprisingly found James to be slow at learning, and resorted to employing the sort of rough discipline by which he himself had probably been forced to learn his subjects. James, having been used to the gentle encouragement of his mother, may well have taken it into his head that he was not going to co-operate with such an unworthy interloper. Learning had been part of his daily diet of interesting things that were offered, accepted and taken on board, and now he was not going to buckle down and take it as though it were being thrust at him like some disgusting swill that he must consume or suffer the consequences: 'Meanwhile the boy was getting to be more venturesome, and needed to be not driven, but led [...] his power of provocation must have been [...] prodigious' (C&G, p. 41).

The tutor's response was brutal; James was 'smitten on the head with a ruler' and had his 'ears pulled till they bled'. In the tutor's defence, the maxim then was 'spare the rod and spoil the child', and so his methods were not quite so outrageous as they seem today. Nevertheless, James did not run to his father telling tales; he bore it all like a man and soldiered on. This gives us an early indication to another side of the boy's character, which he was to retain until his last breath; he could patiently endure anything, even suffering.[5] Had not Christ suffered, and his own mother too? James would have learned from his father the traits of the Scottish Presbyterian character; one must be stoical and endure some pain in life, one must be brave and not complain but carry on. One day, however, James decided he had had enough. Whether he ran off from his lesson or would not come in for it, we cannot tell, but we have the scene in Figure 2.2. We see him in his washtub in the middle of the duck pond,

Figure 2.2 Jemima Wedderburn's illustration of the standoff between the young James and his tutor
James, in the washtub, has taken refuge in the duck pond. His tutor grapples for him with a rake while his young friends, who possibly shared his lessons, look on. Jemima the artist stands on the left while her mother, Mrs Wedderburn, and James' father await the eventual denouement (C&G, p. 42).

with his tutor struggling to bring him to shore using a rake for a boathook. The standoff has been going on for some time, it seems, for already his father and Aunt Isabella are on the scene. His cousin Jemima, the artist who made this record in the late summer or early autumn of 1841 stands on the left; two local lads, James' 'vassals', watch from the far bank. Only the tutor, however, seems at all perturbed!

Although he did not get to know James until shortly after this episode, according to Campbell his torment, both physical and mental, at the hands of his tutor left him with 'a certain hesitation of manner and obliquity of reply', aggravating his slight oddity of character and, undoubtedly, the distress he suffered due to the loss of his mother.

Edinburgh Academy

On the prompting and advice of his sister-in-law Jane Cay, after the duck pond incident John Clerk Maxwell accepted that his son would be better off going to a proper school. He now recognised that James would need more than the sort of rough and ready education that was to be had anywhere nearby. The loss of his mother was part of the problem, and so it was decided that James should not go to a boarding school; he should go to school in Edinburgh, at the new Edinburgh Academy (Plate 2), which was within five minutes' walk from the care of his two aunts, Jane Cay and Isabella Wedderburn. The former lived in a substantial flat at 6 Great Stuart Street while the latter had a town-house of considerable grandeur at 31 Heriot Row (Colour Plates 3 and 4 respectively) which had been John Clerk Maxwell's own home before his marriage. There was space enough there for James to have his own room and he would want for nothing; he would have the company of his

Figure 2.3 (opposite) Nether Corsock, Blackhills and Little Mochrum together with the later addition of Glenlair
See Figure 2.1 for the general location. Nether Corsock, Over and Nether Blackhills, and Little Mochrum were all part of the estate under the Middlebie entail. Glenlair, together with Upper and Nether Glenlair, were brought into the estate later, in 1839, when John Clerk Maxwell purchased them to make a single estate centred on the mansion house that he built in 1831 as 'Nether Corsock'. Only from 1839 onwards was this house and estate referred to as Glenlair, whereafter the name Nether Corsock became attached to the steading a little to its north. (Synthesised from DGA: GGD56/1, *Plan of Nether Corsock by John Gillon*, 1806, and DGA: GGD56/26, *Undated Pencil Plan of Glenlair*, nineteenth century; GGD56/13, *Glenlair Plan for Entailing*, c. 1839, and other maps of the area)

cousin Jemima and a younger cousin Colin Mackenzie. He would have school-friends in the nearby houses and he could visit his aunt Jane whenever he liked; his father would come up on his usual visits to Edinburgh. And so it was decided.

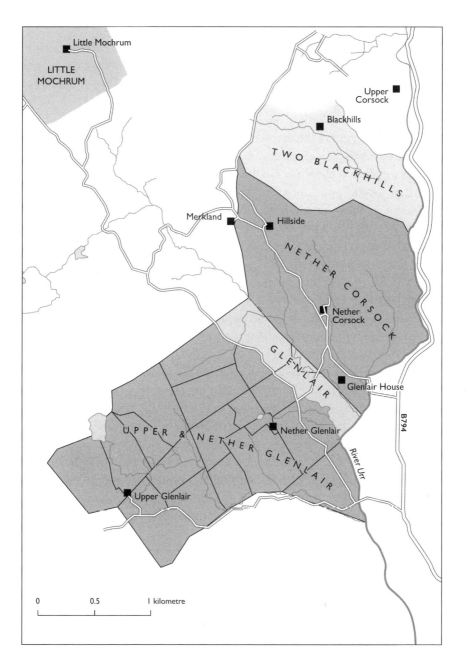

That October, Isabella and Jemima travelled down by carriage to Glenlair, as the recently expanded estate and house at Nether Corsock had now become known (Figure 2.3),[6] and there they stayed a few weeks before commencing the return journey, the top of the coach laden with the trunks carrying all that father and son would need during their sojourn in Edinburgh, and to equip James for school (Wedderburn, c. 1841).

James' first encounter with his new school-mates at Edinburgh Academy was not a happy one. Having joined late in the term, he was not one of the crowd, rather an intruder to be examined and found wanting. If his father and aunts had had the foresight to dress him like any other schoolboy, he might have got away with it, but his plain rustic garb, such as we see in Plate 3, was inappropriate in the new setting. His oddness of manner and Gallowegian brogue only served to confirm first impressions: here was a misfit. Children make few allowances in their judgements, and boys will be boys; the nickname 'Daftie'[7] was handed out to the new lad and the earliest opportunity was taken to show him that he did not fit in. James returned home that afternoon with his clothes bearing full testimony to the treatment he had received in the schoolyard. Nevertheless, according to Campbell he 'was not in the least inwardly perturbed by all this, nor bore any one the slightest malice' (C&G, p. 50). One can only wonder on whose side Campbell had been during James' rough initiation.

It took some time for James to adjust to his new school setting. Even if discipline was more humane and not always directed at himself, his experience under his tutor must have made it difficult for him to accept the dull repetition of learning by rote, while outside the classroom he 'seldom took part in any games [. . .] preferred wandering alone [. . .] sometimes doing queer gymnastics on the few trees that were left [in the schoolyard] [. . .] his heart was at Glenlair' (C&G, p. 51). His hesitancy became more pronounced. He was also frequently absent from school because of childhood illnesses, and when he did attend there were always classmates who would take the opportunity to discomfit him, with the inevitable upshot that his performance in class was not what it might have been. In Campbell's analogy, he was a 'cygnet amongst goslings' (C&G, p. 50). A portrait of James drawn about this time is shown in Plate 4.

Thanks to the extensive library at 'Old 31', as he now called his new home, James managed to engross himself in the works of Swift and Dryden, and later Hobbes and Butler,[8] some of which would have proved difficult reading even for an adult. When his father was in town, they would go for Saturday afternoon walks 'always learning something new, and winning ideas for imagination to feed upon' (C&G, p. 52). On one such occasion he was taken 'to see electro-magnetic machines', for the latest scientific and technical advances had long been of interest to not only his father but also his uncle, John Cay, Sheriff of Linlithgow. John Cay lived just doors away from James, at 11 Heriot Row, and he had been the good friend of John Clerk Maxwell since their days together at the Royal High School. They were the same age, had studied the law together and had qualified as advocates at more or less the same time.[9] Their common interests led to them becoming Fellows of the RSE in 1821; at that time the requirements for election depended on being a gentleman of good reputation with a genuine interest in science or the arts, for which they were both amply qualified. John Cay was to take part in similar excursions with his nephew, for example, to visit William Nicol,[10] who at a later date sent James the gift of a polarising prism.

When his father was not in Edinburgh to divert him, James wrote regularly. But he did not simply observe the customary pleasantries; he took it as an opportunity to let his mind take a creative turn for the benefit of his father's enjoyment. For example, by 'concocting the wildest absurdities, inventing a kind of cypher to communicate some airy nothing, illuminating his letters [...] and adding sketches of school life [...] drawing complicated patterns (C&G, p. 53). Campbell saw this as significant, 'marking the early and spontaneous development of 'the habit of constructing a mental representation of every problem', which was in some degree [...] hereditary.'

By and by, James managed to settle into his new situation and slowly began to progress. By his second year he had managed to get into the top twenty per cent of his class of some seventy boys.[11] His hesitancy, however, still seemed to be something of a problem in Latin and Arithmetic,[12] but by age twelve he could show himself to be top of the class in English and Scripture Biography. In addition, he had managed to prove himself in the schoolyard. A classmate later recounted for Campbell's book:

> On one occasion I remember he turned with tremendous vigour, with a kind of demonic force, on his tormentors. I think he was let alone after that and gradually won the respect even of the most thoughtless of his schoolfellows. (C&G, pp. 55–56)

It was shortly afterwards that James and Lewis Campbell became close friends. Campbell had been in the class since it began, under Mr Carmichael as master, with the intake of 1840. Though James had joined the following year, it was not until about 1843 that their friendship started, and it became all the closer when Campbell's mother remarried and moved into number 27 Heriot Row, practically next door to James:

> We always walked home together, and the talk was incessant, chiefly on Maxwell's side. Some new train of ideas would generally begin just when we reached my mother's door. He would stand there holding the door handle, half in, half out, while, 'Much like a press of people at a door, thronged his inventions', till voices from within complained of the cold draught. (C&G, p. 68)

James had already done some dabbling in geometry on his own, for shortly after his thirteenth birthday, he mentioned in a letter to his father, 'I have made a tetrahedron, a dodecahedron, and two other hedrons, whose names I don't know.' Amazingly, he had not then had any formal lessons in geometry. When work seriously began on the subject in year five (1844–45), however, it began to awaken his interest significantly:

> Our common ground in those days was simple geometry [. . .] but whatever outward rivalry there might be, his companions felt no doubt as to his vast superiority from the first. He seemed to be in the heart of the subject when they were only at the boundary. (C&G, p. 69)

James turned fourteen during the last weeks of that school year, by which time he had had nearly four full years at the Academy; in terms of his steady progress and growing friendships, he seemed completely settled there. When in due course he got the results of the customary end of term examinations, he was able to write proudly to his aunt Jane, who seems to have been out of town, 'I have got the 11th prize for Scholarship, the 1st for English, the prize for English verses, and the Mathematical Medal' (C&G, p. 71).[13]

If winning the Mathematical Medal gives more than ample evidence of progress and branching out, his first prize in English showed he had lost none of his abilities in that direction. The work he submitted was a poem, 'The Gude Schyr James Dowglas' written in the style of an ancient ballad with language drawn from Barbour's *Bruce*. He had researched his topic in several books including Scott's *Marmion*, which he was evidently so well acquainted with that he had to stop himself from falling completely into its style and language when he was writing his poem. Throughout his life, Maxwell would turn to poetry to express his thoughts, or simply for a piece of fun, and a number of his verses were recorded by his friend (C&G, pp. 577–651).[14]

Peter Guthrie Tait belonged to the class below Maxwell and Campbell but was of much the same age. He was already keen on mathematics and took note of Maxwell having won the school medal; in this way he was drawn into Campbell and Maxwell's friendship. All three had an awakening intellect that became the basis of a lasting kinship, while their drive to succeed created the most harmonious of rivalries. If Tait tended to excel in mathematics and Campbell in the classics, then Maxwell tended to fit squarely in between. The autumn term of 1845 would no doubt have seen Maxwell giving Tait a bit of coaching, enjoying expanding on what he himself had learned. From 1844 on, Tait kept a note-book[15] for the purpose of jotting down what they generally referred to as 'props', that is to say, propositions consisting of conjectures, definitions, constructions and proofs. James Clerk Maxwell's name appears under an entry in the notebook headed 'Propositions on the Conical Pendulum'. Although it is apparently signed 'J C Maxwell' and dated 25 May 1847, the writing on the page is Tait's, as is the date.[16] This is evident when it is compared with another entry 'On the Imaginary Roots of Negative Quantities by the Right Reverend Bishop Terrot'[17] signed and dated 27 May by 'P G Tait'. Another entry signed by Tait has, on the opposite side of the page, 'J G dedit' (gave it), while yet another has 'J G fecit' (did it); we may presume 'J G' to be James Gloag, his mathematics master. It seems likely, therefore, that while Tait would copy out things he found of interest, he was careful to attribute them. The entry on the conical pendulum turns out to have been the solution to a 'prop' that Tait had given to Maxwell (see later), which Tait then transcribed into his notebook and honestly attributed to Maxwell.

Later on Tait and Maxwell would produce their own separate manuscripts on some topic of current interest to add to the collection. Two contributions by Maxwell, 'Ovals' and 'Meloid and Apioid', are reproduced in C&G, pp. 91–104; some sheets of the original in Maxwell's own hand are now displayed at the museum at 14 India Street. Another, identified as being in Tait's handwriting, consists of a few pages of propositions on the ellipse.[18] There may have been an element of competition as well as co-operation in this, with one challenging the other to solve a problem. Tait challenged Maxwell to find the cross-section of a torus (a ring) taken in a plane tangent to its inside surface, but Maxwell found the answer. In turn he gave Tait a highly convoluted question about a heavy body rolling on a curve with the 'horizontal component of the force, by which it is actuated [...] to vary as the nth power of the perpendicular upon the axis' (C&G, p. 116). Little wonder that Tait demurred.

His awakening in mathematics and subsequent rapid success no doubt helped James to appreciate all the better the technical aspects of the latest ideas and inventions. His father must also have taken serious note for, when back in Edinburgh during the winter of 1845–46, he redoubled his interest in attending meetings of the RSE and the Royal Society of Arts,[19] this time taking James along with him. Campbell recalled:

> And so it happened that early in his fifteenth year the boy dipped his feet in the current of scientific inquiry [...] he had always something new to tell [...] in February 1846, he called my attention to the glacier-markings on the rocks, and discoursed volubly on this subject, which was then quite recent, and known to comparatively few. (C&G, p. 73)[20]

Through the Royal Society of Arts he came across the work of David Ramsay Hay, who at the time was listed as being a 'decorative painter to the Queen' (EPOD, 1846) and had published not only a volume on the theory of architectural designs but also a reference book of colours:

> Such ideas had a natural fascination for Clerk Maxwell, and he often discoursed on 'egg-and-dart', 'Greek pattern', 'ogive' [...] One of the problems in this department of applied science was how to draw a perfect oval; and [...] [at age fourteen he] became eager to find a true practical solution of this. (C&G, p. 74)

In mentioning a *perfect* oval here, Campbell means one that is truly egg-shaped rather than merely elliptical. James must have seen various forms of decorative ovals, as well as the gamut of decorative stucco embellishments that went with them, on many a ceiling in the fine houses of the New Town. His own school had a prominent oval cupola perched atop the building, and even St Andrew's Church, seen in Plate 5, where he sat in contemplation on many a Sunday morning, was designed in a grand oval. Edinburgh had taken this classical theme to its heart.

Sometime in late 1845, James must have started looking into this seriously. He may have got what he could on the subjects out of books and from the mathematics teacher, Mr Gloag, who seems to have been the sort of amenable master who would have encouraged such an enquiry. By the following February James had experimented with various ideas for generating his ovals geometrically. It is well known that one can draw an ellipse with the aid of a loop of thread placed over two pins stuck into a sheet of drawing paper. An ellipse will be traced out by placing the pencil-point within the loop and pulling it outwards as far away from the pins as it will reach. The shape of the ellipse is varied by moving the pins further apart or closer together, and by changing the length of the loop. Maxwell worked out an ingenious variation of this method, as shown in Appendix 1, which he wrote up in a manuscript of the sort that he and Tait were beginning to put together as part of their joint effort on props. To whose attention it first came is not clear, but it would hardly be surprising if it were not Mr Gloag. Perhaps he mentioned to John

Clerk Maxwell that it was no mere reinvention of something that was already known; or perhaps it was Mr Hay, who although he was by no means a mathematician, probably knew of every practical method for constructing ovals. In the event, John Clerk Maxwell's confidence in James' abilities was such that, in February 1846, he took it to James David Forbes,[21] Professor of Natural Philosophy at Edinburgh University (Grant, 1884, pp. 354–57), for an opinion.

Not only did John Clerk Maxwell know Forbes through the meetings of the RSE (he had probably given the talk on glaciers that James had swallowed up whole), Forbes was a relative[22] and so he could ask his honest opinion without fear of any embarrassment in the event of disappointment. It may well have been this way, for he did not take James' prop directly to the Professor of Mathematics, Philip Kelland, whom he would also have known through the RSE. Forbes indeed seemed to think there was something in it, and it was he who showed it to Kelland for an opinion, and Kelland duly concurred. Forbes wrote back within little over a week of first seeing James' actual paper:

> 3 Park Place, 11th March 1846.
>
> MY DEAR SIR – I am glad to find to-day, from Professor Kelland, that his opinion of your son's paper agrees with mine; namely, that it is most ingenious, most creditable to him [...] I think that the simplicity and elegance of the method would entitle it to be brought before the Royal Society. – Believe me, my dear Sir, yours truly,
>
> JAMES D. FORBES
> (C&G, p. 75)

On 6 April 1846, Professor Forbes duly presented the paper on James' behalf to a meeting of the RSE under the following entry: '1. On the Description of Oval Curves, and those having a plurality of Foci. By Mr. CLERK MAXWELL, junior, with Remarks by Professor FORBES. Communicated by Professor FORBES' (C&G, p. 76). And in due course it was published (Maxwell, 1851),[23] in the same journal that had featured such ground-breaking articles as James Hutton's 'Theory of the Earth' (1788).

John Clerk Maxwell must have been an exceedingly proud man, but there is no indication that it went to James' head in any material way; on the contrary, it would have had the beneficial effect of consolidating the initial boost to his self-confidence from winning the previous year's Mathematical Medal, and it also strengthened his mathematical affinity with Tait. What seems surprising, however, is that Campbell makes no mention whatsoever of any reaction from the Academy; perhaps they considered it a distraction for James, who was not to repeat his previous success when it came to that year's mathematical prize. All the while, he and Tait continued producing props for their little mathematical club. It is Tait who tells us what James contributed: '"The Conical Pendulum", "Descartes' Ovals", "Meloid and Apioid" and "Trifocal Curves". All are drawn up in strict geometrical form, and divided into consecutive propositions' (C&G, p. 86). Since these were written during the course of 1846 and early 1847, it may have been that his paper on oval curves started out as an early attempt to produce such a contribution. What a start!

By the time he had reached the age of fifteen, James had blossomed. He had demonstrated a talent for geometry, a subject that requires a different kind of thought from mere words and numbers; points, lines and curves have to be visualised, and relationships have to be explored by logic and construction rather than by mechanical grinding at facts and formulae. Latin, Greek and English verse might amuse him, but geometry enthralled him. He now made rapid progress in mathematics in general and was to take the final year prize. Given his maturity of mind, even at fifteen he must be regarded as being quite the young man, no longer a mere boy. Considering also his success at the RSE, he must have commanded a degree of respect even from his elders. His father's encouragement, which had been growing since his mathematical medal, must have been even further amplified. Nevertheless, during the late summer of 1846, on the occasion that Campbell made his first visit to Glenlair, we find the pair just being boys, relaxing and enjoying the countryside with few thoughts about geometry and schoolwork.

When they returned to school that autumn for their final year, James' health was not at its best and he missed many days. But this could not have kept him back from his scholarly interests in general and geometry in particular, for it must have been during this phase that he was pursuing extensions to his work on ovals that developed into contributions to his and Tait's mathematical props. Campbell relates that he was now also becoming interested in *physical* phenomena, rather than just abstract mathematical ideas:

> He was certainly more than ever interested in science. The two subjects which most engaged his attention were magnetism and the polarisation of light. He was fond of showing 'Newton's rings', the chromatic effect produced by pressing lenses together, and of watching the changing hues on soap bubbles [...] working with Iceland spar, and twisting his head about to see 'Haidinger's Brushes' [24] in the blue sky with his naked eye. (C&G, p. 84)

This was begun even before he visited William Nicol, who was now approaching eighty, in the spring of 1847. Nicol was a fellow of the RSE and would already have known that James had made something of a mark there in the previous year, and when they met he would have found that James was no one-day wonder. From what we know of James thus far, his knowledge and enthusiasm for the subject must have been obvious to Nicol. His subsequent gift of a polarising prism was perhaps the gesture of an elderly man, one who had made his own mark long time since, doing his little bit to help a rising genius find his way.

According to Campbell, the visit to Nicol

> added a new and important stimulus to his interest [. . .] the phenomena of complementary colours came first, then the composition of white light, then the mixture of colours (not of pigments), then polarisation and the dark lines in the spectrum, then colour-blindness, the yellow spot on the retina, etc. (C&G, p. 84n)

As if that were not enough, shortly after his visit to Nicol, his father took him to a cutler's shop to choose magnets[25] suitable for experimenting with, an activity that he was still pursuing during his autumn vacation at Glenlair. On that occasion, however, John Clerk Maxwell was taking steel to the local smiddy[26] to be made into magnets for his son. It appears that James was eager to get his magnets because a few days later he and Robert Campbell (brother of Lewis) haunted the smiddy until they got them.

Undergraduate at Edinburgh

John Clerk Maxwell had by that summer of 1847 arranged for James to start at Edinburgh University in the following November:

> In deciding not to continue his [son's] classical training, he appears to have been chiefly guided by some disparaging accounts of the condition of the Greek and Latin classes in comparison with those of Logic, Mathematics, and Natural Philosophy. (C&G, p. 90)

Perhaps this is what he gave out to people when they asked him what James was going to be doing by way of preparing for some profession, which was at that time held to be essential even for the sons of the landed gentry. Could he imagine James as a lawyer or a minister of religion? He had the intellect and the learning, but not the personality, and speechifying was not a strong point. A banker? What interest had he or James shown in making money for its own sake? A doctor, perhaps? That would bring demands that made it more of a vocation than a profession. Being a university professor would suit James in intellect, interest and status, and the long vacations were useful if you had a country estate to look after. Furthermore, would he not have recalled his own situation forty years earlier when he studied the law and became an advocate, only to find that, like his grandfather George Clerk Maxwell before him, he had neither the heart nor aptitude for it? In the event he had decided that James must follow the trajectory along which he was already hurtling.

At the age of sixteen, James duly entered what is now called Old College on Edinburgh's South Bridge (Colour Plate 5), enrolling in mathematics, natural philosophy and logic under, respectively, Professors Philip Kelland, James Forbes and Sir William Hamilton.[27] He was sufficiently advanced in his mathematics to be allowed to enter not the first class, but the second.

By the following summer vacation, James had found his niche and was by now hooked on science, as his letters to Campbell on two consecutive days, 5 and 6 July 1848 clearly demonstrate. We give an extract here to give some idea of the flood of things that were going on, all orchestrated in his mind and the whole put into experimental practice under his own inclination and efforts:

> I have regularly set up shop now above the wash-house at the gate, in a garret. I have an old door set on two barrels, and two chairs, of which one is safe, and a skylight above, which will slide up and down. On the door (or table), there is a lot of bowls, jugs, plates, jam pigs, etc., containing water, salt, soda, sulphuric acid, blue vitriol, plumbago ore; also broken glass, iron, and copper wire, copper and zinc plate, bees' wax, sealing wax, clay, rosin, charcoal, a lens, a Smee's Galvanic apparatus [...]
>
> With regard to electro-magnetism, you may tell Bob that I have not begun the machine he speaks of, being occupied with better plans, one of which is rather down cast, however, because the machine when tried went a bit and stuck [...]
>
> July 6. To-day I have set on to the coppering of the jam pig which I polished yesterday. I have stuck in the wires better than ever, and it is going on at a great rate [...]
>
> I bathe regularly every day when dry, and try aquatic experiments. I first made a survey of the pool, and took soundings and marked rocky places well [...] also tried experiments on sound under water, which is very distinct, and I can understand how fishes can be stunned by knocking a stone.

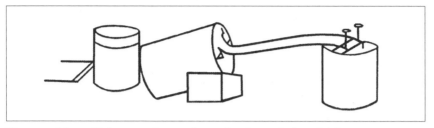

Figure 2.4 Maxwell's homemade experiment for electroplating a 'pig'
The pig here refers to a common type of glazed earthenware pot (C&G, p. 119).

We sometimes get a rope, which I take hold of at one end, and [...] when the water is up, there is sufficient current to keep me up like a kite [...]

I have made regular figures of 14, 26, 32, 38, 62, and 102 sides of cardboard.

Latest intelligence – Electric Telegraph [the problematic electromagnetic machine?]. This is going so as to make a compass spin very much. I must go to see my pig, as it is an hour and half since I left it; so, sir, am your afft. friend, JAMES CLERK MAXWELL. (C&G, pp. 118–120)

Two months later, while still at Glenlair, he revealed that on top of all this, not to mention his diversions with work in the fields, classical reading, time spent in the garden and walking the dogs, he has found time to work on even more 'props':

Then I do props, chiefly on rolling curves, on which subject I have got a great problem divided into Orders, Genera, Species, Varieties, etc.

One curve rolls on another, and with a particular point traces out a third curve on the plane of the first, then the problem is :– Order I. Given any two of these curves, to find the third [and so it goes on]. (C&G, p. 121)

His experimental work now consisted of making bits of unannealed glass of various shapes so that he could examine the pattern of internal stresses within the glass by viewing it with polarised light through a 'crossed' polarising prism, for example, his Nicol prism.[28]

That long summer vacation at Glenlair had been an amazing time of discovery and development. He returned to university in November and entered his second year of study. He kept his 'props' on rolling curves going on the side before eventually submitting the final draft to Professor Kelland for his opinion, for by this time they knew each other well enough, and James was confident enough in his own abilities, to do so. Kelland communicated the paper 'The Theory of Rolling Curves' (Maxwell, 1849)[29] to the RSE in February 1849, three years after James' first paper was read there. Having, it seems, sated himself with curves, he then chose a topic that involved both mathematics and physics, the theory of elastic equilibrium in solids, a subject which perhaps occurred to him while peering at bits of unannealed glass and moulded jelly through crossed polarisers. He started writing this up in March 1850 (C&G, p. 127) and gave it to Professor Forbes, who then put it before the RSE for consideration. The mathematical content was such that they in turn put it out to Professor Kelland for an opinion, and by May James heard back from Forbes that Kelland had read the paper thoroughly and recommended it for publication. But he also warned James that Kelland had complained about 'the great obscurity of several parts, owing to the *abrupt transitions* and want of distinction between what is *assumed* and what is *proved*' (C&G, p. 138). He stressed in no uncertain terms that James had to clarify many such parts before the work could go to print. Nevertheless, he had no doubt that James would be able to do so, for the revised manuscript was to be in Professor Kelland's hands in less than a week! The paper was read shortly thereafter, but it did not appear in print for another three years (Maxwell, 1853).[30] The paper is indeed a tour de force of mathematical physics for someone not quite nineteen years of age. While one can understand his rapid progress with basic experimental physics through learning, enthusiasm and ingenuity, this reveals something deeper, the penetrating power of his analytical mind, visualising the purely abstract stresses and strains and rendering them into three-dimensional differential equations. Nevertheless, it also shows another side that is hardly surprising in one so gifted; in the written word he occasionally found it difficult to express himself sufficiently clearly for even the most capable of readers to be able to follow him.[31] His *Treatise on Electricity and Magnetism* (1873a) left Heinrich Hertz so flummoxed that he gave up on it and worked out Maxwell's electromagnetic theory from the beginning. In his book *Electric Waves* (1893) Hertz commented:

> Many a man has thrown himself with zeal into the study of Maxwell's work, and [...] has nevertheless been compelled to abandon hope of forming for himself a consistent conception of Maxwell's ideas. I have fared no better myself. (p. 20)

> Again, the incompleteness of form referred to renders it more difficult to apply Maxwell's theory to special cases. (p. 196)

On entering his third year at Edinburgh University, Maxwell had completed all the mathematics, natural philosophy and logic that the syllabus offered. He therefore decided to take up moral philosophy under Professor John Wilson (aka Christopher North), chemistry under Professor Gregory, who took the theoretical part, and Professor Kemp, who took the practical work. In addition, he studied German and hung around Professor Forbes' natural philosophy class where he did not formally attend lectures but was afforded the liberty of using the experimental apparatus for his own investigations. These classes may have proved useful, but other than working on his own prop about elastic solids, he was to some extent marking time. There would have been little point in doing a further year just so that he might receive his degree, and so a decision about his immediate future had become pressing by the summer of 1850.

Cambridge

There was talk once more about James becoming an advocate. Family and friends were also asking John Clerk Maxwell about his intentions for James and offering their opinions, but it appears that it was well into the autumn before he was at last listening to their advice. In James' own words:

> He had wished me to be an advocate; but I never attended law classes, as by that time it had already become apparent that my tastes lay in another direction. Moreover, he looked up greatly to James Forbes, and desired that I should be like him. (C&G, pp. 420–421)

> the existence of exclusively scientific men, and in particular of Professor Forbes, convinced my father and myself that a profession was not necessary to a useful life. (Hilts, 1975, p. 59)

He therefore finally accepted that James should go to Cambridge to undergo the rigours of the degree course there, in other words, to set his sights high and aim for the top. The system was that able students would be chosen and coached for the most exacting of final examinations, the mathematical 'tripos'. To get that far required a tremendous amount of effort on the part of both student and coach.

Tait, having got enough from just one year at Edinburgh University, was by then already at Peterhouse College embarking on the course. James D. Forbes voiced a strong preference for Trinity (C&G, pp. 132, 146), but John

Clerk Maxwell, having received many more recommendations for Peterhouse, at length made the decision to send his only son there, some three hundred miles away from home. Now, this would be nothing remarkable today, and most fathers would not hesitate to send their only son to Cambridge if they could afford it. But at that time sickness and death were ever stalking close by, and so it must have been hard for him, now at the age of sixty, to accept that he would not only see much less of James, but might be seeing him for the last time. The mitigating factors were, however, that travel by train was now well established so that Cambridge could be reached from Glenlair within about twenty-four hours; letters travelled just as quickly; and there was even the telegraph for use in an emergency. Not only was Tait already there, a friend and relative called Charles Mackenzie[32] was a lecturer in Caius College.

In October of 1850, James Clerk Maxwell duly enrolled at Cambridge, in St Peter's House of Peterhouse College. Robert Campbell, brother of Lewis, had also enrolled, but at Trinity College. Having achieved so much at Edinburgh, James' first year amongst the freshmen offered him barely enough to hold his attention, and there were the usual distractions of student life. While he did not allow the casual behaviour and occasional merriment of his fellow students to blow him off course, he decided at the end of first year to transfer to the college that Professor Forbes had recommended for him, Trinity.

James, having impressed his tutor William Hopkins, was duly invited to take on the mathematical tripos with Hopkins himself as coach. Hopkins had coached the likes of George Stokes, Arthur Cayley and Phillip Kelland, all of whom became senior wranglers,[33] and also the future Lord Kelvin, who by comparison merely managed to make second wrangler. All four were also Smith's prizemen, a major accolade for Hopkins. Peter Guthrie Tait was already under Hopkins' wing and acquitted himself admirably by becoming senior wrangler and Smith's prizeman at the end of 1852. The coaching primarily consisted of 'mugging up' on typical examination questions from past papers and the like; obscure strategies and techniques that went beyond the bounds of normal practical methods had to be learned, and the work was both difficult and laborious. James, however, seems to have bent only partially to Hopkins' coaching methods, for according to Campbell:

> the pupil to a great extent took his own way, and it may safely be said that no high wrangler of recent years ever entered the Senate House more imperfectly trained [...] But by sheer strength of intellect [...] he obtained the position of Second Wrangler, and [...] equal with the

Senior Wrangler in the higher ordeal of the Smith's Prizes. (C&G, p. 134)

The examination was in January 1854, just a few months after James had suffered a long bout of illness described as a sort of brain fever (C&G, p. 170). It was put down to the strain of his studies, but it could simply have been a virus. Luckily it came in a break between terms but it left him so weakened that he had to be 'careful not to read inordinately hard' in the run up to the final examinations (C&G, p. 171). Nevertheless, if he had been seriously interested in becoming senior wrangler and had focused on doing so by following Hopkins' coaching more assiduously, he may well have achieved it. But his reluctance to buckle down to it seems to recall the brief spell of rebellion against his tutor in his boyhood. James would have accepted what he did achieve as success enough, as would his father, as would everyone else including his old professors, Forbes and Kelland. Tait and the Campbell brothers would have been equally pleased for him, and the difference between senior and second wrangler would have been of no account between Tait and himself.

The important thing was what Maxwell was underneath it all, with his personal traits of self-direction and deeply philosophical enquiry; his pursuance of his own ideas and 'props'; and his insight into mathematical and physical problems. He could not, and would not, have learning drummed into him at the hands of any coach. He had the qualities of genius; and were not his perceived faults, such as Forbes' complaint about his 'obscurity' and 'abrupt transitions', evidence of the same inasmuch as they were simply great leaps of insight? Were his spells of distraction not just episodes of absorption in some 'prop' or other, as his aunt Jane described it (C&G, p. 105)? Even his faltering speech could be put down to a simple failure to find the words adequate to express the ideas in his mind that were not formed from words, but from abstractions. When put on the spot to say his lesson at the Academy it may have simply been a case of nerves, but the problem stayed with him for a long time thereafter. On the other hand, when he was a student we find people saying he could speak eloquently and profoundly on any subject, and yet we find others who described him as saying little, but to the point. Surely, it very much depended on the circumstances, whether he was comfortable with the situation or whether he felt flustered, as the following example in a letter to Lewis Campbell in February 1857, clearly shows:

> Up to the present time I have not even been tempted to mystify anyone [by what I say]. I am glad B—[34] is not here; he would have ruined me.

I once met him. I was as much astonished as he was at the chaotic statements I began to make. But as far as I can learn I have not been misunderstood in anything, and no one has heard a single oracle[35] from my lips. (C&G, pp. 265–266)

Another example was when he stood up at a meeting of the British Association for the Advancement of Science (known as the 'BA'), held in Edinburgh in 1850, to challenge a point made by Sir David Brewster, who had a fair reputation for cruelly demolishing any theory that did not agree with his own. Professor William Swan[36] later recalled the scene:

His utterance then, most likely, would be somewhat spasmodic in character, as it continued to be in later times, his words coming in sudden gushes with notable pauses between; and I can well remember the half-puzzled, half-anxious, and perhaps somewhat incredulous air, with which the president and officers of the section, along with the more conspicuous members who had chosen 'the chief seats' facing the general audience, at first gazed on the raw looking young man who, in broken accents, was addressing them [. . .] But, at all events, he manfully stuck to his text; nor did he sit down before he had gained the respectful attention of his hearers, and had succeeded, as it seemed, in saying all he meant to say. (C&G, p. 489)

These recollections of Campbell say much about James' character, and how he managed to overcome the difficulty he found in speaking under pressure. They may give some clue as to why in later life he was always cautious and measured in what he said, and rarely if ever drew the limelight upon himself.

As soon as his final examinations were over, James was free to revert to his old way of doing things, whereupon he immersed himself in projects that had long interested him and were no doubt constantly brewing in the back of his mind. His mathematical paper 'On the Transformation of Surfaces by Bending' (Maxwell, 1856b)[37] may well have followed directly from his earlier analysis of elastic bodies, helped by the lectures of Professor Stokes. It was presented to the Cambridge Philosophical Society just two months later, in March 1854. He also began investigating how to reproduce mixtures of coloured light in a meaningful quantitative way, picking up on the ideas he had taken from Forbes' demonstrations with a spinning top, which he had doubtless repeated himself.

Perhaps out of curiosity, he made himself a rudimentary ophthalmoscope:[38]

To Miss CAY.
Trin. Coll., Whitsun. Eve, 1854.

> I have made an instrument for seeing into the eye through the pupil [...] and I can see a large part of the back of the eye quite distinctly with the image of the candle on it [...] Dogs' eyes are very beautiful behind, a copper-coloured ground, with glorious bright patches and networks of blue, yellow, and green, with blood-vessels great and small. (C&G, p. 208)

About the same time, however, he discovered something which made an interesting and challenging examination problem (Maxwell, 1854).[39] The idea, suggested to him by the structure of a fish's eye, was a spherical lens[40] which was more refractive (bent light more) at the centre than at the surface. Maxwell showed that if the variation in refractive power took a certain form, then a point of light on the surface of the lens would produce a perfect image of itself at a point diametrically opposite. That is to say, the lens could focus an object at point-blank range. At the time his idea was more abstract than practical,[41] but it gives a clear example of how he was able to think 'outside the box'. This, however, was only part of what he was thinking about at the time, for he went on to carry out a general mathematical analysis of optical instruments to see what was needed to make a perfect image (Maxwell, 1858).[42]

Maxwell, the 'Natural' Natural Philosopher

Little more needs to be said about the origin and route by which James Clerk Maxwell came to be a physicist, and a great one at that. But we must not forget that he had other sides to his character and intellect. He not only thought about things, he thought about them very deeply, at least until he could get 'the *particular* go of it' in his own mind's eye. His lectures in Professor Hamilton's logic and metaphysics classes had stimulated this tendency, for Hamilton was a compelling lecturer on these subjects and James was very much enthralled by them, taking more notes here than in his other classes. At the end of his first year, or perhaps a little later, as an exercise he set himself, he wrote an essay, 'On the Properties of Matter'. The manner in which this essay came to light shows how much it had impressed Hamilton, for it was discovered some years later when he was too ill to continue his classes. His assistant, Professor Baynes, found it tucked away in his private desk drawer.

In order to do it justice, Campbell gives the essay in full (C&G, pp. 109–113). On the age-old problem concerning the nature of a vacuum, Maxwell argued the following:

> If we say it is an accident, those who deny a vacuum challenge us to define it, and say that length, breadth and thickness belong exclusively to matter.
>
> This is not true, for they belong also to geometric figures, which are forms of thought and not of matter; therefore the atomists maintain that empty space is an accident, and has not only a possible but a real existence, and that there is more space empty than full. (C&G, p. 110)

This gives some idea of him mentally reaching out as far as he could possibly go to address, if not actually answer, the basic questions that are central to the understanding of physics, that is to say, natural philosophy. On his appointment to the chair of natural philosophy at Marischal College, Maxwell chose to make this the theme of his inaugural lecture, given on 3 November 1856.[43]

A Scottish Professorship

During the winter of 1854, John Clerk Maxwell took ill while in Edinburgh[44] and James returned to be by his side at 18 India Street,[45] which was now where John's widowed sister Isabella had set up residence; 'Old 31' had become too big for her once her children had grown up and, apart from her son George, moved on. Isabella did not keep the best of health herself, but she would have made sure that her brother had plenty of care and attention; even so James stayed four months and looked after him personally (Riddell, 1930, pp. 28–29). For much of the time he seems to have had little opportunity for anything else, but as soon as his father was on the mend he got back to work. It was during this period that he wrote his first short piece on colour vision, 'Experiments on Colour Vision as Perceived by the Eye', dated 4 January 1855.[46] For this he used a simple spinning top and three discs of coloured paper, each slit along a radius so that they could be interleaved to show any proportions of the three colours he wished, before they were spun by means of the top. In fact, he was photographed about this time with just such a top (Plate 6). Such was the popular interest in this that J. Bryson, the local optician, took to selling the spinning top and coloured papers for the amusement, or scientific interest, of his clientele. Maxwell's conclusion regarding colour vision is accurate, but to the man in the street it sounds like something only a mathematician could come up with: 'the difference between Colour-Blind and ordinary vision is, that colour to the former is a function of two independent variables, but to an ordinary eye, of three' (WDN1, pp. 124).

In this paper, he also introduced the now well-known colour triangle diagram showing how to compose any colour from three primaries. This exploited the fact that changing the intensity of a given colour of light does not affect the colour itself, so that one of the three variables involved can be eliminated by choosing it to be the brightness, making it possible to draw a colour composition diagram in two dimensions rather than three.

'Experiments on Colour, as Perceived by the Eye, with Remarks on Colour-Blindness' (Maxwell, 1855)[47] was subsequently presented to the RSE just two months later by his old chemistry professor, Dr Gregory. In this paper he described several ways of synthesising colour (he had clearly been busy reading up all the information he could on the subject) including his 'colour box',[48] his first attempt at which had been in August 1852, doubtless at Glenlair. At this early stage of his academic career Maxwell was making significant headway simply by studying what interested him; he seems to have had a sixth sense for anything that might be significant. It will come as no surprise, therefore, that another gem or two may be found in that article. First we point out the pearl:

> This result of mixing blue and yellow [light to make a 'pinkish tint'] was, I believe, not previously known. It directly contradicted the received theory of colours, and seemed to be at variance with the fact, that the same blue and yellow paint, when ground together, do make green. (WDN1, p. 146)

Maxwell was therefore one of the first to point out that the mixing of coloured light was not the same as the mixing of pigments (Colour Plate 6), because the former is an additive process and the latter subtractive one.[49] But even before he got that far, he had revealed what we must now recognise as a diamond:

> This theory of colour may be illustrated by a supposed case taken from the art of photography [. . .] Let a plate of red glass be placed before the camera, and an impression taken. The positive of this will be transparent wherever the red light has been abundant in the landscape, and opaque where it has been wanting. Let it now be put in a magic lantern,

along with the red glass, and a red picture will be thrown on the screen. Let this operation be repeated with a green and a violet glass, and, by means of three magic lanterns, let the three images be superimposed on the screen. The colour of any point on the screen will then depend on that of the corresponding point of the landscape; and, by properly adjusting the intensities of the lights, &c., a complete copy of the landscape, as far as visible colour is concerned, will be thrown on the screen. (WDN1, p. 136)

He had simply, off the top of his head, invented colour photography, and never thought to patent it! The first public demonstration took place at a lecture 'On the Theory of Three Primary Colours' (Maxwell, 1861)[50] he gave at the Royal Institution in London on 17 May 1861. The subject he used was a tartan ribbon tied into a rosette (Colour Plate 7(a)). In truth, Maxwell's focus at the time was on developing an accurate theory of colour vision, but the other valuable outcome was his contribution to the understanding of colour blindness.

In October 1855, Maxwell was accepted as a fellow of Trinity College, that is to say he would teach and give lectures, a position which could have been the basis of a lifelong academic career had he wished to pursue it. As an undergraduate he had followed Faraday's experimental researches on electricity and magnetism, and had also taken note of W. Thomson's assertion that there was a close analogy between electrostatics and thermostatics; in conjunction with his mathematical work on elastic deformations, this led him to his first essay in the field of electromagnetics, 'On Faraday's Lines of Force' (Maxwell, 1864a),[51] a considerable body of work read to the Cambridge Philosophical Society in four parts, beginning December 1855. The following February, James Forbes nominated him for fellowship of the RSE despite the fact that he was still only twenty-four years of age.

By this time James was very well settled in Cambridge and had developed a small coterie of close friends; college life being what it is, he felt the loss keenly when any of them had to leave for the sake of following their chosen careers and moving on in life. The only disadvantage was that the academic year was longer than the Scottish one, which had given him the best part of six months in the summer to spend at Glenlair. Nevertheless, he would have been well aware that his father's health had been failing since his long illness in Edinburgh over the previous winter. As well as being nearly sixty-five years old and somewhat overweight, as we have seen, John Clerk Maxwell was not a well man. There may have been some apprehension between him and James that he was nearing his end. No doubt wishing to leave James some reminder of himself, he finally had his portrait painted (from which Stoddart engraved the likeness of him that Campbell used in his book [Plate 1]). Moreover, he wished James to be closer to home. As James himself put it: 'He much wished me to have a Scottish Professorship, that I might have the long vacation free for living at home' (C&G, p. 421).

It perhaps came naturally to a dutiful son like James, who had become so close to his father since the death of his mother, and likewise his father to him, that when the chair of natural philosophy at Marischal College in Aberdeen was advertised, he put his name forward. His father did his bit to solicit suitable references, but sensing that he should once again be by the old man's side, James came home to Glenlair just days before his father passed away, on 3 April 1856. Sadly, John Clerk Maxwell did not see his fondest wish come true. On his father's death, James became the new laird and a man of independent means. He already had the Maxwell name and so the only change under the old entail was that he was now 'Clerk Maxwell of Middlebie'.[52]

James was duly awarded the post just a few weeks after his father's death. In his inaugural lecture, given to his first class in the presence of the assembled dignitaries of the College, Maxwell discussed at length his own concepts of natural philosophy and physics, but it must be said that he did not draw a distinct line between them.[53] At almost the same time, following nomination by James Forbes, he was elected as a fellow of the RSE; it was another plaudit to his son that his father did not live to see. Maxwell took up his chair at Marischal College that November, already in the knowledge that discussions had been ongoing regarding its merger with its neighbour and competitor, the King's College. Had it not been for the rancour between those who had been appointed to negotiate the merger, it could have already been under way. Even although his duties were for the most part fairly routine, as indeed they had been at Cambridge, he did take his teaching seriously. Described as having his fair share of 'misfortunes of the blackboard' (Lamb, 1931), his insights and asides appealed to his more gifted students. While a good deal of teaching and preparation was involved, he still had time for his own researches, continuing with colour vision and electromagnetics and now also embarking on the molecular theory of gases, the theory of optical instruments, and the dynamical top.

As if this were not enough, he had already taken up the challenge of the 1857 Adams Prize Essay, to find an explanation for the rings of Saturn: were they solid, liquid, rubble, or something else entirely? After a good deal of effort on Maxwell's part, his essay 'On the Stability of the Motion of Saturn's Rings' (Maxwell, 1859)[54] was read on 19 April 1858. It took the prize by showing that the rubble theory was the only one that was both plausible and stable. Other competitors for the prize seemed to find the problem simply too much for them, and the Astronomer Royal, Sir George Airy, regarded Maxwell's achievement as one of the most remarkable results of mathematical physics that he had ever seen (WDN1, p. xv).

If the theory of Saturn's rings was Maxwell's first truly great work, his second was only two years in the coming. 'On the Dynamical Theory of Gases' (Maxwell, 1860c) was read at the Meeting of the BA[55] at Aberdeen on 21 September 1859 and subsequently published in full as 'Illustrations of the Dynamical Theory of Gases' (Maxwell, 1860a).[56] Maxwell subscribed to atomic theory, which he later made the topic of a poem poking fun at the president's address to the BA of 1874, of which the following gives the tenor:

How freely he [God] scatters his atoms before the beginning of years;
How he clothes them with force as a garment, those small incompressible
  spheres!
[. . .] Like spherical small British Asses in infinitesimal state [. . .]
First, then, let us honour the atom, so lively, so wise, and so small;
The atomists next let us praise, Epicurus, Lucretius, and all.

While it was hard to conceive of tiny atoms making up tangible materials like solids, the idea that gases consisted of a cloud of atoms had been mooted by Daniel Bernoulli as early as 1738 in his book *Hydrodynamica*.[57] We can call them atoms or molecules, it does not matter which, but the underlying theory is usually referred to as the *molecular* theory of gases. The idea is that in a container filled with a gas, the tiny molecular constituents fly around with considerable velocity. They frequently collide with the walls of the container and the cumulative effect of these collisions acts as a pressure, somewhat like in a hailstorm when a barrage of little icy 'ball-bearings' loudly drums on a roof. While the average velocity of the molecules depends on the temperature of the gas, Maxwell was the first to show that the molecules do not all have the same velocity, rather they have a statistical distribution of velocities, which he duly calculated (WDN1, pp. 380–381). This distribution now bears his name, along with that of Ludwig Boltzmann, who followed up on his groundbreaking work.

There were other important revelations in Maxwell's 1860 paper that helped not only to develop molecular theory but to give it sufficient prominence to get others interested; in addition to Boltzmann, Lord Rayleigh was also impressed by it (Jones, 1973, pp. 65–66) and Sir James Jeans referred to much of Maxwell's original work in his celebrated 'The Dynamical Theory of Gases' (Jeans, 1904), which even the title recalls. Maxwell's other publications during his time at Aberdeen are perhaps overshadowed by these masterpieces on the molecular theory of gases, which as we shall see, could hardly have been appreciated by either his fellow academics or the City Council.

It was in association with the Aberdeen instrument makers Smith and Ramage that Maxwell produced the completed version of his colour mixing box and a fully adjustable version of the dynamical top, a copy of which was provided to his old teacher and mentor back at Edinburgh, James Forbes (Maxwell, 1860b). He also had them make a mechanical model of the behaviour of the rings of Saturn, which he had conceived in December of 1857 (C&G, p. 295) and demonstrated to some acclaim at a lecture given at the RSE (Maxwell, 1857–62) in which he summarised the conclusions of his prize essay.

• • •

The principal at Marischal was the Rev. Daniel Dewar, with whom Maxwell forged a friendship that brought him in contact with Dewar's daughter, Katherine, who, although seven years his senior, was not yet spoken for. Maxwell was invited to stay with the Dewars when they were on their annual vacation near Dunoon in the September of 1857, which must have afforded ample opportunity for Katherine and him to get acquainted. Perhaps Daniel Dewar and his wife had hoped as much in extending the invitation. The following February, James and Katherine were engaged and the wedding took place at Aberdeen in June 1858, whereafter the couple lived with the Dewars at 18 Victoria Street (Jones, 1973, p. 64).

Maxwell seems to have been very happy in his choice of wife, but it was not well received by his family back in Edinburgh. This may well be due to the fact that the senior family members had not been consulted beforehand, or he was going against their advice. John Clerk Maxwell had perhaps thought that the task of seeking a wife for his son could wait until after he had his professorship, but of course it was now too late. Maxwell's aunts Isabella and Jane would have naturally spent some time thinking about such things, and would even have pressed John to do the same, but in the event they were taken by surprise when James simply announced his engagement out of the blue:

18th February 1858

DEAR AUNT [Jane,] This comes to tell you that I am going to have a wife [...] So there is the state of the case. I settled the matter with her [mother], and the rest of them are all conformable [. . .] I hope someday to make you better acquainted [...] For the present you must just take what I say on trust. So good-bye. Your affectionate nephew. (C&G, p. 303)

By what he says elsewhere in the letter, aunt Isabella, his uncle John Cay and Sir George Clerk, the most senior figure on his father's side of the family, effectively received carbon copies of the same. They may therefore have been just as surprised, and perhaps aunt Isabella and Sir George would have been offended that he had not seen fit to consult them, for James should have been thinking about the future of Middlebie, and potentially of Penicuik. James' choice lacked any alliance with a comparable family. Genteel as the Dewars may have been, Daniel Dewar was not landed gentry;[58] he had no estate of his own and had to go to his son-in-law's when it came to summer vacation! For Sir George, the news would have been unwelcome indeed, and he may even have gone so far as to put a shot over the young man's bows.

The issue was that, theoretically at least, James, or any children that he might eventually have, could at some time end up being successors to, or having an interest in, the Penicuik estate; Sir George did by then have grown-up sons and grandsons of his own, but it may be that he was just as upset because James hadn't even considered the wider possibilities. On top of that, there were other legal matters that could be affected, for example, debts owed by John Clerk Maxwell's estate to Sir George. At any rate, on 3 March Sir George served James with a 'summons declarator',[59] a fairly prompt riposte considering James had written to announce his engagement barely a fortnight before. The pretext was that improvements had been made over the years to the Penicuik estate, for which John Clerk Maxwell had had some sort of legal or financial liability. It was all spelled out in detail in the summons, amounting to a total of £6,777 11s 1½d,[60] an enormous sum at the time, probably thirty times a typical professorial salary.

This information has only recently come to light and what happened in consequence is not yet known. Both from the timing, and the fact that it ever took place, seems to reveal the gut reaction of a man of the old school, who simply saw James' behaviour as unacceptable. When James was a boy he had often visited Penicuik House, especially in the Christmas holidays, but whereas there are frequent mentions of various uncles and cousins within his mother's and aunt Isabella's branches of his family, the only mention of Sir George in Campbell and Garnett's biography occurs when James wrote to inform him of his father's death (C&G, p. 254).

There is also scant reference to Katherine Dewar in Campbell and

Garnett, which Jemima Wedderburn put down to Campbell's hands being tied because it was Katherine who had requested him to write it. From their childhood together at 'Old 31' and happy days 'tubbing' at Glenlair, Jemima knew James so well that she could have been his sister; of his marriage, she was to say bluntly in her own memoir:

> This did not give much satisfaction to his friends and relations. The lady was neither pretty, nor healthy, nor agreeable, but much enamoured of him [. . .] and he being of very tender disposition married her out of gratitude. Her mind afterwards became unsettled but he was always most kind to her, and put up with it all. She alienated him from his friends and was of a suspicious and jealous nature [. . .] [after his death, she] published a life of him written by the Revd. Lewis Campbell [. . .] Of course, under these circumstances the history of his married life could not be entered into. (Fairley, 1988, p. 107)

Nevertheless, Campbell did manage to slip into the concluding remarks of his biography: 'Mrs. Wedderburn [aunt Isabella], who had had the care of him during so much of his early life, said on the occasion of his marriage, "James has lived hitherto at the gate of heaven"' (C&G, p. 426). Where she supposed that he may have lived thereafter is left for the reader to imagine.

In consequence of his marriage, Maxwell was no longer able to retain his fellowship at Trinity but, being sufficiently well regarded, he was thereafter made an honorary fellow. In the two years that followed his marriage, the process of college amalgamation at Aberdeen progressed at last under the hand of a royal commission. Since for every two professors of a given subject there would now be only one, the general decision of the commissioners was that the elder man would retire on a pension. This would give the younger man a chance to have a career, give the older man a chance to live out his days in comfort and, by dying the sooner, save on the total expense. There was only one exception; in Maxwell's case, it was the younger man was given the boot (Jones, 1973, p. 66). This has created much speculation that it was the result of him being an inferior lecturer. But according to Niven: 'On the contrary, if we may judge from the number of voluntary students attending his classes in his last College session, he would seem to have been as popular as a professor as he was personally estimable' (WDN1, p. xv).

Another possible reason given by Jones is that Mrs Maxwell interfered because she wanted her husband to be quit of Aberdeen, but this only came as hearsay, originally through Peter Guthrie Tait's wife. It may simply be that, given the affront Katherine must have felt at her husband losing his job to the senior man, she would have preferred people to think that it was they who wanted to get away from the place. But if Maxwell was not a bad teacher, why did the commissioners make the anomalous decision to sack him rather than the older King's College man, Professor David Thomson?

First, the pension was not a consideration for they actually paid him an annuity equal to his average income (Flood et al., 2014, p. 42). Second, if Maxwell had been interviewed at any stage by the commissioners, his quietness and old 'hesitancy' may have resulted in a less than favourable impression. On the other hand, Thomson was robust, well known, influential in the pro-amalgamation campaign and somewhat devious, 'earning himself the soubriquet *crafty* Thomson' (Reid, 2012).[61] Finally, and perhaps equally relevant, could have been the matter of Maxwell's entirely contrasting character, gentle, self-effacing and perhaps too honest for his own good; it is conceivable that he could have offered to go to save the older man! After all, he had independent means, whereas Thomson, who was seventeen years his senior and had a family, may have faced hardship if he lost the post. Indeed, when he had originally applied for the chair at Marischal at Aberdeen, Maxwell had received a request for a reference from Dr William Swan, the attendee at the BA meeting of 1850 who witnessed Maxwell struggling to put a question to David Brewster, who was also a candidate. Rather than demurring, Maxwell granted his request (C&G, p. 252). Interestingly, that Swan asked the comparatively junior Maxwell for a 'good opinion' would seem to indicate that Maxwell had already established something of a reputation for himself.

Since Maxwell knew Professor James Forbes very well, he may also have been aware that Forbes was set to become principal of the United College of St Andrews and the post at Edinburgh would soon be vacant. Edinburgh was a post that would have suited him very well, for the city was a real centre of learning and of learned men, and he would be back amongst his family and old friends. Katherine and he would be able to live there quite comfortably through the winter at 14 India Street, which he now owned, for it would be only a matter of giving notice to the current tenant. With the progress of railways, Glenlair was now only a few hours distant.[62] But it was not to be, for his good friend and rival Peter Guthrie Tait, who had been in Ireland, also took his chance at the chair, and it was he who got it. The *Edinburgh Courant* commented on the chair being awarded to Tait as follows:

> there is another quality which is desirable in a Professor in a University like ours and that is the power of oral exposition proceeding on the supposition of imperfect knowledge or even total ignorance on the part of pupils. (O'Connor & Robertson, 2003)

The implication is that Maxwell was not able to lecture to his students at a sufficiently simple level. His reputation for lofty research seems to have gone against him.

King's College, London

After failing to get the post at Edinburgh, Maxwell subsequently applied for the chair of natural philosophy and astronomy at King's College in London. There were pros and cons, of course. On the one hand he could be nearer other great men of science, especially Faraday, and the possibility of a social life in London may have appealed to Katherine. On the other hand, the academic year was somewhat longer, and it was 300 miles from Glenlair; even so, the journey from London to Glenlair by train could now be done in a day. The chair became his and the couple moved to 8 Palace Gardens Terrace, Kensington, which afforded Maxwell a sizable garret in which to carry out his own experiments. His current version of the colour mixing box was now eight feet long, and so he would have needed ample room! The college was four miles away, but it could be reached by a pretty route, by foot or horse, through parks and along the riverbank.

Things could have taken an altogether different turn, however, for just before the move Maxwell came down with smallpox, which he attributed to visiting the Rood Fair at Dumfries to buy Katherine a horse (C&G, p. 319). The Rood Fair was traditionally held on the last Wednesday of September, and it was soon thereafter that Maxwell came down with this frequently deadly disease that claimed one life in three. He afterwards attributed his survival to Katherine's constant care and devotion, for it was she who nursed him exclusively for fear that any of the servants might catch the disease.

Maxwell gave his inaugural lecture at King's in October 1860 (Maxwell, 1979; Harman, 1990, p. 183). Once settled there, he was asked to join the BA committee for standardising the unit of electrical resistance, designated as the Ohm. While this would seem low-level stuff now, it was then quite fundamental. The experiments took place during 1862 and 1863 at King's College under the general direction of William Thomson, who had become a close friend of Maxwell's. Thomson was knighted for his efforts in 1866 and later became Lord Kelvin, but at this juncture he was simply Professor Thomson, the incumbent of the chair of natural philosophy at Glasgow University since 1846. Like Maxwell, he had been coached by Hopkins at Cambridge, becoming a senior wrangler. There had been connections between the two since about 1850, for in 1849 Maxwell's cousin Jemima had married Hugh

Blackburn, Professor of Mathematics at Glasgow, a lifelong friend of Thomson's who had taken the vacant chair at Glasgow on the death of Thomson's father. Although Thomson was seven years Maxwell's senior, they became close, and Maxwell and his father had often spent a winter holiday at Glasgow, sometimes as guests of the Thomsons, and sometimes as guests of the Blackburns.

The work on the standard Ohm brought Maxwell back in touch with Fleeming Jenkin[63] and afforded him the chance to meet Faraday. In 1861 he was elected a fellow of the Royal Society of London, which gave him an even greater degree of contact with the most important men of science, and there were also frequent meetings at the Royal Institution.

In his next great paper on electromagnetic theory 'On Physical Lines of Force' (Maxwell, 1861–62)[64] he used the description 'physical' in the title in a sense meaning something like 'tangible'. He was attempting to find a rationale for electromagnetic forces in terms of an analogy with a mechanical model in the form of an array of 'molecular vortices'. The paper was published in five separate parts between March 1861 and February 1862. This paper also revealed for the very first time the essential equations of electromagnetic theory, and from it he was able to make the deduction that light was an electromagnetic wave. His critics, however, found his mechanical analogy of molecular vortices too much. The mathematician Cecil Monro, wrote to him:

> The coincidence between the observed velocity of light and your calculated velocity of a transverse vibration in your medium seems a brilliant result. But I must say I think a few such results are wanted before you can get people to think that, every time an electric current is produced, a little file of particles is squeezed along between rows of wheels. (C&G, p. 329)

The critics could not see that his intention in introducing a mechanical model was an analogy and nothing more, a mere stepping stone towards what was really going on. The paper, by its timing, had clearly been on the brew in Maxwell's mind during his last days at Aberdeen, for he had presented the first part of the paper in December of that year, 1860. However, he was probably still correcting proofs for the final parts when he went off again on a new tack, for in December 1861 he wrote to another of his Cambridge friends, H. E. Droop: 'I am trying to form an exact mathematical expression for all that is known about electro-magnetism without the aid of hypothesis' (C&G, p. 330). He had therefore already abandoned his mechanical vortex model and was now postulating in its stead the concept that electrical charges, currents and magnetic dipoles pervade matter and space alike with *electromagnetic fields* governed by equations. While these equations and the conclusions to be drawn from them were little different from those of his previous paper, the key point is that it was an entirely different approach. The eventual paper, 'A Dynamical Theory of the Electromagnetic Field' (Maxwell, 1865)[65] was first read in the December of 1864 and although it proved far more acceptable and added to his growing reputation, in a sense the outcome was academic. He would not live to see the full consequence of his achievement, for that was to take more than twenty years to come to fruition.

On 8 February 1866, not long before leaving King's, Maxwell was awarded the honour of giving the Royal Society of London's Bakerian Lecture. His chosen subject was 'On the Viscosity or Internal Friction of Air and Other Gases' (Maxwell, 1866).[66] This was followed by his major work revisiting the molecular theory of gases, 'On the Dynamical Theory of Gases' (Maxwell, 1867b).[67] One of Maxwell's important and memorable contributions to this theory is a paradox that he dreamt up, now called Maxwell's Demon. He first described it to Peter Guthrie Tait in a letter of 11 December 1867. Maxwell mentioned no demons as such, rather 'a very neat and light-fingered being'; it was actually Kelvin who later dubbed it a 'demon'. Maxwell imagined the 'demon' operating a shutter between two sealed compartments both filled with the same gas. It was the demon's task to allow molecules through the shutter according to whether they were moving slower or faster than average. Now, the molecules in a cold gas move slower than those in a hot gas, and therefore by instructing the demon to let all the slower molecules into one chamber and all the faster ones into the other, in a manner analogous to sorting out white and black sheep at a gate, he could end up with one chamber full of hot gas and the other full of cold, which means that it would be possible to get both heating and cooling for free. But we all know that in practice it takes work (energy) to create both these things – or we would not have any energy bills to pay. Even in theory, this result appears to be a paradox. Maxwell's point was to demonstrate that the results of a molecular gas theory are entirely statistical, and as such their physical interpretation has to be properly qualified.[68]

According to Niven, 'the mental strain involved in the production of so much valuable work, combined with the duties of his professorship [...] [for] nine months of the year' (WDN1, p. xvi) led him to resign his chair and retire to Glenlair. For one who knew him so well Campbell makes little comment on the matter. In contrast, he takes up time with Katherine's devotion to her husband during his two major illnesses of that period, and so it is fair to ask whether Katherine was part of the reason. But if this was the case, Maxwell would never have divulged it.

45

Certainly, over the years Maxwell's abilities as a lecturer have come under scrutiny. Firstly there was speculation that he had not been a good lecturer at Aberdeen, and then there was the reportage as to why Tait had been preferred over him for the chair at Edinburgh. As to King's, Hearnshaw (1929, pp. 247–248) informs us that in October 1863, an assistant lecturer had to be appointed to help Maxwell control his unruly students, but this was to little avail and he was effectively sacked February 1865, with his post being awarded instead to the assistant, Mr Grylls Adams. Randall (1963, pp. 19–21) points out that while there was reliable hearsay evidence, the records for the period in question are missing.

Amongst Cay family papers, a note by Maxwell's cousin William Dyce Cay (1887) mentions that Maxwell had told him in 1864, before the eventual demise, 'the students did not care for instruction except in engineering or practical mathematical subjects'. Maxwell clearly understood the situation, but whether he offered his resignation or it was asked for is an entirely moot point; either way, he was struggling in the lecture theatre. While he could readily teach those students who wanted to learn, it would seem those that could not follow him simply resorted to classroom rebellion.

Retreat to Glenlair

On finally leaving King's at the Easter of 1866,[69] it was Maxwell's intention to carry on working independently at Glenlair. He could carry out his experiments there just as before, he could correspond by post that would arrive the following day, he could quite easily be in Edinburgh or Glasgow when he wished to, and he could manage very well financially without a professorial chair. Best of all, he would not have all those annually repeating demands on his time and energy.

Late that summer when Mr and Mrs Maxwell were settled back at Glenlair, he had his second close escape from a life-threatening disease, and it was Katherine who once more saw him through it. Although erysipelas is an infection of the skin, it makes the patient extremely unwell. Nowadays, a course of antibiotics would normally see it off in a few days, but at that time there was no effective medical treatment for any such thing. James had caught it from a simple scratch on the face inflicted by a low branch, and within a couple of days he was brought very low, to the point that: 'her quiet reading of their usual portion of Scripture every evening, was the utmost mental effort which he could bear' (C&G p. 320). His eventual recovery was to take several weeks.

At Glenlair, Maxwell could occupy himself as of old, engaging freely in his experiments, ideas and writing, and he was far from inactive. Until 1862, however, Glenlair had lacked a proper access from the main road,[70] which ran on the other side of the River Urr. He had therefore engaged his cousin William Dyce Cay, a civil engineer, to design and supervise the building of a suitable stone bridge to remedy the problem. This turned out to be crucial, for Maxwell's correspondence was by now so voluminous that the new access road was a great benefit in transporting it. Evidently, he been much concerned about the extra burden that his scientific work was creating for the poor postman, for he had his own post-box installed at the far side of the bridge. It was then just a brief walk, for him rather than the postman, to take his mail back and forward from the post-box, and he did so every day, in fair weather and in foul.

Maxwell was also able to get the benefit of other improvements he had made at Glenlair during his time at King's. He had already contributed generously to the endowment of Corsock Church and the building of its manse. Parton Church, however, would always be dear to his memory, for it was within the ruined walls of the old kirk there that his father and mother were laid to rest. Then he turned his thoughts to his own dwelling house.

His father had built Glenlair House on the estate in 1831 at a place originally called Nether Corsock, with Walter Newall as architect to implement his fairly functional concept of what he required as a fledgling country laird cum farmer. Apparently he had often toyed with plans for houses, and so when it came to the building of farm offices and outhouses in 1841–42, he took charge himself and they were built to his own plans. No doubt he would have taken note of James' hopes for the garret space. By that time, however, he had completed his plans for the estate by selling off some of his more distant and isolated farms and buying the neighbouring farm of Glenlair. Nevertheless, because of the entail, he was still John Clerk Maxwell of Middlebie; likewise, in due course his son was James Clerk Maxwell of Middlebie rather than of Glenlair.[71] The consolidation of the Glenlair estate should have been a happy occasion, but the year had been 1839 and Maxwell's mother, who must have shared in the same dream, had not lived to see the reality.

Following a tour in Italy in 1867, it was the young laird's turn to take a hand in shaping the estate. He decided to realise his father's dream of one day remodelling Glenlair House in a way more fitting to the needs of a country gentleman (Colour Plate 8). Referring to some of his father's old sketches as to what shape it might take, he worked on a suitable design. His father's original plain building was to be kept and extended from its westward gable.

It would have two storeys running crosswise with the original, a lofty roof and windows, and a new entrance vestibule. One of Newall's apprentices, James Barbour, was engaged as architect and he produced the working plans of 1868 that incorporated Maxwell's ideas, both for the overall scheme and some of the decorative touches.[72]

Even while the building work was going on, Maxwell was working on scientific ideas and turning them into papers. Between 1868 and 1869, he followed up on his finding in 'On Physical Lines of Force' and 'A Dynamical Theory of the Electromagnetic Field' that the speed of light, υ, in any medium was to be given by a simple formula containing only two measurable constants, ε and μ.[73] He had previously obtained these constants from existing data, but now he undertook a new measurement of his own which gave a result of υ = 288,000 km/s, just 3 per cent lower than Foucault's latest and fairly accurate value for the speed of light, 298,000 km/s. This was another important step towards confirming that light and electromagnetic waves were the same sort of thing.

Another paper, and on a completely different tangent, was 'On Governors' (Maxwell, 1868b)[74] in which he gave a mathematical analysis of different types of governors and distinguished them from regulators.[75] It was the origin of control theory, an important branch of applied mathematics that is now used in mechanical and hydraulic machinery, for example, in the control of robotic arms and manipulators, and in the power steering and brakes on vehicles. It is also of major importance in electronic circuits, as in the case of various types of amplifiers, to increase their frequency range and fidelity. Modern aircraft employ sophisticated control systems that make them stable in flight, something that, for various design objectives, their airframes are not capable of achieving on their own.

In this period he also wrote three papers that were directly applicable to civil engineering. They were extensions of an earlier paper, 'On Reciprocal Figures and Diagrams of Forces' (Maxwell, 1864b) which had possibly been inspired by the sort of structural calculations that would have been of interest to his cousin, William Dyce Cay. These papers were 'On the Theory of Diagrams of Forces as Applied to Roofs and Bridges' (Maxwell, 1867a); 'On Reciprocal Diagrams in Space and their Relation to Airy's Function of Stress' (Maxwell, 1869b); and 'On Reciprocal Figures, Frames, and Diagrams of Forces' (Maxwell, 1872).[76] The last of these papers, submitted on 17 December 1869, was read before the RSE on 7 February 1870 and subsequently won for Maxwell the Society's Keith Medal for 1871.[77]

On top of this, Maxwell was working on two major books, one of which was his famous *Treatise on Electricity and Magnetism*, in two volumes

(Maxwell, 1873a). Although his *Theory of Heat* (Maxwell, 1870), is now much less well known, it ran to ten editions by 1891 and was revised by Lord Rayleigh for a fresh edition in 1902.

Return to the Fray

All the while, the Maxwells made visits to London and James attended the annual meetings of the BA at various places around the country. He would also spend some winter weeks in Cambridge with the mathematical tripos. In 1868 he was tempted to apply for principal of St Andrews, but thought better of it, no doubt because it would have involved more administration than scientific work. But it must have been in the back of his mind that he would like to get back into the thick of things, as spending the best part of each year at Glenlair involved a fair degree of isolation, and while it meant that he could do work there aplenty, he was out of the mainstream and had no form of assistance with his paperwork or experiments other than what Katherine could provide.

In the meantime, the chancellor of Cambridge University, William Cavendish, 7th Duke of Devonshire, had offered funds for a new laboratory there for research into heat, electricity and magnetism. Although this was extremely generous, the Duke had an ulterior motive; Henry Cavendish FRS (1731–1810), the scientist best known for his discovery of hydrogen gas, had been a grandson of the second duke, and the present duke wanted to do something in his memory. Keen to bring this to fruition, in February of 1871, the university senate established a chair of experimental physics. But the laboratory was only part of what the Duke had in mind; Henry Cavendish had spent nearly fifty years conducting his researches, and while he had kept copious records he had published little of them. The string to be attached to the new chair was that the successful applicant would have to take on the burden of the task that had been left undone. Worthy though that task might have been, for a man at the cutting edge, as Maxwell was, it would be a significant diversion of his talents.

Maxwell's old friend, now Sir William Thomson, appears to have been the first choice for the new chair, but he would not put his name forward. Thoughts then turned to Maxwell who, at first also reticent, let himself be persuaded, but only on condition he would be able to step down after a year if he so wished (Riddell, 1930, p. 39). Now, the job was a fantastic opportunity: finances were in place; the location was in one of the country's greatest scientific centres; it was on Maxwell's home turf, as it were; and there was a brand new laboratory to be designed, built and equipped from first principles. On the negative side was the workload that this would initially involve, on top of which there was the chore of examining and editing Henry Cavendish's research papers. At least Cavendish had been very highly regarded, and the task would have been worthwhile doing; it could not be turned down out of hand. But as he was no stranger to work, it is unlikely that the workload in itself would have put Maxwell off, and so why did he stipulate from the outset that he should effectively be allowed to try it for a year before committing to stay longer?

There must have been some difficulty in the back of his mind. Certainly, we must think back to the circumstances of him leaving King's, and with the object of retiring rather than finding a more suitable post. The prospect of doing battle once more with unruly, unwilling students must have been something of a consideration, but he would have doubtless been reassured that he would have only the best students. From another angle, it has been hinted that Katherine was of a somewhat nervy and needy disposition, and recalling Jemima Blackburn's words, 'Her mind afterwards became unsettled [. . .] alienated him from his friends and was of a suspicious and jealous nature.' Whether from this or from some other cause, Campbell (p. 372) also says that eventually 'The last few years of Maxwell's life were saddened by the serious and protracted illness of Mrs. Maxwell.' If Katherine's health and disposition had indeed been the concern behind his reticence, upon receiving the requisite assurances that he could leave after a year if he wished to, he finally accepted the post and the Maxwells moved to Cambridge.

Maxwell gave his inaugural lecture (Maxwell, 1871a) on 25 October 1871, and when the year had passed the couple were still at Cambridge and living in a house he had leased at 11 Scroope Terrace.[78] Busy though he was, during this period he found time to publish some novel ideas in 'Remarks on the Mathematical Classification of Physical Quantities' (Maxwell, 1871b).[79] The first was dimensional analysis, a simple device that is now a standard for checking physical formulas, and the second was the naming and visual representation of some important mathematical concepts that occur regularly in electromagnetic theory and fluid flow. His diagrams, Figure 2.5, show at once the idea.

The laboratory itself was then duly completed and handed over by June 1874. Unfortunately, the money for equipment did not stretch as far as had been hoped and there was a temporary shortage for some time. During this initial period of planning, organising and building, he had also completed *Treatise on Electricity and Magnetism*, and so from 1874 onwards he was able to return anew to his research. Moreover, he now had some able students to

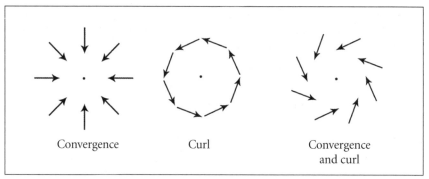

Figure 2.5 Maxwell's legacy to the terminology of mathematics
The terms curl and divergence (negative of 'convergence') are in regular use today. Maxwell realised that each takes a distinct mathematical form and his diagrams depicting them allude to patterns of fluid flow that we would easily recognise when draining a bath. He instigated these ideas based on Hamilton and Tait's quaternions when their practical application was still in its formative stages (WDN2, p. 266).

work under him on his experimental work, and by all accounts he was a very patient mentor, unstinting with his time and advice. One such able student was William Garnett who, having graduated in the year before as a fifth wrangler, was appointed by Maxwell as his first demonstrator in experimental physics at the Cavendish. In November 1874, Garnett was elected a fellow of St John's College and he went on to become a close friend of Maxwell's. It was he who, together with Lewis Campbell, wrote Maxwell's biography that we refer to so frequently here as 'C&G'; Campbell wrote his life and Garnett wrote his works.

In this period Maxwell wrote a textbook on mechanics, *Matter and Motion* (Maxwell, 1876a) and many articles and reviews in various publications including *Reports of the British Association*, *Nature* and *Encyclopaedia Britannica*. Amongst the interesting titles are 'On the Solution of Electrical Problems by the Transformation of Conjugate Functions' (Maxwell, 1876b)[80] which pointed out a way of solving a given problem in electrostatics by transforming it to a geometrically related problem for which the solution was already known. This elegant mathematical idea based on 'conformal mapping' has since been referred to in almost every standard textbook on electricity and magnetism.

While Maxwell also went so far as to write on the protection of buildings from lightning strikes, his note on the subject (Maxwell, 1876c)[81] does not refer to lightning conductor design as much as it does to his opinion that a lighting conductor might only serve to increase the probability of a strike:

> It appears to me that these arrangements are calculated rather for the benefit of the surrounding country and for the relief of clouds labouring under an accumulation of electricity, than for the protection of the building on which the conductor is erected.

To which he adds, as a tongue-in-cheek conclusion: 'It is hardly necessary to add, that it is not advisable, during a thunderstorm, to stand on the roof of a house so protected, or to stand on the ground outside and lean against the wall.' Nevertheless, a lightning conductor was eventually installed at Glenlair (Plate 7), and so we may adduce that it was more likely to have been the work of his successor.

The editing of *The Electrical Researches of the Hon. Henry Cavendish from 1771 to 1781* proceeded alongside all else that was going on, and at least the work had an interesting and fruitful side to it. Maxwell was able to uncover important advances made by Cavendish such as his anticipation of Ohm's law, the measurement of capacitance and inductance, and the exactness of the inverse square law of electric force. The last of these is so fundamental that Maxwell organised a repeat of Cavendish's experiment. Cavendish had shown that the power of the law could not differ from two by more than 1.7 per cent, but while it may be hard to see the point of redoing the experiment, Maxwell had his reasons and took great pains to make sure that the apparatus for the new measurement would be of the very best quality. The experiment was carried out at Cavendish Laboratory in the summer of 1877 by Donald MacAlister,[82] with the result that the power of two in the inverse square was found to be correct to one part in a million (Maxwell, 1880). Maxwell may have been relieved to find that the outcome was so conclusive, for the inverse square law is highly essential to electromagnetic theory.[83]

In one of his last papers, 'On Stresses in Rarefied Gases Arising from Inequalities of Temperature' (Maxwell, 1879),[84] Maxwell explained one of the key physical mechanisms behind the radiometer, a curiosity that was invented in 1873 by the chemist Sir William Crooke.[85] The device in question used to be seen on many a window ledge, spinning in the sunshine seemingly of its own will.[86]

The End

The 'Cavendish Papers', having been published in 1879, saw Maxwell start on a second edition of his *Treatise on Electricity and Magnetism*; but unhappily it was left to his friend William Thomson to complete. In the spring of 1879

Maxwell realised that he was quite ill; he had been suffering from dyspepsia for about two years and had begun self-medicating with sodium carbonate, but it had now worsened and he had grown very weak. It is characteristic of the man that his first mention of it to his physician, Dr George Paget, was made only as an aside in the course of writing to him about Katherine's health.

The Maxwells returned to Glenlair at the end of term in the hope that rest and fresh air would restore James' health. From his letters back to Cambridge, no-one would have guessed that this was not having the desired effect, and so when William Garnett came with his new wife to visit the Maxwells that September, he was shocked to see just how his friend's health had plummeted; he was now so fragile 'he could not bear the shaking of the carriage' (C&G, p. 408). Nevertheless, his mind was still alert, his conviviality undiminished, and the Garnetts were treated with Maxwell's typical generosity and hospitality.

Maxwell's Edinburgh physician, Professor Sanders, was an old friend, and it was he who broke the news to James that he had stomach cancer, the same disease that had taken his mother's life at the very same age, and that he had about a month to live. Maxwell took the news with typical resolve and stoicism, and his only concern seems to have been for the well-being of his wife:

> unable to nurse him as of old, she seemed more than ever dependent on his care. To the last, he regularly gave the orders that were necessary for her comfort, and endeavoured to see that they were carried out. (C&G, p. 409)

Just as had been the case when his birth was imminent, Glenlair was deemed to be too far away from medical expertise. Dr Lorraine, the local doctor at Castle Douglas, evidently concurred, for already Professor Sanders had been asked to come all the way from Edinburgh to see the patient. But Edinburgh would not do, for aunt Isabella and aunt Jane were now gone, and James would not have wanted to be a burden to any of his cousins. In early October it was therefore decided that they would return to Cambridge, where he could be looked after by Dr Paget, and there they would await the inevitable.

Maxwell was attended by Dr Lorraine on the long journey, which he barely managed; his younger cousin Colin Mackenzie, with whom he had played by the shore at Silverknowe on the banks of the Forth nearly forty years before, came to be by his side during his last days. Colin had qualified as a Writer to the Signet and frequently acted as Maxwell's legal advisor; Colin and he were close, and he may well have been summoned to make sure that his last wishes were duly observed. His parting words to Lewis Campbell during the final days of his illness record his selflessness in the face of adversity: 'I have been thinking how very gently I have been always dealt with. I have never had a violent shove in all my life' (C&G p. 421).

When he died on 5 November 1879, his thoughts had once again been for Katherine rather than for himself. His cousin Jemima was to say of him, 'He [...] had the sweetest temper of any man I ever knew. I do not think I ever saw him angry or heard him say a word against any one' (Fairley, 1988, p. 109). He was succeeded by his cousin Andrew Wedderburn, who thereby became Wedderburn Maxwell of Middlebie. Although Katherine may have been given the liferent of Glenlair (see Chapter 14), it was no longer hers and how often she chose to retire there we can only guess at. From the address given in public records after her death,[87] she appears to have stayed on at Scroope Terrace. James was buried with his mother and father in the family lair within the walls of the old kirk at Parton (Colour Plate 9), and Katherine followed him there after her own death on 12 December 1886.

In his early days at school in Edinburgh, when he was awkward and hesitant, being treated as an oddball and labelled as 'Daftie' by his schoolfellows, Lewis Campbell graciously said that James Clerk Maxwell 'was a cygnet among geese'. Not only did our cygnet transform into a swan, he had soared even higher than his parents could ever have dreamt of. Yet, like our native swans, he preferred grace and silence to clamour, and never once did he cry 'Look at me!' Maxwell was so far out of the limelight as to be in the shadow of his good friend William Thomson who, through his discoveries and inventions, had succeeded in making himself not only famous but wealthy, on top of which he was also made a peer, and became Lord Kelvin.

But Maxwell was made of different stuff and did not seek such things; he had gone his own way and had done what he did just for its own sake. Although he won many academic prizes and medals, a number of which are listed in Appendix 2, these were the due recognition of his scientific peers rather than the nation. Such recognition came to others, in various walks of life, simply because what they did thrust them into the public eye, for example, Sir Walter Scott and Robert Louis Stevenson, writers of poems and novels; James Young Simpson, pioneer of anaesthesia; David Livingstone, discoverer of Victoria Falls. Only in November 2008, 129 years after his death, was a statue erected to commemorate Maxwell in the city of his birthplace, Edinburgh (Plate 8).[88]

In Appendix 3, we have attempted to give some idea of the true significance of Maxwell's scientific legacy. For most of the 150 years that have passed since Maxwell gave to the world a substantially complete basis for electromagnetic theory, his name has been known only to the few who have plied their trade in academia or engineering. Even the majority of those who did know his name knew little more of him than that he had been a nineteenth-century Scottish physicist. More recently, the dawning of the wireless age has increased interest in the man who, as Basil Mahon (2003) put it, 'changed everything'. The internet has given us ready access to many books and articles about Maxwell and his works, and a simple search for 'James Clerk Maxwell' throws up page after page of results. His Wikipedia entry gives immediate access to a considerable body of information about him, and many sites are even dedicated to just him.

However, Maxwell still has a long way to go. The most recently available (2002) BBC poll conducted to find out those regarded as being Britain's greatest ever people[89] shows that James Clerk Maxwell did make the list, but only at number ninety-three. After stripping it down to only those people who were scientists or engineers, we are left with just seventeen names,[90] but even so Maxwell is third from the bottom, whereas Faraday is number six. It simply reflects the fact that people in general still know little about Maxwell and what he did, and what is even more depressing is that they do not actually think a great deal about science in general. For example, in the full poll, even Michael Faraday came behind the comedy actor Michael Crawford. On the positive side, many things have been done to commemorate Maxwell, and the most noteworthy of them are to be found in Appendix 2.

Maxwell, as a principled Christian, would have been the first to recognise that due rewards do not always fall to those who most deserve them. The recognition that an individual receives for their achievements is quite a separate thing from the achievements themselves. There are those who will grab for glory, and there are those such as James Clerk Maxwell, for whom glory did not matter. For him, it was enough to do what he did for its own sake, or as he would have put it, for the sake of God. The final words on him must go to his friend Lewis Campbell (C&G, p. 431):

> He never sought for fame, but with sacred devotion continued in mature life the labours which had been his spontaneous delight since boyhood.

• • •

From the life and achievements of James Clerk Maxwell we now turn to the lives of his forebears and of some of their better known contemporaries, most of whom hailed from southern Scotland or northern England. A statistically unusual number of them showed the same sort of traits that their remarkable descendant held in abundance. Admittedly, many others showed traits that lay in diametrically opposite directions. All the same, they were not the less noteworthy. As his family name suggests, amongst them were the Clerks and the Maxwells, but who were they? And who were the others whose names did not get passed on? It is only right that we should find the answer, for these were the human ingredients from whom the genius of James Clerk Maxwell was forged.

Peter Guthrie Tait and Lewis Campbell

We can hardly move on from the life of James Clerk Maxwell without saying something about the lives of his two closest friends from their schooldays together at Edinburgh Academy. Campbell had taken the prize in mathematics and had been school dux in 1846; while Tait was class dux and both he and Maxwell took prizes in mathematics, neither had ever been school dux. And so, in the way of young people, they were the best of friends and at the same time the best of rivals. There were no grudges between them, even later in life when it came to Tait being preferred over Maxwell for the chair of natural philosophy at Edinburgh. While Tait helped influence Maxwell's work and career, Campbell seems to have been purely and simply his friend, the greatest significance being that as a classicist rather than a scientist, he could write a memorial of James' life and work without being drawn into the scientific debate; he gives us James the boy and Maxwell the man, and no more.

Just two months older than James Clerk Maxwell, Peter Guthrie Tait[91] was born in Dalkeith, a county town about ten miles to the south-east of Edinburgh. In 1837, not long after beginning his education at Dalkeith Grammar, where long before some of the young Clerks had studied, his mother Mary was widowed. She, Peter and his two sisters then moved to Edinburgh to live with their uncle, a Mr John Ronaldson, who was a 'writer' and banker. From about 1845 they lived at Somerset House (EPOD, 1845), now the Raeburn Hotel, on the west side of Stockbridge. The location would have been convenient for Peter attending Edinburgh Academy, which he began in 1841, for it is just a ten-minute walk away. In 1847 he matriculated at Edinburgh University along with his friend James, but on completing first year he left for Peterhouse College, Cambridge, where he took his BA in 1852 and became both Smith's prizeman and the youngest senior wrangler then on record.

From 1854 Tait was Professor of Mathematics at Queen's University Belfast, where he took up an interest in quaternions, the invention of William Rowan Hamilton. Quaternions proved to be the foundation of vector algebra, and Tait did much to promote their use and later published a treatise on them (Tait, 1867). Not long after taking up the chair of natural philosophy at Edinburgh in 1860, he formed the idea of writing a magnum opus on mathematical physics, and having written most of the first volume he was joined in his efforts by Sir William Thomson, later Lord Kelvin, who was well known to both Tait and Maxwell. On the halfpenny postcards that they exchanged when they had some point of physics to discuss, the trio referred to themselves as respectively T (Tait), T′ (Thomson) and dp/dt (Maxwell). The last of these refers to a thermodynamic formula, $dp/dt = JCM$! They clearly still shared a schoolboy sense of fun. It took Tait and Kelvin several years to get the first, and only, published volume into print (Thomson & Tait, 1867). Tait died in 1901 and was buried in his family plot in the south-east corner of St John's churchyard at Lothian Road, Edinburgh.

Peter Guthrie Tait had four main influences on James Clerk Maxwell. When they were at school, he was a great encouragement to the fledgling mathematician in Maxwell. By rivalry, they both pushed themselves higher; by collaboration, they learned much each from the other. Second, all through their adult lives, they corresponded with one another, carrying on a briefer but perhaps more focused and equally rewarding dialogue. Third, Tait specifically persuaded Maxwell to incorporate quaternion equations, that is to say, a proper vector formalism, in his *Treatise on Electricity and Magnetism*. To be fair, Maxwell did not abandon the original form of his original equations; the quaternion form was introduced as an alternative without comment as to what Maxwell thought about them. Finally, when they were both looking for suitable academic posts in 1860, Tait sealed Maxwell's fate by beating him to the Edinburgh chair. Had things been otherwise, Maxwell would never have been at King's, he possibly would not have retired for a spell to Glenlair, and latterly may have been too settled at Edinburgh to take on the Cavendish.

While it was his old school friend Lewis Campbell who wrote Maxwell's life, herein 'C&G', it was Tait who was asked by the RSE to write his obituary (Tait, 1880). As the following extract shows, as opposed to the usual form of flattering valedictory one might expect to find in such a piece, he spoke openly and honestly about Maxwell's traits:

> But the rapidity of his thinking, which he could not control, was such as to destroy, except for the very highest class of students, the value of his lectures. His books and his written addresses [. . .] are models of clear and precise exposition; but his extempore lectures exhibited, in a

manner most aggravating to the listener, the extraordinary fertility of his imagination.

We may therefore take his final words on Maxwell as nothing less than he thought to be the very truth:

> Scotland may well be proud of the galaxy of grand scientific men whom she numbers among her own recently lost ones; yet [. . .] she will assign a place in the very front rank to James Clerk Maxwell.

• • •

Lewis Campbell was born on 3 September 1830, at 13 Howard Place, Edinburgh. From *The Life of James Clerk Maxwell* we know that Campbell's mother was actually a Mrs Morrieson: 'a journal kept by my mother, then Mrs Morrieson, in which she occasionally noted matters relating to her sons' friends' (C&G, p. 132). We also find in Huxley's *Memorials* of Campbell (1914) that she had been born Eliza Constantia Pryce (1796–1864), was Welsh, and was well regarded for her poetical works (Roberts, 1908). Her first husband had been Capt. Robert Campbell RN, governor of Ascension Island from 1819 to 1823, but he had died in 1832 leaving her with two very young sons, Lewis and Robert, to fend for. She worked hard to put her sons to school, latterly at the Edinburgh Academy, which Lewis attended from 1840. It may be presumed that Mrs Campbell, as she then was, put her talents as a writer to good use in this endeavour, for in 1833 she published *Welsh Tales*, followed by a separate Scottish edition in 1837. It was in 1844 that she remarried to Lt Col Hugh Morrieson of the 20th (Bengal) Native Infantry, who had retired from service in the East India Company in 1841 (Asiatic Journal and Monthly Register, 1841).

As to where they lived, Lewis tells us:

> somewhere in 1843 or 1844 [. . .] my own closer intimacy and lifelong friendship with James Clerk Maxwell began. Shortly after this we became near neighbours, my mother's new domicile being 27 Heriot Row, and we were continually together for about three years. (C&G, p. 55)

Coincidence of coincidences, 27 Heriot Row was the very house owned by James Clerk Maxwell's great-uncle Lord Newton, and in which his great-aunt Bethinia had lived until 1842 (EPOD, 1806–42).

Lewis Campbell was already in Mr Carmichael's class at Edinburgh Academy when James Clerk Maxwell joined it in November 1841. His brother,

Robert, two years younger, was also at Edinburgh Academy and a friend of James too, for he visited Glenlair in 1847 and also ended up at Trinity College, Cambridge (Henderson & Grierson, 1914, pp. 122–123). Lewis was a frequent visitor to Glenlair (Huxley, 1914, p. vii), and it is clear that James was close friends indeed with both Campbell boys.

Having excelled at Edinburgh Academy, as did his friends, Lewis went to the University of Glasgow. Although he and they were now going their separate ways, they continued to keep in touch; Lewis and James exchanged many letters from which we get glimpses into their lives, and Lewis continued to visit James at Glenlair during further summer vacations (C&G, pp. 154, 174). At Glasgow, Lewis read classics and went on to Oxford where he won a Snell exhibition scholarship to Balliol College. His academic successes continued with a first in 'Greats', that is to say classics, in 1853. He became a fellow of Queen's College and was ordained as an Anglican priest.

When Campbell married Frances Pitt in May 1858, Maxwell made the long trip down from Aberdeen to Milford-on-Sea, Hampshire, to be his best man. The newlyweds were soon to follow James' journey back to Aberdeen to attend James' own wedding to Katherine Dewar. While at Milford, Campbell coached students for university entrance, one of whom was Charles Hope Cay (1841–1869), son of Robert Dundas and Isabella (Dyce) Cay (Chapter 14). It is likely that the recommendation to study with Lewis Campbell would have come from Maxwell himself. It produced the desired result, for in 1860 Cay won a classics scholarship to Caius College Cambridge.

A significant change came for Campbell when in 1863 he was elected to the chair of Greek at the University of St Andrews where James D. Forbes, relative and former mentor of Maxwell, was principal. After Maxwell's death in 1879, Campbell began the book on his life that we have made such frequent reference to here. Not being grounded in science, he concentrated on Maxwell's life, while his chosen co-author, William Garnett, who had been Maxwell's demonstrator at the Cavendish, wrote on his scientific work. The first edition was published in 1882 with the second following in 1884.

In 1898 the Campbells moved to Alassio on the Italian Riviera, calling their new home Sant' Andrea. He died aged seventy-eight at Lake Maggiore. In the time that had passed since the death of his dear old friend James, he must have had much pleasure in hearing of the discovery of electromagnetic waves and of the emergence of wireless telegraphy, for he wrote: 'yet his electromagnetic theory of light was undoubtedly the precursor of the Rontgen rays [. . .] and wireless telegraphy is an application of ideas on which he used to discourse to me' (Huxley, 1914, p. 386). How Campbell's admiration for his old friend must have turned to the profoundest sense of wonder!

CHAPTER 3
The Early Clerks of Penicuik

Our history of the Clerks of Penicuik from the time of Mary, Queen of Scots to the middle of the eighteenth century is contained in the original manuscript of a memoir written by Baron Sir John Clerk, 2nd Baronet of Penicuik, which bore the admonition: 'This book may be read by all my friends in the House of Pennicuik, but is never to be lent or carried out of the House.'
    Nevertheless, about 1890 the Dowager Lady Clerk gave her permission for it to be published by the Scottish History Society (Clerk & Gray, 1892, hereinafter referred to as BJC). The Clerk history recorded by the Baron was briefly recapitulated and updated by Miss Isabella Clerk, a daughter of Sir George Clerk, 6th Baronet of Penicuik, for the use of Lewis Campbell in his 1882 *Life of James Clerk Maxwell*. Campbell also added material from a book of autograph letters which was then in the possession of James Clerk Maxwell's widow, Katherine Dewar. Finally, local Penicuik historian John J. Wilson provided a history of the Clerk family in his *Annals of Penicuik* (1891, pp. 150–164). While he gives no specific attribution for his sources, Wilson was involved with the Scottish History Society in gaining access to the Baron's manuscript. Beyond this, the archives held at NAS help to fill in some of the gaps and also provide some interesting details.

The Origins of the Clerks

According to the personal memoir of Baron Sir John Clerk, the Clerks of Penicuik are descended from John Clerk of Killiehuntly (1511–1574). Killiehuntly and the Dell of Killiehuntly lie about two miles to the east of Kingussie, by the banks of the River Tromie not far where it meets the mighty Spey. This is in the very heart of Scotland in a region that was in ancient times called Badenoch, the purlieu of the Duke of Gordon. However, that was not their original seat, for Clerk is not a Highland name. Douglas (1798) suggests that the Clerks had been prominent citizens of Montrose on the east coast of

Scotland since the fourteenth century. They later gained lands in Fettercairn, about ten miles to the north-east of Montrose, and those of Killiehuntly, which even by the shortest route is some ninety miles distant and over mountains. In being a loyal supporter of Mary, Queen of Scots, John had gone against his feudal superior the Duke of Gordon, and so had to abandon his Badenoch estates in 1568 when the ill-fated Queen eventually quit her kingdom never to return. Things had gone badly for her and her adherents, and they would get worse.

Despite this setback, John's son, William Clerk, was a prosperous merchant in Montrose during the latter part of the sixteenth and early part of the seventeenth centuries. Of him we know little more other than that he died in 1620.

The First Clerk of Penicuik

William Clerk's son, John (1611–1674), was a merchant like his father and proved to be a very able one indeed. Having moved to Paris in 1634 in the hope of making his fortune, by good commercial prowess he managed to succeed in his goal to the extent that, when he returned to Scotland in 1646, he was able to purchase, piece by piece, the extinct baronetcy of Penicuik lying on the River North Esk some ten miles south of Edinburgh. Nowadays Penicuik also refers to the dormitory town that serves that city. Having been a mere hamlet when the first Clerk arrived on the scene, by the middle of the nineteenth century it had grown into a busy little town with three paper mills and a coal mine. The Penicuik estate itself lies astride the River North Esk between the Peebles and Carlops roads, just to the south-west of the town.

John Clerk also purchased the ancient lands of Wrychtishousis, or Wright's Houses (Wilson, 1886, p. 432), then a substantial property just outside Edinburgh, on the road south to Penicuik where it met the Burgh Muir (now Bruntsfield Links). He married Mary Gray, daughter of Sir William Gray of Pittendrum, whose grandmother had been a lady-in-waiting to Mary, Queen of Scots (Family Tree 1), and who had left her some pieces of gold jewellery that was given to her by that ill-fated queen before her execution in 1587.[1]

John Clerk left his Penicuik estates to his first son, also named John, who became 1st Baronet of Penicuik, and to his second son James he gave Wright's Houses. A third son, Robert, became a surgeon in Edinburgh; he saved the life of his nephew, the 1st Baronet's eldest son, without whom this story would never have taken place, and it was also from this Robert Clerk that descended another prominent branch of the family, the Clerks of Listonshiels and Rattray.

Of John Clerk's daughters, Margaret married William Aikman of Cairnie;[2]

their son, William Aikman (Robinson et al., 1798), was the well-known artist who painted the portraits of the 2nd Baronet and of his friend, the poet Allan Ramsay. John Clerk's youngest daughter Catherine, born in 1662, married Sir David Forbes of Newhall,[3] not far from Penicuik, where William Clerk, John Clerk's grandson, and Allan Ramsay were frequent guests; they will enter our story in Chapter 5.

The First Baronet of Penicuik

John Clerk of Penicuik's eldest son was also called John (1649–1722). This John Clerk had both the wealth and good fortune to be created a Baronet of Nova Scotia by King Charles II at the age of thirty. Since he is the third John Clerk encountered thus far, it will be helpful to differentiate them as being respectively John Clerk of Killiehuntly, John Clerk of Penicuik, and John Clerk 1st Baronet of Penicuik. This particular type of baronetcy had been instituted by Charles' grandfather James VI[4] for the purpose of settling the colony of Nova Scotia. The new baronets were required to sponsor settlers there, or to pay an equivalent amount of money to the Crown, and so a wealthy landed gentleman who was lacking a hereditary title could become a baronet provided that he was a loyal subject, had a decent lineage, and was prepared to make the necessary financial contribution to the exchequer. In this manner, therefore, the existing baronetcy of Penicuik was created. One of the ancient and seemingly arcane duties of the previous line of Penicuik baronets was that whenever the king went hunting on Edinburgh's Burgh Muir (which name is recalled by the present-day district of Boroughmuir), he had to sit on the Buckstone (which likewise gives its name to an Edinburgh suburb) and 'wind three blasts of a horn'. After 1603, visits to Scotland by the British monarch were very rare indeed, and by the time that this situation was set to rights, most of the Burgh Muir was houses and parkland. It is therefore unlikely that any of the baronets was ever called upon to do this for real. Nevertheless, it is still part of the Clerk motto, 'Free for a Blast'.

The 1st Baronet did, however, take his other duties as a landowner seriously, for he became the member representing Edinburgh in the Scottish Parliament, and a lieutenant-colonel in the Earl of Lauderdale's regiment. Lauderdale had been for the Covenanter cause,[5] but when he saw the monarchy under threat, he effectively changed sides and became a Royalist. Having tried in vain to save Charles I from defeat, he was imprisoned and had his estates confiscated by Cromwell. At the Restoration, however, his fortunes were once more in the ascendant and he was one of those included in the party that went to Breda in 1660 to bring Charles II back to his 'rightful place'. As for John Clerk, Charles II would have granted him a baronet's letters patent only if he were certain of his loyalty, for in 1679 the Covenanting troubles in Scotland still dragged on. The battle of Rullion Green had taken place in 1666 on John's own back doorstep, and the uprising at Bothwell Bridge in 1679, though further afield, occurred *in the very year of his grant*. We must surely therefore see John Clerk as leaning towards the Royalist cause rather than that of the rebellious Covenanters. Having sympathy with the Stuarts would be consistent with the loyalist leanings of his great-grandfather and, indeed, his connection with Lauderdale serves to confirm it. Although he belonged to the Church of Scotland rather than the pro-Royalist Episcopal Church, this only serves to indicate that he, like many others at that time, kept his political loyalties separate from his religious ones. For the Covenanters, however, religious convictions outweighed all else, and many paid the ultimate price for it. The Baronet therefore had to tread very carefully, for had the tide turned so overwhelmingly against the present King as it had done against Charles I, his father, the Baronet could have found himself in great difficulty. As it turned out, the ousting of James VII in 1689 in favour of William and Mary somewhat settled religious tensions and the Covenanting troubles eventually came to an end.

The Baronet saw action of a different kind when sixteen robbers attacked his residence at Newbiggin[6] on the night of 17 December 1692. He managed to fend off his assailants until some of his tenants were able to come to his aid. Having made off with a few pieces silverware, the robbers were eventually caught and prosecuted.[7] It is hard to imagine that, being an officer in Lauderdale's regiment, he never took part in any action against the Covenanters. Perhaps some of them who were still beyond the law were minded to settle old scores? Was the assault on Penicuik House such a case? It seems to have been a large party for the assailants to have been simply robbers, and more likely they were out to make some such trouble.

The Baronet married Elizabeth Henderson in 1674. She was the daughter of Henry Henderson, or Henryson, of Elvingston near Haddington, and they had three sons and four daughters, who were:

John (1676–1755), who was his successor;
Elizabeth;
Henry (–1715), 'bred to sea', and close friend of his brother John;[8]
Barbara;
William (1681–1723), an advocate, who it was that ended up playing the crucial role in the Clerk Maxwell succession;

Sophia;
Mary, who married the Rev. Alexander Moncrieff of Culfargie,
   minister of Abernethy, in 1722.

Elizabeth Henderson died at the age of twenty-five (BJC, p. 8). If so, she must have been about eighteen at the time of her marriage and had had a child almost every year. From the same source we hear that after about ten years as a widower the Baronet married again in September 1692, to Christian Kilpatrick, who bore him eight more children, making fifteen in all:

James, general inspector of the outposts of North Britain;
Catherine;
Christian;
Robert (1702–1761), who was from 1732 a commissary in Edinburgh;[9]
Margaret;
David;
Hugh (1709–1750) a merchant who traded with North Carolina;
Alexander, a painter (–1775).

With such a large family Sir John decided that it would be fitting to build a family mausoleum at St Mungo's in Penicuik; although there is no longer any access, it still stands.[10] Following his death on 10 March 1722 he too was buried there. Around the time of his daughter Mary's wedding he had been beginning to feel quite weak, which in these days was not in itself remarkable considering he was then about seventy-three years of age. Leaving his family to enjoy the rest of the wedding celebration, he retired to his bed early to read by the light of a candle, and for a time did so. All the while, his wife and daughters came and went from the room, always taking care not to disturb him. When his wife finally came to bed, however, she found him apparently still reading, but not so, for Sir John Clerk, the 1st Baronet of Penicuik, had passed away some hours before, in complete tranquillity, without anyone being aware that he had gone (BJC, p. 108–109). The peaceful manner of his passing was something that 'he always wisht for, ευθανασια' (BJC, p. 9).

The line of descent from the 1st Baronet of Penicuik down to James Clerk Maxwell is to be found in Family Tree 2.

CHAPTER 4

The Baron

Sir John, the 1st Baronet, died peacefully on 10 March 1722 and was succeeded by his eldest son of his first marriage, John (1676–1755). By the time of his succession at the age of forty-six, the 2nd Baronet was a scholar and an illustrious yet principled man of many talents and achievements. We must be thankful that he also wrote on a wide variety of subjects, although largely for his own interest rather than publication. His propensity for it led him to write the fascinating memoir published by the Scottish History Society from which the bulk of our information about the early Clerks, and indeed for this chapter, has been directly drawn. It gives us a rare insight into the life of someone who was not only lucky enough to have been born in times of great change, but also helped to see things through and reap the due rewards. While he had a comfortable life, received the best education, did the grand tour, achieved high office, and had the time and opportunity to pursue his own intellectual interests, it must be remembered that all this took place nearly three centuries ago when severe illnesses, family deaths, financial ruin, even rebellions, were frequent visitors.

Early Life

There is no sign that the young John Clerk's early scholarly achievements were the result of parental pressure. His mother having died even before he went to school, there was nobody to shield him from the strict parental discipline that a son would normally have expected to receive if and when he failed to make sufficient progress in the serious matter of his education. However, it seems that he was fortunate enough to need little of this sort of 'encouragement' and always spoke of his father treating him fairly and kindly. He clearly had a facility for learning, and at the same time possessed the self-discipline to apply himself to the subject in hand, whatever it might be.

He recalled few troubles with his early education at the local school in

Penicuik, where he would have studied along with sons of other country gentlemen under a 'dominie' who would teach them everything from arithmetic to Latin and Greek. One difficulty, however, was not only the strictness of one such teacher, but the severe discipline he meted out to transgressors. Although this did not hold John back, he took it as a lesson in life that children should never be treated in such a manner. Secondly, he admits to having had very poor handwriting, because he and his classmates were regularly made to copy out the minister's weekly sermons — which meant having to write so quickly that he became careless. It is also evident that he regretted not learning English, by which he meant the refined language of the English gentry, for the Lowland Scots gentry and professional people still adhered to the old vernacular as spoken on the streets and in the shops and taverns. As any follower of our national poet, Robert Burns, will know, there were many differences from standard English; there was a separate lexicography of words, some of which are still common. While grammar was essentially the same, for those who could read and write, spelling was not only irregular but variable. The courtroom speech of another John Clerk, Lord Eldin, who was the grandson of the 2nd Baronet amply demonstrates that the use of such language was not the result of any lack of education, for even at the start of the nineteenth century it was fairly common (C&G, pp. 19, 21 and note 3 thereto).

By the start of the eighteenth century, however, the Scottish gentility had been beginning to think that they would have to acquire the more genteel language of their southern cousins in order to take full advantage of political union. While the writings of our present John Clerk are comparatively refined, they still include a number of Scottish words and a variety of spellings, one of the reasons he gave for not letting the manuscripts be seen outside the immediate family.

University

After his Penicuik schooling, in 1691 the young Clerk went off to Glasgow University to apply himself to the study of logic and metaphysics. This he ably did for two years, after which he came to the conclusion that what he was learning was of no use to him, and in fact, he felt it was harming him. His father must have known his son well, for he was allowed to withdraw and to proceed to Leiden in Holland,[1] as did many young Scotsmen of the time, to study civil law. When he set off in October 1694, England was at war with France, and so his ship departed as part of a convoy of ninety merchantmen at Queensferry that was bound for Rotterdam under the escort of two Dutch frigates. After a rough passage and a hot pursuit by French privateers that ended in near disaster, he eventually got to his destination.

As well as the civil law, he studied mathematics, philosophy and music. While his father had made it plain that he must succeed in the law, he found mathematics and philosophy to be so interesting as to be a distraction, even to the point of neglecting all else. He realised, however, that this was doing him no good and so he learned to keep his fascination with these subjects in check. On the other hand, another diversion may have been beneficial, for becoming a master on the harpsichord not only provided him a useful social accomplishment but also therapy in the form of relief from his academic studies. His lectures were mostly conducted in Latin, in which at least he had a grounding from his schooldays, whereas outside the university normal conversation was in Dutch, which was yet another subject for him to master.

The overall burden of his studies and pursuits must have been considerable indeed. All the same, he was still eager to stretch himself even further by adding to them rhetoric, the art of drawing, history (conducted in Dutch) and ecclesiastical history. Naturally enough, fitting all these things into the available time did not work out very well, and so it took him an additional year to qualify, which he finally did in May 1697 along with a dozen or so other Scots.

While his father wanted his son back home in Penicuik, there is little mention of his reproaching him for the fact that his time abroad had been extended, and he generously gave his son leave to do a grand tour of Europe, albeit with minimal funds.

The Grand Tour

The May of 1697 must have been a critical time for the young John Clerk. Having reached a high of successfully completing his degree and then anticipating his grand tour, things went badly awry when he came down with smallpox, which then had a fatality rate close to 30 per cent. Luckily for him, he had made firm friends with a fellow scholar, Herman Boerhaave (1668–1738), a Dutchman who had studied medicine and was keen to try a new remedy. It seemed to work, the skin lesions cleared up in days and John was able to set off with all speed two days before the graduation ceremony was to take place, and with no thought as to a travelling companion to guide him. With great hopes and a mere £100 from his father, he simply decided to trust in providence. Boerhaave, on the other hand, stayed on at Leiden and began to make it famous as a centre of medical learning.

John travelled through Germany to Nuremberg with such companions as he found on the way. He then took a boat down the Danube to Vienna, where the British Ambassador, Lord Lexington, introduced him at the court of Leopold I. This gained him access not only to Viennese society, but to some of the finest visual and performing arts in Europe, not to mention libraries full of the choicest books.

But Vienna was only one such stop on his grand tour, for he also had designs on Venice and Rome. We may readily imagine the society, the arts and the antiquities that he sought, found and devoured in these places. When he reached Rome in the late autumn of 1697, however, the smallpox that his friend Boerhaave had 'cured' returned with a vengeance, and he became very seriously ill. Having placed his faith in providence, it came to his aid in the shape of Fr Cosimo, a Franciscan monk, who turned out not only to be the son of a Scottish advocate[2] but was actually a distant relative (BJC, p. 26).

Fr Cosimo, having only just met John Clerk, tended his sick young acquaintance with the assistance of the nuns of the Tor di Specchi, who nursed him until he was at last restored to health. From then on, Cosimo kept close to John for more or less the remainder of his time in Italy. Once recovered, John found that his endeavours with the harpsichord were highly beneficial, for his musical abilities gained him access to the cream of society. He was also able to study music under some of the best teachers of the day, Pasquini and Corelli, and gained the favour of an antiquarian and philosopher by the name of Chaprigni, who kept a busy social calendar of musical and literary assemblies at his own house. He had also formed a close friendship with a young English nobleman, Wriothesley Russell[3] with whom he visited Naples. Nevertheless, and perhaps to please his father, he also took up his studies of the law again under a judge, Monsignor Caprara, whom he assisted with the reading of legal papers.

Things continued in this way until he could no longer put off his father's entreaties that he return to Penicuik. He set off on his journey home in December 1698 after having spent fifteen months in Rome and having had the time of his life. Monsignor Chaprigni had died before John's departure, but he had thought highly of his young Scottish protégé and bequeathed him several of his antiquities, amongst them a head of Cicero, a bust of Otho and a little statue of Diana, all of which were packed off to Penicuik House. Despite the urgings of his father, John was in no mood to rush directly home, for he had not yet seen Florence. With Fr Cosimo as his companion, he was accepted at Florence by the Grand Duke of Tuscany, who not only entertained him well but so took to the young Scot that he gave him the run of his libraries and even went so far as to bestow honours on him. Finally heeding his father's call, John set off for Genoa to find a ship to take him to Marseilles, stopping en route at Pisa and Leghorn, where he finally parted company with Fr Cosimo. Having benefited from the Father's company and diligent attention for about eighteen months, John must have found it a difficult adieu.

From Genoa, John eventually got to Marseilles only after another perilous sea voyage. He then headed straight for Paris and the court of Louis IV. But what he saw at Versailles seemed to him a mere repetition of what he had seen in Rome, and it disappointed him. The music, the operas and the comedies were uninspiring, and only the dancing and conversation pleased him. Despite this, and despite the further exhortations of his father to come on home, he lingered in Paris, taking time to renew his interest in mathematics and philosophy, until the early summer.

Having been fully two years on his grand tour, he now set off for Brussels and Antwerp, where he visited the churches and viewed the paintings of Rubens, still finding nothing that compared with Rome. Ever reluctant to hurry, he put off three more months visiting Dr Boerhaave and other friends at Leiden. Though Holland suited him very well and he was reluctant to depart, he eventually took his leave and got back to London, but only after diverting via Harwich as his voyage was once again beset by storms. In London he was met by his uncle, William Clerk, who may well have been sent there to collect him, if not actually to bring him back from Holland. They travelled back north together, with John reaching Penicuik House by 2 November 1699, after an overall absence of five years and a few days, and an expenditure well in excess of what his father had allowed him.[4]

A Brief Marriage

Once home, John completed his studies in civil law and was admitted as an advocate in Edinburgh on 20 July 1700. It was an opportune time, for things were finally settling down and there was even talk of political union with England. His next step, however, was to find a wife, a matter that his father was already working on. John did not at all take to the first proposed match, was too late for the second, but was lucky with the third. He had met Lady Margaret Stuart, daughter of Alexander Stuart, 3rd Earl of Galloway, and his wife Lady Mary Douglas. Although it was an excellent match approved by all concerned, his father balked at putting enough money into the marriage settlement to secure the Earl's agreement. A*mor vincit omnia*, as they say, for Lady Margaret persuaded her father to take Clerk's offer, such as it was, and allow the marriage.

John Clerk and Margaret Stuart were married on 6 March 1701 and took a house in Edinburgh[5] so that John could attend court. Much to their happiness, a child was soon on the way and a son was safely delivered on 20 December of that year. But from his joy, John was plunged into despair by his wife's rapid decline and death. At first her life was not feared for but somehow she had got it into her mind that she would die, and so the best doctors were brought in. They intervened only because they found it impossible to dissuade her from this morbid fixation, and ironically it was their intervention that led to her death. While John believed the doctors had acted for the best, his own uncle Robert amongst them, he could not help thinking that things would have turned out differently had they simply done nothing. It seems to have left him with a long-standing belief that doctors only made matters worse.[6]

Margaret died the following day and was buried in the family vault at Penicuik. Her son, named John after his father and forefathers, was placed in the care of an aunt, Lady Aikman of Cairnie (see page 61). Margaret, however, was a cousin of the Duke of Queensberry, a very influential statesman, and because of the calamitous turn of events he was keen to take John under his wing and do what he could for him. John's father-in-law, the Earl of Galloway, though perhaps less influential, was of a similar mind, and in 1703 helped to get him elected to the Scottish parliament as member for Whithorn in the south-west of Galloway. In that same year, John was appointed to the commission enquiring into the public accounts and national debt of Scotland, this time through the Duke's influence, and was even entrusted with drawing up its report. Despite his despair at the loss of his wife, things were improving in other directions; now the Duke of Argyll was nominating him to a council of enquiry into the trade and commerce of the Scottish nation. This sort of work suited John better than the bar, and within a few years he had gained a 'thorough acquaintance with all the Finances of Scotland, and the whole management of the Lords of the Treasury, and Exchequer' (BJC, p. 58).

Part of the Union

When barely thirty, John was nominated by Queensberry as one of the thirty-one Scottish commissioners for negotiating the political union of Scotland and England. The two countries had joined crowns under James VI in 1603, making him James I of England, but they had retained their separate parliaments in Edinburgh and in London. Possibly thinking himself insufficient for the role, or just as likely that it would end like previous attempts, in failure, he at first demurred, causing Queensberry to threaten withdrawing not only his patronage but also his friendship. It seems Queensberry's efforts on John's behalf were not purely altruistic, for in pressing John to accept he no doubt knew that he needed not only someone he could trust, but also someone of exceptional abilities, who would see what was going on and who would manage all the complexities, and would keep him informed.

Clerk had then little choice but to accept the appointment and so became one of four commissioners whose duty it was to confer daily with their English counterparts. He focused on the financial aspects of the treaty and by 'assiduous application, to the study of the momentous questions' (Somerville, 1798, p. x) he managed to establish himself as a diligent and capable public servant. The negotiations were in the end successful and the Articles of Union were concluded on 22 July 1706. On the following day John Clerk took his place when the commissioners presented the signed articles to Queen Anne, filing in two by two, Scottish and English side by side (House of Lords, 2007b). The articles were duly passed by the Scottish parliament on 16 January 1707 and thereafter by the English parliament on the 6 March (House of Lords, 2007a). Following a great procession up the High Street to the Parliament House, the final acts to dissolve the Scottish parliament were passed on 25 March.

In the meantime, however, an incident that was both horrible and bizarre was unfolding at the other end of the Royal Mile. Queensberry House[7] had been left just about deserted as more or less the whole household went up to the parliament as either participants or spectators. Alone in the house was a kitchen boy, left to turn the roast for the evening's dinner on a spit before the kitchen fire. That is to say, he should have been alone. In a separate wing of the house, the Duke's eldest son, an imbecile, was kept locked up, not only for his own safety but for that of the household at large; he was extremely strong and of a violent disposition. In the absence of supervision, he managed to escape from his confinement and find the kitchen. Once there, he murdered the poor kitchen boy, and according to Chambers (1980, p. 337), placed his corpse on the spit and proceeded to roast it.[8]

For a job done well, John Clerk was duly appointed as one of the five Barons of the Scottish Exchequer that was set up pursuant to the Treaty of Union. This John Clerk is frequently referred to as 'the Baron' simply to distinguish him from all the other John Clerks of his line, and it will be useful for us to do the same here. His portrait by his cousin William Aikman (Plate 9) shows him in his robes office.

Remarriage

On 15 February 1709, the Baron resumed the role of family man by remarrying. His new wife was Janet Inglis, a daughter of Sir James Inglis of Cramond and Anne Houston, of the Houston of Houston family, who bore him a further nine sons and seven daughters. Their sons were (BJC, p. 2):

James, born on 2 December 1709, eight years to the month after his half-brother John;
George, born on 31 October 1715;
Henry and Patrick, twins born on 5 October 1718;
John, born on 10 December 1728 and named after his now deceased half-brother (BJC, p. 135 & n3);
Mathew, born on 15 March 1732;
Adam, born in May 1737.

A previous son named Henry or 'Hary' died aged three (BJC, p. 85) and there had also been a William who died before reaching his first birthday (BJC, p. 114). Of the surviving sons, James became the 3rd Baronet (Chapter 10), while George (Chapter 8) became not only George Clerk of Dumcrieff but also George Clerk Maxwell of Middlebie and, for a brief time, the 4th Baronet of Penicuik. Their younger brother John did not succeed to any family estates but nevertheless did succeed in making a name for himself as John Clerk of Eldin (C&G, pp. 19–21),[9] an engraver of some distinction. Of the twins, Patrick became a lieutenant in the army and died in 1741 at the siege of Cartagena in South America, while Henry became a lieutenant in the navy and died on active service somewhere in East Indies during 1745.[10] Mathew also went into the army to become an engineer and met his death in July 1758 at the Siege of Ticonderoga.[11] Their last son, Adam, was born, much to the joy and very great surprise of the Baron and his wife Janet when she was about fifty-one years old! (BJC, p. 146). Of Adam we know little, only that it appears he had once been a crew member on his uncle Hugh Clerk's North Carolina trading ship and may have thereafter joined the navy. He died about 1757 'abroad in the service of his country'.[12]

The daughters of John Clerk and Janet Inglis who survived infancy were:

Anne, born on 4 June 1712, died unmarried in November 1755;[13]
Elizabeth, or Betty, born on 10 August 1713. She married Robert Pringle of the Stichel family and who was later the judge Lord Edgefield (BJC, p. 145);

Jean, or Jen, born on 5 February 1717 and married James Smollett
of Bonhill, a cousin of the writer Tobias Smollett, on 17 January
1740 (BJC, p. 155);

Johanna, or Joanna, born on 10 March 1724 and died unmarried
(very probably in November 1781, for details of which see
Chapter 8);

Barbara, or Babie, born on 17 October 1725. She never married and in
later life lived with or nearby her sister-in-law Lady Mary Clerk
(see Chapter 9);

Janet or Jennet, born on 10 August 1727. She married James Carmichael
of Hailes but died without issue in 1784 (BJC, p. 222).

After his remarriage, the Baron lived the life of a conventional country gentleman. While he would regularly commute on horseback to Edinburgh when the Exchequer Court was sitting, he would otherwise be engaged in developing his estates, researching antiquities or studying the classics. The family would also spend vacations at Moffat in Dumfriesshire or Lawers in Perthshire, where they could indulge in fishing and shooting, or taking the goat's milk or spa water cures that were then in vogue. Almost everything that he found interesting was recorded for posterity and his own enjoyment.

The Jacobite Rising of 1715

This seemingly idyllic existence was interrupted by the Jacobite Rebellion of 1715 in support of James Francis Edward Stuart, better known today as the 'Old Pretender'. He had been just a year old in 1689 when his father, James VII, had forfeited his crown after fleeing to Europe. The Baron would have had cause to remember that year, for it was then that he had the dreadful accident that lamed him for life. When, on 6 September 1715, the Jacobite standard had been raised by the Earl of Mar and the rebels under Brigadier Mackintosh came south from Perth to threaten Edinburgh, the city was already well prepared for the event. About 400 able-bodied citizens had been called to arms and, ever the loyal servant of his country, the Baron was amongst them. The rebels had balked at a direct assault on the city itself and instead attempted to infiltrate the castle with an advance party of about fifty men on the night of 8 September. Unfortunately for them, word of the intended assault reached Edinburgh and the garrison was on the alert.

Those of the raiding party who had managed to escape fled to the half-ruined Citadel of Leith,[14] which the main Jacobite army had just taken after crossing the River Forth. Meanwhile the Duke of Argyll had been despatched to Edinburgh to take overall charge of its defences, whereupon he sacked the garrison commander for having taken the impending threat too lightly. After taking personal command, he confronted the Jacobites on 15 October at the Citadel. But when Argyll returned to the castle to fetch some artillery (possibly a feint), the rebel army slipped away during the night (Grant, 1850, pp. 213–218; Arnot, 1779, pp. 107–108). For the Baron, therefore, it had been but a brief fling. In the meantime, back home at Penicuik, Janet Inglis awaited not only news of her husband, but the imminent birth of her son George, who was born a fortnight later.

As to the 'Old Pretender' himself, he did not set foot on Scottish soil until after the battle of Sheriffmuir.[15] He eventually landed at Peterhead in December 1715 but on failing to impress the leaders of his rebel army, he took ill and went home to France. Somewhat ironically, he departed from Montrose, the Angus town to which John Clerk of Killiehuntly had retreated in the wake of a previous Stuart failure. Back in France he fared no better, for he was shunned by his French backers, to whom he was now a mere embarrassment.

Cammo, Mavisbank and Dumcrieff

After his remarriage and presumably some improvement in his finances, the Baron purchased Cammo, an estate adjacent to his father-in-law's at Cramond.[16] Cramond lies by the east side of the River Almond where it flows into the Forth Estuary while Cammo is a little further south. Cramond, now best known as a picturesque seaside village and the site of some significant Roman remains, was purchased by the Inglis family in 1622. They had lived in the old four-storey tower house until about 1680 when James Inglis built the adjacent Cramond House.[17]

Cammo obviously had the advantages of being near both his new wife's parents and the Roman site, while its distance from Edinburgh was only half of what he had to travel from Penicuik: 'by this purchess I had a very agreable retirement and abundance of Exercise in riding between Edin. and my House in the Country' (BJC, p. 78). Furthermore, the Baron would have known there was a family connection with Cammo, for the Mowbrays of Cammo were ancestors of his; Giles Mowbray, maidservant to Mary, Queen of Scots, had been the grandmother of Mary Gray, his own grandmother (Family Tree 1).

At Cammo, one of the first things the Baron did was to start a plantation, an activity he pursued both there and at Penicuik:

> My plantations at Pennicuik no ways hindered me from improving the Lands and Gardens of Cammo, for I did a great deal about it, all the Plantations except a few Firrs on the East side of the House having been made by me. (BJC, p. 84)

He mentions that he managed to keep on planting at Cammo even during the turbulent year 1715 and 'his long residence at Cammo, and his connection with Sir John Inglis [his brother-in-law] were the means of enriching his museum with innumerable [...] remains [...] found at Cramond' (Wood, 1794).

The Penicuik and Lasswade estates having fallen to him on the death of his father in March 1722, in the following winter the new baronet, now himself Sir John Clerk, reluctantly decided to sell Cammo and return to Penicuik. He 'resolved to build a small house at Mavisbank, under the Town of Lonhead, which my Father inclined frequently to have done, because his Coal works there [...] required his frequent attendance' (BJC, p. 113).

The Baron had helped his ageing father run his lucrative coal mines in the North Esk valley. Being the owners of the land, the Clerks had the right to exploit the coal seams lying beneath its surface, and one might ask whether John Clerk had designs on the coal rights when he bought the Penicuik estate. At any rate, it must have been a major source of income. The multitude of coal mines that were dotted all around the North Esk valley, from the Industrial Revolution to fairly recent times, gave rise to many of the present-day towns in the area, such as Loanhead itself. The main problem with the exploitation of the coal was getting access to the seams, but where they met the deep valley it was possible to dig into them horizontally, from the steeply sloping surface. We therefore hear the Baron talk of beginning new 'levels' to keep his coal revenues flowing.

The years 1722 and 1723 were full of sadness for the Baron: his father had died in March 1722 and his son, being far from well even then, died in August of a prolonged 'hectique fever'[18] that may have ended in pneumonia. By the end of the year his brother William was also unwell with a similar condition that endured until the following April, when he too died of his illness (see Chapter 5).

Cammo was eventually sold in the summer of 1724 for £4,300, bringing him some £1,500 profit.

• • •

Mavisbank (Historic Scotland, 1971; Glendinning et al., 1996) was built just to the north of Loanhead[19] on the sloping side of the valley of the North Esk

River, which flows down from the Pentland Hills, through Penicuik, past Loanhead and thence to the Firth of Forth at Musselburgh. Based on the Baron's own design, the building commenced in 1723 under the supervision of the architect William Adam (1689–1748)[20] and the Baron's regular stonemason, John Baxter (Skempton, 2002; DSA, 2014). The exterior was completed in the summer of 1724, which date is carved on one of the lintels, but he did not live in the house until the early summer of 1727 (BJC, p. 132). He now carried on with the finishing touches and adding outbuildings, gardens and enclosures.

At the previously desolate moor of Loanhead nearby, he also built houses and created enclosures and an avenue. In Loanhead itself he built a Town Council house and installed timber pipes to bring in a water supply. In the midst of this he was expanding his coal workings by constructing a 'level' running '300 fathoms under ground' (BJC, pp. 233–234).

• • •

As already mentioned, the Baron was in the habit of taking a late summer vacation at Moffat,[21] where he and his family could benefit from the spas and enjoy the local shooting and fishing, just as his father had done for many years before. Being unconnected with any family there that might offer them hospitality, it was necessary to take lodgings, which at times were neither easy to find nor convenient to his needs. In 'summer, 1727' he therefore purchased the estate of Dumcrieff from his friend and mentor, Charles Duke of Queensbury (BJC, p. 130).[22]

Originally called Drumcreich, it was a small estate with a farm and dwelling house (BJC, pp. 249–250). Lying about one and a half miles southeast of Moffat, it had been in the possession of the Murray family until 1717 when it was sold to the younger brother of the Duke of Queensberry, then just sixteen years old. The property fell to the Duke himself in 1726 and, presumably being surplus to his needs, he sold it to the Baron at the end of that year, a little in advance of the Baron's own recollection of events, but at any rate it was indeed the Baron's property by 1727.

As it stood, however, the existing house at Dumcrieff was not suitable and so he and his family carried on staying at lodgings in Moffat until such time as the house could be rebuilt. In the summer of 1728, they were lodged in an old tower house called Frenchland[23] where they had the dining room and two bedrooms, but had to bathe in the kitchen because a decent tub could not be fitted through the old house's narrow doorways. By that time the Baron had made plans for creating a park enclosed by drystane dykes,[24] which were then relatively unknown. Later that year he improved the parklands by exchanging part of his land for half the acreage adjacent belonging to his neighbour, Tod of Craigieburn.[25] In August 1729, the Clerks once again stayed at Frenchland,[26] while the work continued until 1733, when a final deadline had been set for Dumcrieff to be ready for occupation by the Clerks during their August vacation. Dickson, the Baron's factor, wrote: 'I hope the hous will be in order when you come [...] you may eather bring your coach or chaise, [if] your honour will give orders to John Black to put Doars on the barn' (Prevost, 1977).

Even so, work continued on the parks with Robert Clerk, most likely the Baron's half-brother, acting as the factor. He wrote to the Baron a year later, on 21 September 1734, to tell him the parks would at last be finished by the end of November, but he updated that during the following week, complaining of lack of progress due to foul weather and the concomitant flooding. Even worse, there was a great deluge later that night that caused a good deal of damage round about. Robert therefore wrote to the Baron in the morning, 1 October 1734, to tell him the bad news, 'Old Herbert here says the lyke flood was never seen in Moffat water in his dayes' (Prevost, 1977).

He went on to say that he would like to see the back of the place, and the only time he ever wished to see it again was from the next world! Alas, we do not know whether he first departed from Dumcrieff or this world, for this was the last of this particular episode in the Baron's affairs.

Dumcrieff, upon which the Baron must have expended an enormous amount of time and money for the land; building and furnishing the house; and creating the amenities such as the parks, gardens, drystane enclosures and plantations, took a very long time to complete, some six or seven years. Possibly because of the expense involved, he could not have afforded to do it any faster. Nevertheless, Dumcrieff was not just a holiday home for the Baron and his family; it was to become of enormous significance to the first Clerk Maxwells.

Public Offices, Interests and Later Life

In the years following the building of Mavisbank, the Baron, now entering his fifties, continued with his duties at the Exchequer[27] and the management of his own coal mines and estates as usual. Nevertheless, he took on some new public offices and still found time to broaden his personal pursuits. In 1726, his mentor the Duke of Queensberry made him one of the trustees acting in his affairs. This occasioned trips to the Duke's family seat at

Drumlanrig, which no doubt had helped sow the seeds of his idea of building himself a place near Moffat, which although not quite en route was less than twenty miles distant.

On a visit to Carlisle in 1724, the Baron had met the antiquarian William Gilpin of Scaleby Castle, 'I had all the pleasure imaginable in his Company', and afterwards corresponded with him (BJC, pp. xxiv, 119). Although the correspondence was cut short by Gilpin's death at the end of that year, it is worth mentioning that his sister Dorothy married Jabez Cay who was the first Cay of Charlton Hall (see Chapter 14). In late 1726, the Baron formed a correspondence 'upon Greek and Latine Literature, and particularly upon Antiquities', which he had ever been interested in, with some like-minded English gentlemen: Roger Gale, Esq, excise commissioner; Dr William Stukeley, physician turned clergyman; and the Earl of Pembroke. This resulted in his acceptance as member of the Society of Antiquaries in London, whose president was then the Earl of Hertford.[28]

In the spring of the following year he travelled to London where he visited many friends and useful contacts. One in particular, his new friend and correspondent, Roger Gale, took him to visit the Royal Society of London at Crane Court, off Fleet Street. The Baron had just shortly before visited its recently elected president, Sir Hans Sloane, a great collector, who showed him a greater 'Treasure of Valouable Antiquities, jewels, and medals, and Gold, silver, and copper' than he had ever seen before. But it was not until around November of 1729 when at last he was informed, 'by a letter from Roger Gale, my good English friend and constant correspondent, that I was chosen a Member of the Royal Society [of London, on 16 October]' (BJC, p. 126).

In the following year, 1727, the Baron was appointed one of the Trustees for Manufactures in Scotland,[29] a commission that had newly been set up following representations by the Baron himself, his cousin Duncan Forbes of Culloden,[30] Charles Erskine[31] and others, to strengthen provisions under the Articles of Union for boosting the economic situation in Scotland. He was a man very suitable for the position, for he was not only skilful in the law, politics and negotiation, he was also of an inquisitive mind and certainly knew a great deal about mining.

The year 1731 saw two of the Baron's sons leaving home to continue their education. For George, it was a trip no further than Penrith, to a boarding school at Lowther, while for James it was a case of following in his father's footsteps and setting off first for London and thence to Leiden, where George was to join him some five years later. George returned in 1737, but it was two more years before James came home after an extended grand tour. Like his father, he had a fascination for art, music and Europe.

The Baron's interest in antiquities and philosophy led to him becoming a founder member of the Philosophical Society of Edinburgh, of which he was one of the original co-vice-presidents.[32] He was already a Fellow of the Royal Society of London, but he now had access to a stimulating environment where he could meet with local *illuminati*, gentlemen such as himself, the number of whom was now swelling rapidly. Indeed, he soon presented a paper on the early languages of Britain and indulged his interest in astronomy. He noted, sometimes in great detail, comets and eclipses of the Sun, Moon and Jupiter.[33] He had wondered, as the ancients had done, whether the brilliant comet of 1744 heralded an impending calamity; indeed, Bonnie Prince Charlie had already set sail on an exploratory venture with the French marshal, Count de Saxe.

Following his son George's rather early marriage to his cousin Dorothea in 1735, the Baron busied himself in sorting out their rather complicated affairs (see Chapter 8). His efforts in building Dumcrieff had been partly for their benefit and when it was eventually finished he turned his attention back to Mavisbank, to put a 'finishing hand' to it, which he succeeded in doing by the summer of 1739. He continued to live there to be near his coal workings at Loanhead, otherwise living at his hereditary seat of Newbiggin, at Penicuik, which he eventually decided was the more fitting of the two for a baronet.

When the second Jacobite rebellion eventually came in the summer of 1745, the Baron was nearly seventy and too old to take up the call to arms as he had done thirty years before. On receiving the news that Bonnie Prince Charlie's rebel army had entered Edinburgh, he quite sensibly preferred to be out of the way rather than in the action. Heading for England with his wife and eldest daughter, he left it to his eldest sons James and George to defend the family honour, as one might expect, from the government side. This rebellion was a much bigger affair than 'the 1715', with Edinburgh and environs under rebel occupation for about six weeks, but when the rebels eventually marched south, the Baron and his family headed home from the comparative safety of Durham. Once back at Newbiggin, they discovered that sixteen to twenty of the Highlanders had billeted themselves there, in the course of which they had made free with his stores of hay and meal and helped themselves to his best horses.

Although this was to be the Baron's last great physical upheaval, the following years saw his health decline, which he valiantly tried not to show. Nevertheless, at the Exchequer Court he was now involved in managing those estates in Scotland that had been forfeited by rebel lairds in the aftermath of the '45. He was still able to make the journey to Dumcrieff and Drumlanrig, and even ventured north, to Perthshire, to take the goat whey near Lawers and to see the progress made in getting the 'Highlanders' to take up linen manufacturing for a living. Walking any distance got ever more difficult, but even when he was just two years short of eighty, a great age then, he still managed to ride well. This we know because he had the misfortune to be thrown off his horse at Penicuik bridge. Although he ended up no worse than badly bruised, the experience must have revived unpleasant memories of his boyhood accident on a horse, which had nearly killed him and left him with a permanent limp.

Still enjoying the thrill of learning something new, the Baron spent time poring over his beloved books and writing his manuscripts. Even if he was now allowing his son and heir James to do some of the work, he was still the instigator of many improvements in the village and about his estate; for example, the Terregles Tower on Knight's Law was built when he was about seventy-five years old. Over the years he had accomplished many things at Penicuik (Wilson, 1891, pp. 10, 153–154; BJC, p. 233), and most of the things he built are still there today as a memorial to a man of many talents. Ever the skilled administrator and clever manager, he made rapid progress, and he possessed energy and erudition in equal measure. In October 1755, the *Scots Magazine* announced his death to the nation: 'Octr. 4. At his seat of Pennycuik, Sir John Clerk of Pennycuik, one of the Barons of the Exchequer. He had been Baron since the union in 1707. He is succeeded in estate, and the title of Baronet, by his eldest son, James.'

His only enduring regret was his unrefined manner of speech and writing, which he always felt placed him at a considerable disadvantage to his English cousins. He consequently felt that his manuscripts were inferior; he frequently revised them, and would even forbid them to be seen except by friends and family. He earnestly desired his sons to do much better in this respect, so that they could truly become gentlemen, but a key piece of advice that he pressed on them was, 'You have nothing else to depend on but your being a scholar and behaving well' (BJC, p. 138, n1).

It was a motto that he himself had lived by. His wife Janet survived him by just over four years.

CHAPTER 5
William Clerk and Agnes Maxwell

The 1st Baronet of Penicuik had three sons by his first marriage. The first was his heir, John, whose story we have just recounted; the second was Henry, a favourite brother of John's who took to the sea and died unmarried; the third was William, born on 12 July 1681 and named after his great-grandfather. Although the direct line of the Clerk succession passed him by, it was William who played the pivotal role in our story by marrying Agnes, a daughter of one of the many minor Maxwell lairds with properties in and around Dumfries, for it was through this marriage that the Clerk Maxwell line was begun.

Although from his portrait painted in the early 1700s (Plate 10) William looked very much like his elder brother John, they were altogether different characters. William was less inclined to apply himself and rather more inclined to be wayward. In fact, it seems his father had his difficulties bringing William to heel and even wished him to sign a bond for his good behaviour.[1] He wanted William to prepare for life in the same manner as his older brother, and so he was first put to Glasgow University, where his best efforts were such that he was expelled for having an affair with a servant, Janet Steinstone (Stevenson), with whom he fathered a son, Henry Clerk.[2]

William was consequently sent away to Leiden just before his nineteenth birthday to continue his studies there, but he was back in Edinburgh again within a couple of years. In all probability things were such that his father would never have contemplated letting him go on any grand tour, and could have prevented it by denying him the letters of introduction that he would have needed.

Clerk and Ramsay

Despite his wayward tendencies, by early 1704 William managed to qualify as an advocate (Grant, 1944, p. 37). By no means does this imply a complete reform on his part, for he was ever the spendthrift[3] and his restless spirit earned him the nickname 'Wandering Willie': 'From his liking to visit, and shift about, from house to house; among his companions, he got the name of "Wandering Willie"' (Ramsay, 1808, p. 640n). The footnote goes on: 'When at New Hall House, he slept in one of the garret rooms, adjoining to those of Allan Ramsay, and Mr Tytler.'[4]

Until 1714, Newhall near Carlops belonged to Sir David Forbes,[5] who was William's uncle by marriage on his mother's side (Chapter 3), and after that date it passed to his son John (1683–1735) who, like his cousin William, was an advocate. The Allan Ramsay being referred to here was the well-known early Scottish poet[6] and friend of the Baron and his family. He and William Clerk were friends, possibly from as early as about 1710. Feckless though he may have been, William was not untalented; he drew, composed songs and wrote poems, such as *Letter in Verse from Mr WILLIAM CLERK, Advocate to Dr ALEXANDER PENNECUIK of New Hall, May 1714* (Ramsay, 1808, pp. 640–641). William, Ramsay and Dr Pennecuik had something in common; their fondness for wandering the environs of Penicuik, no doubt often ending up in each other's company, either at Newhall or at one of the many local 'howffs'.[7]

When he was about fourteen, the young Allan Ramsay had been sent by his stepfather to Edinburgh, some fifty miles distant, to learn a trade. The reason was simple enough: the rural economic conditions of the time were poor and Ramsay would have to be able to fend for himself in the city. Indeed, his older brother Robert had been sent off on the same journey five years earlier. After he had settled in Edinburgh and been accepted as an apprentice wigmaker in 1704, Allan Ramsay became close friends with William Clerk in spite of their social backgrounds being entirely different. This was not unusual; for in the Scottish society of the time there was a much smaller gap between the landed gentry and those that worked for a living: a laird such as the Baron might manage a coal mine and a country lad like Ramsay might become a scholar. In addition, there were some common factors: Ramsay's father had been a mining manager at Leadhills, and William and John Clerk were both keen on poetry, something the young Ramsay was showing a talent for.

But how did William Clerk and Allan Ramsay first become acquainted? We cannot say but, like William, Ramsay seems to have been fond of wandering and Newhall may have been a social nucleus that attracted them both. This is evident from Ramsay's poetic works, for his pastoral comedy, *The Gentle Shepherd* (Ramsay, 1808), is set at Habbie's Howe, a noted beauty spot on the Newhall estate. Their acquaintance may well have begun with nothing more than a chance encounter at one of the local howffs,[8] for they were both fond of ale.

As Ramsay was a wigmaker, their acquaintance could also have been occasioned in an entirely different way, by William, the advocate, coming to have his wig dressed at Ramsay's shop near Edinburgh's law courts. It is, moreover, possible that they could both have been members of one of the many Edinburgh clubs that were so popular at the time.

The Easy Club

Allan Ramsay became one of the early members of the Easy Club in 1712, when his career as a poet was only just beginning (Ramsay & Robertson, c. 1886, p. xxxii). The club initially comprised just eight 'young' men, all allegedly unmarried and in or about their twenties; on the face of it, it seems to have been just the sort of social and literary enterprise that both Allan Ramsay and William Clerk would have readily joined in with. The club's stated literary objective was to emulate English literary ideas and values; for example, the members would read and discuss Addison's *Spectator*. They even chose English literary pseudonyms, of which more shortly. The description Easy referred to a main tenet of the club, that its members should be 'easy' in each other's company, in particular, differences of either a religious or political nature were to be strictly out of bounds.

The use of pseudonyms suggests that the members did have certain reputations that they needed to put aside. It would have allowed them to forget who they were for a while and so be able to enjoy the company of their fellow enthusiasts without falling into a quarrel. An alternative view, of course, is that they used pseudonyms to allow themselves the freedom to express certain opinions that they could not give voice to in their own name. The description of themselves as being 'unmarried, and in or about their twenties' (Rogers, 1884, p. 358), seems so far from the mark as to suggest there was indeed some element of intentional obfuscation: Archibald Pitcairne was nearly sixty; Thomas Ruddiman was nearly forty; and James Ross, Allan Ramsay's father-in-law, could hardly have been in his twenties.

But if they were trying to be obscure, what was the reason? The year 1712 was a time of rising Jacobite sentiment, and even if an uprising was not actually imminent, talk of one was beginning to circulate. Some of the Easy Club's members were staunch Jacobites.[9] Moreover, there was an entirely separate current of nationalistic sentiment; those involved wanted the clock turned back to before the Union of 1707. Of course, there were those that saw a Jacobite rising as a good solution to both of these issues. For example, members such as Archibald Pitcairne and Thomas Ruddiman were not only

Jacobite sympathisers but had been involved in disputes over it. Pitcairne was particularly disputatious and had even been suspended from the Royal College of Physicians of Edinburgh for pushing his opinions too far (Henderson, 1896). Hepburn of Keith was also a staunch Jacobite, and was so well-known for it that on the outbreak of the 1715 uprising troops were immediately sent to arrest him, with the unfortunate result that Hepburn's son became the first casualty of the conflict. Another, John Fergus,[10] was a staunch nationalist, for his rantings on the subject were recorded in the club's minutes (Andrews, 2002, p. 2), and he alone refused to take an English pseudonym, taking instead the name of George Buchanan from the outset. The problem for the nationalists was that they were taken for Jacobites irrespective of who they thought should be the rightful monarch.

If men such as these did ever leave their factions at the club door, it did not last long and the club's sentiments began to swing more overtly towards nationalism. Indeed, all of the original English pseudonyms were soon replaced by Scottish ones. Ramsay, who had styled himself as Isaac Bickerstaffe,[11] switched to 'Gavin Douglas', the sixteenth-century Bishop of Dunkeld who had been a poet.

Ramsay became a burgess of the city in 1710 by right of being by this time a master of his trade. Nevertheless, he would not have found it easy to join a club of select gentlemen. Furthermore, as an early member he must have already been in very good standing with at least one of the other members for whom the question of status was not a problem. It is said that the person in question was a lawyer called James Ross (Smeaton, 1896, pp. 46–47), his future father-in-law. Ross was well enough off but did not belong to the upper echelons of his profession. Nevertheless, he cultivated friends at all levels in society and his house in Blair's Close off Castlehill[12] (Smeaton, 1896, p. 10) was a focal point for Edinburgh's literati. But, as the story of Ramsay's wooing of Ross's ward Christian Ross[13] suggests, Ramsay was both persistent and persuasive (Ramsay & Robertson, c. 1886); to be persuaded to allow the marriage, Ross may have been reasonably impressed by Ramsay's prospects as a master wigmaker, but he was probably susceptible to the fact that Ramsay's literary potential could eventually serve to boost his own standing amongst those that attended his soirees. At any rate, he vouched for Ramsay at the Easy Club and persuaded him to produce his earliest known poem as his introductory address, in effect launching the aspiring young poet's literary career.

While Allan Ramsay's membership of the Easy Club is well established, the social writer Rev. Charles Rogers (1884, pp. 356–358) has it that the Baron, William's brother, was also a member, in fact its secretary. However, if any of the Clerk brothers was involved with the Easy Club, it was far more likely to have been William. The true identity is obscured by incomplete knowledge of the real names behind the pseudonyms of each of the (approximately) twelve members. Even the accuracy of those identities that are supposed to be known is in doubt, for at least two have been disputed (Martin, 1931, p. 26; Gibson, 1927). We therefore cannot be certain, we can only argue that it would have been uncharacteristic of the well-regulated Baron, while his more self-indulgent brother William would have been more likely to join in the nationalistic banter of the Easy Club.

From what we know of John Clerk, he would have been disinclined even to appear to depart from the establishment line; he did Queensberry's bidding and duly took part in the negotiation of the Union; and he also came out for King George in the Jacobite rising of 1715. If neither Jacobite nor nationalist, he would have been quick to distance himself from the Easy Club at the first whiff of any such leanings. Nevertheless, Rogers does state that he was the secretary, and that it was he who took the pseudonym George Buchanan.

In 1713, Ramsay wrote an elegy on the death of the disputatious Jacobite, Dr Pitcairne. Although it does not appear in any of Ramsay's own publications or later collections, it was read to the Easy Club and printed by them for circulation, making it Ramsay's first published work. It opens with a couplet originating from Gavin Douglas' *Aeneid*:[14]

Some yonder bene for ready gold in hand,
Sold and betrayed their native realm and land. (Pittock, 2006, p. 154)

This, and the nationalistic tenor of the poem in general, is unmistakeable, and indeed it is the reason that Ramsay never published it in his own collections. 'George Buchanan', however, wrote a response in praise of Ramsay's elegy that, by reflecting these very lines, could not conceivably have come from John Clerk,

And then at last by whom [their darling liberty] was basely sold
For the dire thirst and love of English gold (Rogers, 1884, p. 368)

If so, he would only have been damning himself.

Although Smeaton (1896), who saw the original club minutes, was able to identify about half of the members, John Clerk was not amongst them. Nor does Andrews (2002) mention Clerk, rather he proposes that George Buchanan was actually the name taken by John Fergus. But even if Rogers was wrong in some respects, it is curious how he came to assert that John

Clerk was a member.[15] But if John Clerk was not a member, and there is any grain of truth in Roger's assertion, it could well be the case that he was thinking of another Clerk with poetic inclinations, William,[16] who was hardly so cautious as his brother. He was a friend of Ramsay's, a fellow poet and of a similar age, about thirty to Ramsay's twenty-six when the club opened, and if the Baron had been for the Union he may well have chosen to take the opposite stance.

Furthermore, Ramsay went on in the poem to point out that none of the Easy Club's members had tainted themselves in such a manner (Pittock, 2006, pp. 154–156); he could not, in all seriousness, have said that of John Clerk, whose position as a baron of the exchequer and a little monetary 'compensation' had been his reward for the part he played in bringing about the Union. At any rate, Ramsay did not publish the poem and, by the time the uprising was actually imminent, he no doubt realised the real danger that these politically imprudent verses posed – anything that carried the slightest whiff of disloyalty of any sort would have been seen as seditious. The other members no doubt began to see things in pretty much the same light and so the club was soon disbanded.

Despite being on the opposite side of the fence from the Baron as far as the 'Union' was concerned, Ramsay became a firm friend of the Clerk family as a whole. Grant, however (BJC, p. 229n1), found cause to remark that, good friend and frequent visitor to Penicuik House though Allan Ramsay was, he is mentioned only once in the Baron's memoir. On the other hand, we are also told that the Baron had a portrait of Ramsay, painted by his cousin William Aikman (see Chapter 3) inscribed on the back with a comic verse bearing the line 'By Aikman's hand is Ramsay's snout [. . .]' signed 'J. C, Pennicuik, 5 May 1723'. It is therefore likely that the Baron commissioned the portrait but, either way, it signifies that by that date at least there was a strong friendship between him and the poet.

The date, however, may have some significance, for it was a month to the day after the death of William. Other evidence of the friendship of Allan Ramsay is to be found in his poem 'To Sir John Clerk: On the Death of his Son John', in 1722. The Baron in addition helped Ramsay's son, also called Allan, when he was trying to establish himself as a portrait painter in Edinburgh. He also advised the young artist on the design of a house that he was building for his father, the 'Goose-Pie' that still looks down on the New Town from its vantage point on the north side of Castlehill. Further, he helped him get introductions to the artistic community in Italy (Ingamells, 2004). In turn, Ramsay the poet wrote a poem for the Baron's second son, James, shortly before the Baron's death, and there is more besides, including letters written in familiar terms to the Baron's son George, and the obelisk on the Penicuik estate raised in 1759 to the memory of the poet.

Making the Maxwell Connection

While it is known that William Clerk married the heiress Agnes Maxwell (C&G, p. 17; BJC, p. 114), he had previous designs of marriage to at least two other women. The first was Henrietta Porteous (1678?–1762) of Hawkshaw[17] which lies in the Tweed valley close by the principal route from Edinburgh to Moffat. While William was known for wandering, Hawkshaw was some distance from Penicuik; but it did lie only ten miles north of Moffat, a place frequented by members of the Clerk family in the late summer months (see Chapter 4), at least as early as 1706 (BJC, p. 63). Since William's return from Leiden was about 1702 and Henrietta married Michael Anderson of Tushielaw in April 1706[18] it seems the affair would have taken place sometime during 1703–05. However, his father would not give his permission for the match, perhaps because Henrietta's family had a certain reputation owing to a remarkable atrocity that occurred during the civil war of 1641–50 (Pennecuik, 1815, p. 244n) in which sixteen of Cromwell's horsemen were captured and slaughtered in cold blood by a Porteous laird. If that was not enough to put off Sir John Clerk, Hawkshaw had been put up as surety against debts (Buchan & Paton, 1925, vol. 3, pp. 381–415). On William's side, he had a record of irresponsible behaviour and as yet could have had no great income. It would have therefore been clear in his father's mind that William needed some considerable time to prove himself ready for marriage.

Henrietta was widowed in 1719 but by this time William had moved on in his search for a wife. He secondly pursued Helen Smollett, daughter of Sir James Smollett of Bonhill, the grandfather of the author Tobias Smollett. Sir James was a director of the Bank of Scotland and so Helen could have proved a very good catch indeed. The match having doubtless been scotched by both sets of parents,[19] in an interesting twist to the story, William's niece Jean Clerk was to marry Helen's nephew, the advocate James Smollett of Bonhill.[20]

William's eye soon afterwards alighted on Agnes Maxwell (Plate 11), daughter of John Maxwell of Middlebie in the county of Dumfriesshire (now Dumfries and Galloway).[21] Dumfries was and is an ancient royal burgh, the principal town in the county of the same name. As such, it was home to all the main markets, courts and kirks for miles around. It was also the most important town between Edinburgh (or for that matter Glasgow) and Carlisle. Lying about seventeen miles east of Dumfries itself, the village of

Middlebie is some considerable distance from Penicuik, in fact nearly seventy miles. It may be that William came about Middlebie or Dumfries on one of his wanderings, or equally well, Agnes could have gone to Moffat for pretty much the same reasons that the Clerks did, for the spas, or she may even have been visiting family in Edinburgh.

What we know for certain, however, is that in September 1718 William wrote a letter to his beloved 'A', begging her to accept his proposal of marriage.[22] He could not come to see her because his father was desperately ill, which was at least true,[23] and he recalled to her that when he had last seen her at Dumfries, his father had horses sent to Moffat, to bring him back in all haste. However, he had found the time to spare for a visit with one of 'A's aunts. Poetically, he signed himself 'Damon'. We do not know the reply, but in January of 1719 he wrote again, this time with a song, extolling the virtues of his 'Celia'.[24]

> *When Celia gracefully walks through the plain,*
> *she's admir'd by each shepherd, adored by each swain.*

If he were being true to Agnes, then she was his Celia. But how do we know for sure? For the answer we must turn to literature of the time, in which the Celia of the song, and the Damon of the previous letter, are the archetypal characters of pastoral poetry, for example, as employed by George Farquhar (1677?–1707) in *Thus Damon knocked at Celia's door* (Farquhar, 1700). The same names were still current when Ramsay chose them for the principal characters in *The Morning Interview* (Ramsay, 1719–22).[25] They would have been understood by almost everyone of that era and therefore it required no great leap of imagination for William, the country lad, and Agnes, his true love, to see themselves as the Damon and Celia in his song.

By the time of the next letter,[26] the couple were married and expecting their first child. William wrote it from his father's house at Penicuik, where it seems that Agnes was not welcome. He was asking her to be patient and reassured her that in time things would sort themselves out. This certainly suggests that William had married Agnes against his father's wishes, and his reasons were no doubt much the same as before. We may guess that the marriage had taken place sometime after September 1718, for his letter from that time shows him to be still in the process of wooing Agnes. No record of the marriage has been discovered in the old parish records on Scotland's People, but the same is true for many Clerk marriages. There is a record, however, of the birth of their only child, a daughter Dorothea, who was born in Edinburgh on 28 August 1720.[27]

In 1722 we find William writing to Agnes on two further occasions when he was at Penicuik House and she at Castlehill,[28] which is therefore most likely to be where they both lived in Edinburgh. The later of the two letters is dated 9 March 1722, when he had been attending the wedding of his sister Mary and the Rev. Alexander Moncrieff of Culfargie[29] minister at Abernethy. William mentions that he needed to stay on because of his father's poor health, and indeed Sir John, having been unwell for a couple of days, died peacefully the very next night (Chapter 3). Note again, that Agnes was not with him at Penicuik.

The Baron's youngest son William, aged one, took ill in November 1722 and died in the early days of January 1723, to be followed by his two-year-old sister Mary, who died in March. The Baron's wife was then also very ill and 'was reduced to a meer shadow'. It therefore seems that some serious illness was doing the rounds of the Clerk family, for William Clerk himself had been ill since at least December 1722. From about February 1723 he was being medicated on a regular basis with such things as 'pectoral pills', 'anti-febrile decoctions', 'spirit of hartshorn' (ammonia solution) and laudanum.[30] At last, on 5 April several things were tried, but the very last was for Agnes, '2 ounces of the spirit of wine and camphire', to fortify her in her grief; clearly, William had died that very day. Aged forty-one[31] he left behind a widow who was still in her early twenties, and Dorothea,[32] the only child of their marriage and just over two and a half years old.

From the funeral bill[33] we learn that William was buried at Penicuik on 12 April 1723. The funeral procession was no mean affair, for the journey from Edinburgh required five vehicles, including two mourning coaches and two ordinary coaches, each drawn by six horses!

William left a number of debts, most of which Agnes had to settle.[34] Gentlemen's bills often went for years without being paid, as so aptly described in some of Allan Ramsay's poems,[35] and William was no exception. His personal bills and bonds dating back as far as March 1720 went unsettled, and to these were added his medical bills and funeral expenses. Having sold William's horse and books to raise some money, Agnes was left with a balance of nearly £100 to pay off.[36] This she may have been able to do with the help of the man she was to marry, of whom we will hear later, in Chapter 7.

William was the Baron's only surviving full brother, but in his memoirs the most he could say of him was that he had been an advocate! As to his sister-in-law Agnes Clerk and his niece Dorothea, he says little more than that he and Agnes became joint tutors of the child. Nevertheless, Dorothea Clerk was in due course to become central to the Baron's plans.

CHAPTER 6

The Maxwells

The Origins of Maxwell

The Maxwells were a powerful and noble family with two principal branches, one with lands in and around Dumfries and its seat at Caerlaverock Castle, and the other based in the lower Clyde valley with its seat at Pollok on the south-western outskirts of Glasgow. The original form of the name is sometimes given as Maccuswell, which had the variations Maccusville and Maccusweil, both evident in Maxwheel, the present name of a pool on the River Tweed near Kelso Bridge, around which the village of Maxwellheugh developed. King David I, who reigned from 1124 to 1153, had granted land there to Maccus (Douglas, 1899) and the pool Maccus' Weil would have been his.[1]

As a result of growing Norman influence in southern Scotland, Herbert, a descendant of Maccus, was styled as Herbert de Maccusville. His son's name, however, is generally given as John de Maccuswell. John flourished in the early thirteenth century and was the sheriff of Roxburgh. It was he who acquired the barony of Caerlaverock under Alexander II and in 1231 became Great Chamberlain of Scotland. He had obviously become a powerful man, and in time this place, rather than Kelso, became the main focus of the Maxwells (Figure 2.1).

Sir John was succeeded by his son Eumerus, variants of which name, Aymer and Homer, crop up in later generations. Through giving lands in Renfrewshire to John, his second son, Eumerus established the line of the Maxwells of Pollok. His eldest son Herbert de Maxeswell succeeded him at Caerlaverock, and by the end of the century it had gone to his son of the same name, but now given as Herbert Maxwell (Burke, 1834–38, vol. 1, pp. 325–326). This then, is how the Maxwell name evolved over a period of less than 200 years.

The 1st Lord Maxwell of Caerlaverock was created sometime between 1440 and 1445 (Paul, 1909, pp. 474–475). Lord Herbert Maxwell was a direct descendant of the Sir Herbert Maxwell mentioned above, and including him the succession of Maxwell lords was:

Herbert	d. before 1453	
Robert	d. 1485–86	
John	d. 1484	Also called Master of Maxwell because his father, having merely resigned in his favour in 1478, actually survived him
John	aft. 1454–1513	Died at the battle of Flodden
Robert	c. 1494–1546	Lord Provost of Edinburgh in 1528 and 1535–37
Robert	d. 1552	His widow Beatrix was mentally ill during 1562
Robert	c. 1550–1553	Died an infant
John	1553–1593	Born after the death of his father, he succeeded his infant brother Robert. Created Earl of Morton in 1581, but this was rescinded in 1585. Killed by James Johnstone at Dryfe Sands
John	bef. 1583–1613	Attainted and executed at Edinburgh on 21 May 1613 for the murder of James Johnstone
Robert	d. 1646	Succeeded his brother John when the attainder was lifted. Created Earl of Nithsdale in 1621 in lieu of the Earldom of Morton

Up to and including the first Earl of Nithsdale, there were therefore ten Lords Maxwell, spanning eight generations. Some writers do not credit John, Master of Maxwell, as being one of the lords, so there is the difficulty of having two different ways of numbering them. Our numbering here is both consistent with Paul (1909) and Andrew Wedderburn-Maxwell's genealogical notes.[2]

Going now only so far back as the eighth lord, John Maxwell was born on 24 April 1553, when Scotland was ruled by James Hamilton the Earl of Arran, regent for Mary, Queen of Scots, who was then just ten years old. After her return from France in 1561, Mary reigned in her own right but was forced to abdicate in 1567 following the scandal of her husband Lord Darley's murder and her precipitate remarriage to the Earl of Bothwell. With Mary's subsequent imprisonment, regents once again had to be appointed, for her son, James VI, was only a year old. The last of the four regents of that troubled era was James Douglas, Earl of Morton, who had been appointed in 1572, and although he was successful in quelling civil strife, he was eventually accused of complicity in Darnley's murder. It is easily overlooked that Darnley was the young King James' father, and if Morton had indeed been a party to the murder, he was heading for trouble when James became old enough to do something about it. It is likely that James' other advisers plotted against Morton and planted the seeds in James' mind, and in June 1581 Morton was tried, convicted and duly executed, whereupon his titles and estates were escheat by the Crown.

Perhaps because he was still within the manipulation of his curators, in the following October James bestowed the forfeited earldom of Morton to John, 8th Lord Maxwell, who was not then thirty years old; having recently purchased the barony of Carlyle, he was now styled John, Earl of Morton and Lord Carlyle and Eskdaill.

Unfortunately, in a few years' time the pendulum swung back in the other direction, and King James revoked John Maxwell's earldom and restored it to Regent Morton's nephew, Archibald Douglas, Earl of Angus. Perhaps James had never believed in Morton's complicity, or perhaps he was still being influenced by his advisors, but it was a remarkable thing that the earldom should be first given and then taken away so soon. Lord Maxwell was nevertheless unabashed and continued to refer to himself, as his followers generally did, as the Earl of Morton.

And so it continued until 1593 when our 'Earl of Morton' was killed in a battle at the Dryfe Sands[3] near Lockerbie (Figure 2.1). These were still very troubled times, especially in the border regions (Fraser, 1971) and while it had long been the case, the source of the problem had generally been tit-for-tat raids and concomitant feuding between families of all ranks on opposing sides of the border, where the ordinary rule of law was unknown. As a measure intended to prevent these cross-border feuds escalating into a state of perpetual war, the entire region had been divided into Marches, three on each side, each under the wardenship of one of the most powerful regional barons who then effectively ruled them in their own right. Hot pursuit and retribution for wrongdoing had to be done according to strict March Law enforced by the wardens, often to their own advantage (Scott, 1838, pp. 23–24). Despite their 'Jeddart justice' being rough in the extreme, it brought no lasting peace, for the border reivers were no more than bandits out for their own gain. March Law was simply a means of keeping the lid on a situation that was beyond the control of the respective monarchies.

Towards the end of the sixteenth century, the 8th Lord Maxwell, 'Earl of Morton', was warden of the Scottish Western March. Seeking to put an end to a long-running feud between the Maxwells and the Johnstones in his own favour, he obtained a commission from James VI to raise an army of about

2,000 men with the object of taking Sir James Johnstone of Dunskellie and his henchmen by force and bringing them to Edinburgh for trial as outlaws. The size of this army gives some indication of how difficult he thought the task might be, and so he also offered a bounty for Johnstone's 'hand or head'; he clearly had little intention of letting him get as far as Edinburgh, let alone stand trial. Johnstone, however, got wind of the impending raid and quickly managed to get a few hundred men together for the purpose of making his own pre-emptive strike against Lord Maxwell. In this he entirely succeeded, with the Maxwell army being routed at Dryfe Sands on 6 or 7 December 1593. Sir James, having himself offered a bounty for Maxwell, paid it to William Johnstone of Wamphray who had succeeded in cutting off Lord Maxwell's hand as he fled. Having parted company with his hand, Lord Maxwell did not escape very far and he was soon slain. Of his kinsmen who managed to survive, a great number were slashed on the head, leaving them with a permanent scar that became known as a 'Lockerbie Lick'.[4]

If the Maxwell–Johnstone feud had been bad under the 8th Lord Maxwell, things did not improve under his son John. While there was no further battle on the scale of Dryfe Sands, John was a hothead whose conduct seemed even more out of control than was usual amongst these fearsome people.[5] He sought to quarrel with the re-established Earl of Morton by trying to assert his own claims to that earldom, for which he was imprisoned in Edinburgh Castle in 1607. But he escaped and was soon back in his old haunts around Dumfries, still showing complete disregard for King James' authority. Perhaps he realised that he could not carry on living as an outlaw forever, for he then tried to negotiate reconciliation with Johnstone, but the outcome of an arranged private meeting on 6 April 1608 with his old family foe was that Johnstone ended up dead, shot twice in the back. Whether the unstable Maxwell simply could not keep himself under control or he had actually been engaging in a ruse to avenge his father cannot be said for sure. Either way, in the Crown's view he had 'treasonably murdered' Johnstone.

Lord Maxwell was tried[6] in 1609 but, as on previous occasions, he managed to escape, this time fleeing to France in order to lie low and let the heat die down. But, impetuous as ever, he returned too soon. Arrested in Caithness, he was brought back to Edinburgh and executed according to his original sentence at 4pm on 21 May 1613, his estates and titles being forfeit to the Crown. He seemingly went to his death more penitent than he had ever been while he lived.[7]

John, 9th Lord Maxwell, died without leaving a male heir and he was therefore succeeded by his brother Robert. From 1616 on, his family's former rights and titles were restored in stages, culminating in Robert being created

1st Earl of Nithsdale in 1620, with all due precedence reinstated back to 1581, that is to say, when his father received his short-lived earldom. This honour, and the fact that he lived till 1646, demonstrates that the 10th Lord Maxwell was much more circumspect and peaceable than either his father or his older brother had been. He had, like all his forefathers, remained a staunch Roman Catholic[8] throughout the reformation years and the covenanting times. This eventually went against him during the Scottish Civil War of 1644–45, because it cast him on the Royalist side, just as his father had been in the days of Mary, Queen of Scots. Along with his son, he consequently joined from the outset Montrose's campaign in defence of Charles I (McDowall, 1873, pp. 358–359);[9] being on the losing side he had to flee to the Isle of Man and his estates and titles were forfeit. He died in Edinburgh in 1646.

The Maxwells of Middlebie

We draw a line here on Maxwell lords because the next step in the Clerk Maxwell line came about through another John Maxwell, a bastard of the 8th Lord Maxwell who died at Dryfe Sands. He was therefore a half-brother to the 9th and 10th Lords, John and Robert. In May 1624 Robert, then Earl of Nithsdale, granted his half-brother John the Twenty Merk Land of Middlebie to the east of Dumfries (Figure 2.1), making him the first John Maxwell of Middlebie. Much of the information about his line of succession is shown in Family Tree 3, our second excerpt from John Clerk Maxwell's draft, but here we give a more detailed, and hopefully more legible, reconstruction. Each generation is numbered 1, 2, 3, etc., with appended roman numerals distinguishing siblings:

1. **John Maxwell of Middlebie** married **Marion Maxwell**, daughter of **Homer Maxwell of Speddoch**, see below, in 1628, and had sons, Robert, William, Homer and James. Homer and James, together with a sister Elspet, died within two months of each other in the April and June of 1636 (DGFHS, 2008), presumably as a result of some epidemic disease. John died in 1639, supposedly murdered by John and Thomas Blake,[10] whereafter by September 1641 Marion had remarried to Andrew Stewart (Burke, 1894, p. 1365) who was a legal writer.[11] In 1637, she appeared before Dumfries Kirk session for failure to attend communion (Bryson & Veitch, 1825, p. 347), the implication being that she preferred to adhere to Roman Catholicism. Andrew Stewart died in 1662 and she survived him until 4 January 1674 (DGFHS, 2008).

2. **Robert Maxwell of Middlebie** succeeded his father in 1639 and, unless one of the following names is erroneous, he married twice. They are **Miss Dacre**[12] and **Miss Maxwell of Kilbean**.[13] One was named Mary, for in 1649 Lady Mary of Middlebie was upbraided in the Kirk of Dumfries for lack of attendance at communion, just as her mother-in law Marion had been in 1637 (*Literary Gazette*, 1825, p. 460). Robert died before 1679 and was succeeded by his son John 3(i).

3. (i) **John Maxwell of Middlebie**, referred to as the 'Entailer', married **Dorothea Bellenden** or **Ballantine** in 1680. It was he who executed the crucial entail of the lands of Middlebie in 1722. Contrary to what is given in earlier histories, in addition to the daughter Agnes who eventually became the heiress of Middlebie, he had two sons, John and Robert, and an elder daughter Anna.

    (ii) **Robert McLellan**, who lived in Glasgow.[14] The different surname could imply either that he was a half-brother, or had changed his name as a condition for inheriting property (his nephew Robert Maxwell 4(ii) below, is a further example of this – see note 17 thereto).

4. (i) **John Maxwell of Middlebie** the '**Younger**' or '**Last**', was confined in May 1715 in the Tolbooth of Dumfries for three days following a breach of the peace, threatening behaviour, abusing his parents and family, and burning their house over their heads – and it was not his first offence. Liberated by consent of his father, he was again jailed on 15 May 1722 for a further breach of the peace and 'making a riot' in Chappell's legal chambers. John had many debts, and it is likely that he had gone to these chambers as a result of his father drafting an entail, presumably aimed at preventing him from selling Middlebie. Indeed, the entail was executed just a few weeks later on 9 June. As to his debts, John the Younger had a string of them, for which he was imprisoned during 1724 on several occasions. He married Grace Smith[15] but died without issue sometime about January 1728.[16]

    (ii) **Robert Maxwell**, who changed his name to, and bore the arms of, McMillane.[17] We must assume that in doing so he was debarred from the succession of Middlebie, otherwise he must have died without issue before 1722 when the entail was drawn up. He is also mentioned along with his father in 1730 concerning a debt,[18] but such things often went back years and so the year 1730 is no indication that he was then alive.

    (iii) **Anna Maxwell**, eldest daughter of John Maxwell of Middlebie 3(i).[19]

    (iv) **Agnes Maxwell, heiress of Middlebie** was born on 9 February 1697.[20] She first married, in 1719 or early 1720, **William Clerk**, the

Baron's younger brother, whom we met in Chapter 5. However, contrary to what is often said, he did not take the surname of Maxwell; his own signature and all contemporary references to him are in the form of William Clerk. Furthermore, he predeceased Agnes in 1723, before she became heir to Middlebie, for both her father and her elder brother were still then alive. William and Agnes had a daughter, Dorothea, born on 28 August 1720.[21]

Agnes secondly married, on 26 July 1725,[22] a naturalised Huguenot émigré, **Major James Le Blanc**, who predeceased her in 1727 (see Chapter 7). She died in March 1728, just one or two months after her elder brother, leaving Dorothea in the care of her brother-in-law Baron Sir John Clerk. Consequently, she did not survive long enough to succeed to Middlebie except in name only, and the estate was then held in trust during her daughter's minority.

5. **Dorothea Clerk**, having been left an orphan at the age of seven became heiress to Middlebie and was the ward of her uncle, Baron Sir John Clerk, 2nd Baronet of Penicuik. In 1735, she 'privately' married her cousin, George Clerk, the Baron's second son. Although she was barely fifteen years old at the time, the legal age for marriage for females was then twelve. A portrait believed to be of Dorothea is shown in Plate 12.

What we know of the marriage of William Clerk and Agnes Maxwell has already been recounted in Chapter 5. This fusion of the Clerk and Maxwell lines was subsequently to be reinforced by the marriage of their daughter Dorothea, heiress of Middlebie, to George Clerk, the Baron's second surviving son, in 1735. As will soon be discovered, the name Clerk Maxwell did not come about until some three years thereafter, and we will now go on to uncover the reason for this.

The Middlebie Entail

The name Clerk Maxwell owes its origin to a deed called the Middlebie entail. Some explanation of the idea behind an entail, and the specific conditions of the Middlebie entail, is therefore in order.

John Maxwell of Middlebie and Homer Maxwell of Speddoch are just two examples of the old custom whereby landowners of significant heritable properties were known by the name of their property. Not only did they derive status from this, they also had a very personal attachment to their properties. It can therefore be seen that a landowner would do all that he possibly could to hold onto the land for himself and his successors in circumstances where two critical factors prevailed, uncertain finances and high rates of mortality. The first of these meant that if he had to borrow a significant amount of money then the property would be at risk if he became unable to repay the debt. The second meant that the property might fall into the hands of female heirs, for which the rules were different, so that the estate could be split up. It could also mean that if he had no children at all to pass the property on to, it could end up with some distant relative. Having considered all this, the landowner could opt to entail either all or at least a significant part of the property, particularly the part that would continue to carry the name. By this apparently simple means, he could try to ensure that it would be neither exposed to debt nor split amongst portioners. If he so wished he could also divert the natural line of succession and impose mandatory conditions on heirs.

The entail itself took the form of a legally binding deed for the purpose. One of the sundry conditions which was typical in entails is often mentioned as being the root of the Clerk Maxwell name: 'The next important clause [...] is an obligation to assume the name and to bear the arms of the Entailer' (Committee on Scotch Entails, 12/5/1828, p. 8).

Once the entail had been signed, registered and recorded it was fully binding and after the death of the entailer himself, no alteration could be made to its stipulations except by a petition to Parliament. A private Act had to be passed in order to allow the entail to be broken. Failing such an Act, an heir could not break the conditions of the entail without putting themselves in jeopardy of having their succession declared null and void.

While an entail involved conditions to prevent the estate from being encumbered by debt, this could not prevent the person in possession of an entailed estate from falling into serious personal debt. Moreover, on a person's death, their debts were included in their estate, the value of which could then be in effect negative! It is natural that a deceased person's creditors would try to get their money back out of the estate before the heir walked away with it, and they could always go to court to do so. But what they could not do was get their hands on an entailed property for this purpose. Unfortunately, neither could the heir sell the property; they would either have to find some other means of paying off the debts or step aside and let the next heir in line take his or her chance at it. It is therefore clear that entails did as much to create problems for heirs as they did to safeguard their heritage.

The Middlebie entail is a prime example of both the execution of the entail[23] and the 1738 Act of Parliament (11º Georgii II, Private Acts, 16) procured by petition, which allowed part of the property to be sold off. The important part of the entail is summarised in the Act itself, and with the key parts emphasised for clarity, we find:

> Whereas **John Maxwell of Middlebie** being feized in Fee or poffeffed of the **Twenty Merk-land of Middlebie** [...] difpone and provide Premifes **to himfelf for Life;**
> which failing to **John Maxwell his eldeft lawful son**, and to the Heirs Male lawfully to be procreate of his Body, and to the Heirs Male of their Bodies;
> which failing to **the Heirs Male lawfully to be procreated of his own body** [i.e. any other sons], and to the Heirs Male of their Bodies;
> which failing, to **Agnes Maxwell his Daughter, Spoufe to Mr. William Clerk, Advocate,** and to the Heirs Male lawfully to be procreate of her Body, in that or any other marriage;
> which failing to **the Heirs Female lawfully to be procreated of the faid John Maxwell Younger** his body;
> which failing, **to the faid Agnes Maxwell and to the Heirs Female which should be procreate of her Body**, in that or any other Marriage;
> which failing, to **the Heirs Male of the Body of Robert Maxwell of Schalloch,** his Brother-german, and the Heirs Male of their Bodies;
> which failing, to the **Heirs Male of the Body of Robert Maxwell Younger of Arkland;**
> which failing, to the **Heirs Male of the Body of Homer Maxwell, his Brother-german,** and the Heirs Male of their Bodies [...]

Following which, it would go to the heirs of Grizzel Grierson, who was a descendant of Homer Maxwell of Speddoch (see note 31 of Chapter 4). It then continued:

> and it should not be in the Power of the faid John Maxwell Younger, or any of the Heirs of Tailzie to alter or change the Order of Succeffion thereby eftablifhed, or to fell or mortgage any Part of the faid Lands and others aforefaid, nor to contract Debt, whereby any Part thereof might be charged or incumbered.

From this, therefore, it is quite clear the route by which Dorothea Clerk came to be heiress to the Middlebie estate on her mother Agnes' death in March 1728; John Maxwell of Middlebie the Entailer and John Maxwell of Middlebie the Younger were both then dead.

As to the stipulation that the name Maxwell had to be used, the entail itself states:[24] '[they] shall be obliged to assume bear and use the surname arms and designation of Maxwell of Midlbie [sic] as their proper arms surname and designation in all time hereafter'. But it also goes on to say: 'the said heirs female who shall succeed shall be obliged to marie gentlemen or adequate matches who shall use and weir the said name, arms and designation', and, furthermore, anyone who did not adhere to these stipulations would forfeit their right to succeed.

Without the entail, Dorothea Clerk may not have been the sole successor to Middlebie. It specified that it would go to her notwithstanding that she was a female heir. It is not easy to say whether this would have been the outcome without the entail, but the entail certainly made certain of it. It also ensured that the property of Middlebie that she became heir to could not be sold off and split up. The creditors of John Maxwell of Middlebie and his son John Maxwell the Younger would have to wait for a further ten years after Agnes' death to see any of their money. Finally, the provisions of this entail created the Clerk Maxwell name which had to be taken by Dorothea's husband, George Clerk, at the time she eventually inherited her property in 1738.

Maxwell of Speddoch Line

Before leaving the Maxwells, we now briefly return to the origins of Marion Maxwell, the wife of the first John Maxwell of Middlebie. The story is of more than passing interest because it reveals the relationship between the Maxwells, the Church they followed, and the Church they had to live with.

The first John Maxwell of Middlebie married Marion Maxwell, the youngest daughter of Homer Maxwell of Speddoch,[25] and their marriage took place in 1628 (see Family Tree 3). Being one of many lesser cadets of the Maxwells, it is hard to find anything more than mere fragments about them. Nevertheless, we do know that a previous Homer Maxwell was a burgess of Dumfries and, in 1543, a Baillie (Johnstone, 1889, p. 200).
By 1569, the Abbey of Hollywood near Dumfries had been secularised, and its rights and lands returned to the Crown (Barbour, 1884–85, pp. 64–65). Homer Maxwell, son of the earlier baillie of Dumfries,[26] was one of the Maxwells who benefited from this for he was granted a sasine on part of the Abbey lands about seven miles to the north-west of Dumfries. This place was called Speddoch (Figure 2.1), whence Homer gained the appendage to his name. It is rather likely that Homer had been a hereditary tacksman there before he received the grant of the land (Aitken, 1887–88, pp. 113–114), but at

any rate McDowall (1873, pp. 208, 272, 774) informs us that by 1584 he was a commissary of Dumfries and by 1593, its provost. He was also a baillie in 1585 and a commissioner (representative) for the burgh at a convention of the Scottish parliament. As a henchman of the 8th Lord Maxwell, he was present at the battle of Dryfe Sands in 1593:

> *The town Dumfries two hundred sent,*
> *All picked and chosen every one;*
> *With them their Provost, Maxwell, went,*
> *A bold, intrepid, daring man.* (McDowall, 1873, p. 272)

Homer died *c.* 1611 and was succeeded by his son who, unsurprisingly, was another Homer Maxwell of Speddoch (Aitken, 1887–88, p. 120).

Now, we may guess that in profiting from the demise of the abbey, Commissary Homer Maxwell of Speddoch had converted to the Protestant Church. Furthermore, the provost and baillies had to attend the kirk for the Sunday service.[27] But if so, any such apparent conversion was unlikely to be genuine because, even as late as 1588, the Reformation was struggling to take hold in and around Dumfries. In that particular year a special General Assembly had to be convened to deal with the problem (McDowall, 1873, p. 235), and our Commissary Homer Maxwell appears in a list of those found guilty of attending illegal masses conducted by an itinerant Jesuit priest:

> the Lord Hereis [kinsman of 8th Lord Maxwell] [. . .] *Mr. Homer Maxwell, commissar*, John Mackgie, commissar clerk [. . .] My Ladie Hereis, elder and younger, my Ladie Morton [Lord Maxwell's wife] [. . .] the Lady Tweddail, Papists, apostats, interteaners, and professed favourers of Jesuits. (McDowall, 1873, p. 236)

There are also numerous examples in Dumfries of people being called before the kirk session to be denounced for the sort of nonconformities that marked them out as Catholics.[28]

It is clear that the Maxwell lords and their kinsmen served two masters: the Roman Catholic Church, on the one hand, and the Crown, which espoused the Protestant Church, on the other. While they could not serve both masters simultaneously, they would serve whichever master happened to be pressing on them most urgently at the time, that is to say, when they deigned to serve either of them. A typical example how people in general behaved is that of Maister Ninian Dalzell, who in 1579 was a licensed minister and head of the Protestant grammar school in Dumfries – in spite of which his pupils were being taught the Roman catechism (Aitken, 1887–88, p. 120)! Maister Ninian, however, was eventually found out and duly sacked by the General Assembly. For most people it would have been simply a case of doing whatever was necessary to get by, and such dissimulation was simply a means to an end; only a person with totally inflexible principles would have thought to do otherwise.

The second and last of our Homer Maxwells of Speddoch married Nicolas, possibly also a Maxwell, who died in January 1641 (DGFHS, 2008); this is about as much as we know about her, but it does corroborate Family Tree 3, taken from John Clerk Maxwell. Homer and Nicolas had a daughter, Marion, and it was she who married back into Lord Maxwell's family in 1628 by marrying John Maxwell of Middlebie, the first of the Maxwells of that line. After she joined that line, we hear no more of the Maxwells of Speddoch.

Summary of the Maxwell of Speddoch Line

1. **Homer Maxwell**, a burgess of Dumfries who was also a baillie in 1543. He, or his son, acquired the lands of Speddoch from Lord Maxwell. In 1547 he married **Janet Gordon**, and had a son, Homer.
2. **Homer Maxwell of Speddoch**, baillie and commissary of Dumfries, provost 1593–94, who was at the battle of Dryfe Sands with Lord Maxwell in 1593. He died in 1611, whereupon he was succeeded by his son, also Homer.
3. **Homer Maxwell of Speddoch**, who married **Nicolas** (probably also a Maxwell) and died in 1641 and whose youngest daughter was Marion.
4. **Marion Maxwell**, who firstly married **John Maxwell**, the first **Maxwell of Middlebie** in 1628. After his murder in 1639, she married **Andrew Stewart**. She adhered to Roman Catholicism and died on 4 January 1674.

CHAPTER 7
Agnes Maxwell and James Le Blanc

Who was James Le Blanc?

After William Clerk's death in 1723, his widow Agnes Maxwell remarried on 26 July 1725 to Major James Le Blanc, a French Protestant. Somewhat surprisingly, there were a handful of Frenchmen of that name in and around Edinburgh in the early 1700s, long before the town decided on a plan to bring in weavers from Picardie (Dobson, 2005, p. 61). If Major Le Blanc had come directly from France, he must have decided at some point to anglicise his Christian name. Many such French Protestants, generally referred to as Huguenots, had fled France following the persecution in the counter-reformationary upheavals of the late sixteenth century. Several Huguenot massacres were perpetrated by the Roman Catholic majority, notably in Paris on St Bartholomew's Day, 1572.

If the Edict of Nantes in 1598 was intended to alleviate these troubles, it was only partially successful and over time it came to be disregarded. Emigration consequently continued, with many Huguenots coming to England and Ireland, some of them eventually finding their way to Scotland. It is likely that James Le Blanc came by way of Ireland, for in 1694 he appears to have been a lieutenant in Lord John Murray's Regiment of foot (Dalton, 1896, p. 386) which had been engaged in the aftermath of James VII's last stand at the battle of the Boyne of 1690. Although his forename is not given in the regimental rolls, in a petition to the Scottish parliament James Le Blanc stated that he was indeed a lieutenant in that very regiment.[1]

It is clear from the foregoing that James Le Blanc, onetime soldier, was probably born around or before 1675, and that he was from a relatively well-off family, otherwise he would not have been a commissioned officer in 1694. This setting is also typical of many Huguenot merchant families who fled France and managed to set up business very successfully in other parts of Europe. From his petition of 1705, we find that Le Blanc was in partnership with a Mr William Scott in the business of manufacturing and working glass.

He ground and polished glass panes making windows in carriages and sedan chairs (Turnbull, 2001, pp. 191–198). Le Blanc's side of the trade was the production of mirrors which were either framed on their own or fitted into furniture.[2] Nevertheless he supplied and even hired out other glass products, for example, lighting sconces for use at funerals, and furniture, including a fine 'Scritour' (writing desk) supplied to the young Laird of Grant.[3]

The business, however, was not without its problems and in 1705 he and Scott had to petition the Scottish parliament to try to get the duty lifted on the rough glass plates they imported from France.[4] While they, or more specifically Scott, had taken part in an earlier enterprise to set up a glass factory at Morison's Haven in Prestongrange (Turnbull, 2001, pp. 189–204), it had not been entirely successful, and a competitor by the name of Sarah Dalrymple held the monopoly in Edinburgh on the sort of glass plates wanted by Scott and Le Blanc; naturally she wanted them to use her product rather than import them. After petition and counter petition[5] some sort of compromise was reached, which Chambers somewhat inaccurately alludes to in his *Domestic Annals of Scotland* (1861, pp. 154–155).

James Le Blanc married a 'gentlewoman of this countrey', Elizabeth Houston, daughter of the deceased John Houston of Wester Southbarr, a cadet of the Houstons of Houston, and in 1707 he was one of a number of other immigrants who had done well for themselves and their adopted country, and were accordingly naturalised in the run-up to the Union of the Parliaments.[6] In the Act of naturalisation, James Le Blanc's entry is paired with Daniel Lasagette, both given as merchants. The link to Lasagette is interesting, firstly inasmuch as the records for Newhailes House (Rock, 2013a) give a broader idea of the type of merchandise he dealt in, and secondly because Daniel Lasagette was known to the Clerks of Penicuik, and even wanted to marry one of the Baron's sisters. He was still in contact with the Baron until around 1730.[7]

James Le Blanc and the Baron

The next we hear of James Le Blanc seems to be at first sight a simple coincidence. In his memoirs the Baron tells us, '[In June and July 1710] I resolved to take a journey [. . .] [to Bath to take the waters.] [. . .] I traveled with two friends, one Major Leblanc, a French man, and one Mr. Robert Clerk' (BJC, p. 76).

How the Baron first met James Le Blanc can only be conjectured, but the Baron was interested in all manner of things, and the manufacture of glass and mirrors could well have been one of them. From a different perspective, however, we can also reveal a family connection. Here we recall that the

Baron's second wife was Janet Inglis, whose mother was Anne Houston. She was the daughter of Sir Patrick Houston of Houston. As we know, James Le Blanc had married, in 1699, Elizabeth Houston, daughter of John Houston.[8] The two Houston families were highly likely to have been connected, as Family Tree 4(a) demonstrates.

If the Baron and James Le Blanc became acquainted because both their wives had a Houston connection, it would have happened between 1709 and 1710, consistent with the time of the Baron's second marriage and his first mention of Le Blanc in his memoir. At any rate, the paths of the Baron and James Le Blanc crossed again in 1715, when Le Blanc rejoined the army in the face of the first Jacobite rebellion. Having already been a serving army officer, he was able to join the garrison of Edinburgh Castle with the rank of captain. His name features in army accounts for 1715–16: 'Capt. James Le Blanc, for subsistence of his party which marched from Edinburgh to Berwick to convoy money, 9*l*. 5*s*. 6*d*.' (Lincoln, 1715–16). On 23 January 1716, during the final days of the uprising, he wrote to the Baron informing him of progress. This, however, is just one of half a dozen letters that are preserved from the period from 1716 to 1725 confirming that their relationship was one of more than just nodding acquaintance.[9] From these we find that later in 1716 Le Blanc was in London, still in the army and hoping for promotion, and indeed he was made up to the rank of major. He stayed on with the castle garrison under Lieutenant-General George Preston, for in 1720 he was in charge of distributing pay to the garrison, and in 1723 a sasine is recorded 'in favour of Major James Le Blanc of the Castle of Edinburgh'.[10]

The records further show that in September 1724 the Major was accused by the then Lord Provost, John Campbell, of false muster, a charge implying that he had neglected to account properly for the number of men in his charge, a circumstance no doubt related to the fact that he was the one paying out their wages. The Major consulted the Baron about the charges, for there is a copy of the complaint against him in the Clerk Papers.[11] He was duly court-martialled by Brigadier-General Preston, and wrote to the Baron about the case.[12] However, he seems to have got off, perhaps on condition he pay back some money, for it was left to Brigadier-General Preston to claim money the Major owed him from his heirs.[13]

Remarriage, Debt and Inheritance

Perhaps through his friendship with the Baron, or perhaps through being a known figure at Edinburgh Castle, the Major became acquainted with Agnes

Maxwell, who had lived close by on Castlehill during her marriage to William Clerk, now deceased. Remarriage would have been providential for Agnes, for she had been left with William's debts to clear. Mr Arthur, William Clerk's tailor, had to wait eighteen months for his bill to be paid, but Mr Stoddart of Lasswade had to wait over four years for repayment of money he had lent to William.[14] Since her marriage to William was not sanctioned by the Clerk family, it is unlikely that Agnes had the provisions of a marriage contract to fall back on. She may have had some property, but she was not yet heiress to Middlebie for both her father and elder brother were then still alive. And even if she were to inherit the estate, it was heavily burdened with debt. On the other hand, given that the Baron was one of her daughter Dorothea's tutors, she would at least have had her basic needs provided for.

Major James Le Blanc and Agnes Maxwell were married regularly in Edinburgh on 26 August 1725[15] and, as a serving member of the garrison, he would have had the privilege of marrying in St Margaret's chapel in the castle. How things went until the Major's death on 27 July 1727, the day following their second wedding anniversary, we do not know. His medical bills reveal an illness for which he had been treated since the previous December by John and David Knox.[16] James Le Blanc's death must have been a severe shock for Agnes and her young daughter Dorothea. Her first husband, William, had died in 1723, followed by her father John Maxwell 'the Entailer' in October 1725; now this.

One of the things that we may surmise about the Major is that shortly after his marriage he was concerned that he might be found liable for some of William Clerk's debts. It would seem that he addressed the issue by compiling a list of these debts, which he then had endorsed both by the Baron and his brother Robert, certifying that the bills had been paid by Agnes herself out of what little William had left her.[17] There was an additional detail, however, in that William's funeral bill had been receipted in favour of the 'the Barron Clerk' rather than 'Mrs Clerk'. To remedy this, there is a further endorsement in the Baron's own hand: 'I doe acknowledge that this monie said to be payed by me did belong to Mrs Clerk spouse to my deceased brother Mr William Clerk. John Clerk.' Unfortunately, we cannot fathom what really lay behind these actions, but it does seem that the Major went to some pains to make clear that all such debt was repaid and accounted for.

As a substantial loan to James Clephan and his wife[18] would seem to indicate, the Major had been wealthy. Certainly, the Baron (BJC, p. 134) tells us that he left a fair bit of money to Agnes and some friends, and he also bequeathed a substantial legacy to Dorothea, who was then about seven years old. In addition, in the days preceding his death the Major had made out a disposition in favour of Agnes herself, of 'All & Sundry Lands Heretages Tennements Annualrent & Hereditaments Whatsoever now pertaining and belonging to me'.[19] But Agnes herself was to die barely eight months later, in March 1728, leaving Dorothea, then still only seven years old, an orphan under the care and tutorship of her uncle the Baron, who presumably took her into his own family.

Sorting out Agnes' estate proved to be highly complex. Firstly, her late husband the Major had died not long before her. He had left a considerable sum of money but was also owed money by some and in turn he owed money to others. This was a common situation in an age when people borrowed from other people rather than from banks. Besides the money, the Major also had heritable property which normally had to be inherited down the blood line and could not be simply passed to his widow or step-daughter. To circumvent this, he had made a disposition to Agnes before his death, but the validity of such a scheme would be open to challenge, and of this we will soon hear more. The other complicating factor was that Agnes' father John Maxwell, the 'Entailer' of Middlebie, having died in the autumn of 1725, left an estate heavily burdened with debts. Her elder brother, John Maxwell of Middlebie 'the Last' or 'Younger', having succeeded to this sorry situation, did not live long enough to do much about it. He died in January 1728, just two years after his father, leaving Agnes as the next heir of entail (see Chapter 6). Since Agnes' own death followed just two months later, her succession to Middlebie was never legally formalised. The more considerable consequence of the entail was that her death left her child, Dorothea, as the new heiress of Middlebie. As a minor, the process of her legally inheriting the property would take some time, during which the estate could hopefully be sorted out.

To understand the full implications of the situation, however, it will be necessary to investigate the legal complexities and background surrounding Agnes Maxwell's inheritance in more detail. The laws of entail required that the estate concerned should not be encumbered with debt but, if that was ever strictly true, by 1728 it was certainly not the case for Middlebie. Such a condition could easily be subverted, for example by getting someone else to assume responsibility for the debt, or pretending that there were sufficient other assets to cover it. But the creditors would still want their loans repaid, irrespective of whether or not the debtor's only remaining asset was an entailed property.

Creditors began to pile up actions against Dorothea both as heir to her mother's personal debts, and as heir to the debts against the Middlebie estate. One would have thought that since Agnes had been left a lot of money by the Major, this would have been immediately employed to clear these off. But the money was now Dorothea's, and she was under the wardship of her uncle, the Baron, who endeavoured to keep the creditors at bay until she legally succeeded to her estate many years later. In 1728 the Major's creditors, headed by a Dr Thomas Young of Killicanty,[20] filed a petition to the King asking to be appointed as legatees. This Dr Young was the Major's brother-in-law, for he had married Elizabeth Houston's younger sister, Agnes. The Crown was, and still is, the heir of last resort, and so they hoped that the King would grant them the rights to the Major's heritable property to allow them to recover whatever they could.[21] This plea was passed on to the Barons of the Exchequer (including the Baron himself) whose recommendation to the King on 10 December 1728 was: 'As no heir appears who can by the laws of Scotland succeed to the property (his relations being Roman Catholics and debarred from the succession of heritable subjects) [...] they are of opinion that it will be proper [...] to pass a signature for granting to Dr Thomas Young the real estate of the Major'(Redington, 1889). The Crown's decision was given a month later: 'granting to Dr Thomas Young, physician in Edinburgh, as trustee [...] by the disposition made by the said Major [Le Blanc] of date 1727, July 14, [to his wife Agnes] all lands, heritages, tenements, annual rents, &c [...] *there being none to succeed within the tenth degree*' (Shaw, 1897, pp. 1–13).

The remarkable thing here, of course, is the fact that the Major's disposition of the heritable property in Agnes' favour was considered as being null and void. Given the Major was a Huguenot, for the most part his relations would have been Protestant, and if there had been even one still alive, in Britain at least, they would have provided an heir. But the reading of the Barons' wording, 'his relations being Roman Catholics', is perfectly clear. To prevent a loophole in the 1701 Act for Preventing the Grouth of Popery,[22] the law regarding dispositions of property had been made much the same as for inheritance. The Act stated, inter alia:

> no [Roman Catholic] [...] shall be capable to purchase and enjoy, by any voluntary disposition or deed that shall be made to them [...] any lands, houses, tenements, annualrents [etc., (...) and] *the said voluntary dispositions and deeds* [...] *shall themselves become void and null*.

Everything that transpired is consistent with Agnes Maxwell having been outed as a Roman Catholic. The Maxwell Lords had all been steadfast in their faith as Roman Catholics, and their kinsmen, the Maxwells of Middlebie, had been more or less duty bound to follow their example (Chapter 6). However, as we have already pointed out, because the repression they suffered went so far, people survived by bending with the wind. It was a matter of keeping the authorities and kirk session at bay while all along maintaining their Roman Catholic faith, if need be in secret. There must also have been those who tried in good faith to convert to a Protestant religion but changed their mind at a later stage, particularly in the face of death.

It is fairly certain that Agnes would have been raised as a Roman Catholic. She and her first husband William Clerk, a Protestant, would already have encountered all the obstacles that a marriage across the religious divide would have presented at a time, not long after the first Jacobite rebellion, when anti Catholic sentiment would have been considerable. Sir John Clerk, the 1st Baronet, refusing to welcome Agnes into his home (Chapter 5) provides some indication that this may well have been the case, that is to say, not only had his son married against his wishes, he had married a Roman Catholic to boot.

Even if Agnes had made every appearance of converting to the Protestant faith, which she did at the very latest before her marriage to the Major, there would always have been suspicions that it was a sham. Agnes was a religious woman[23] who would not readily give up her true faith, and if she had offended God by doing so, the death of a father, brother and two husbands in the space of a few years may well have occasioned much soul searching on her part. Surely, when her own time in the temporal world was running out, reaffirming her true faith would have been uppermost in her mind. If she had sent for a priest to make her confession, it is likely that it would not have gone unnoticed. Even the mere suspicion of it would have been enough to set her husband's creditors scuttling to the law courts.

There is no evidence that William Clerk ever converted to Roman Catholicism. When William died, the Baron brought his body back to Penicuik for burial, and by the size of the cortege it was a very public affair. If William had been a Roman Catholic, the burial would have been a low-key affair somewhere out of the way. Penicuik had very few Roman Catholics at the time, and there were no Roman Catholic churches for miles around (Wilson, 1891). The evidence therefore points to William having been, at his baptism and death at least, a Protestant.

James Le Blanc, a Protestant of Huguenot extraction and a man of some wealth and standing in the city of Edinburgh, would not have been likely to marry an overt Roman Catholic. Quite probably, whether out of love or practical necessity, Agnes went through the motions of becoming a Protestant to marry William. After William's death the Baron, as Dorothea's tutor, may well have advised her of the negative consequences should she consider reversion to Roman Catholicism, notably regarding the Middlebie succession and, later, also regarding James Le Blanc's disposition to her of his heritable property. When it came to Agnes' turn to face death, however, it would seem natural that she would have been more concerned to save her soul than her fortune.

We are now left to ponder as to whether Dorothea was raised a Protestant. While it is probable, at least for outward appearance's sake, what sort of religious guidance she received at home can only be guessed at. The law at least recognised that a child could not profess any religion. As far as Dorothea's succession to Middlebie was concerned, it could be deferred for up to ten years (see Chapter 8) but, when the time did come, the 1701 Act stated that in order to be retoured heir, unless she was manifestly Protestant, she would have to renounce Roman Catholicism under oath, for which a strict formula was provided. In the meantime the succession was let to lie, and John Maxwell of Middlebie's creditors simply had to 'whistle for their money' for the next ten years. As to Dr Young, as late as 1735 he was still digging into the affairs of the Major's first wife, Elizabeth Houston, but to a purpose unknown.[24]

CHAPTER 8

The First Clerk Maxwells

The origins of the first Clerk Maxwells, George and Dorothea, have already been briefly mentioned. George Clerk was the third son of the Baron, Sir John Clerk, and his second wife Janet Inglis (Chapter 4) while Dorothea Clerk (Chapter 5) was the only child of the Baron's younger brother William Clerk and his wife Agnes Maxwell, heiress of Middlebie (Chapter 7). George and Dorothea were therefore first cousins. How they came to be Clerk Maxwells will now be revealed in this first detailed account of George's life and career. On Dorothea's part, the fact that she brought the estate of Middlebie into the Clerk family was to be a major influence in shaping the lives of the second Clerk Maxwell, her grandson John, and in turn his son James, who was the third and last of the line.

George's Early Life

In Scotland, until fairly recent times, children were usually named after family members from earlier generations. This was certainly the case amongst the Clerks of Penicuik, but the naming of the Baron and Janet Inglis' second surviving son was a departure from this protocol: 'it pleased God to make up the loss of my son Hary by the birth of another son, whom I christned George, after the patron of the cause which I had espoused during the Rebellion. He was born 31 octob^r 1715' (BJC, p. 96). The significance of the date and name was later clarified by Lord Eldin:[1] 'the day on which the [Jacobite] Rebels under Brigadier McIntosh advanced to Jock's Lodge with an intention to surprize the City. He was named George after the King as testimony of loyalty'.

The eagerness of the Baron to call his son after the King was no doubt presaged by him already having called his second son James, as was customary, in honour of his wife's father; the name of the Old Pretender was not at all popular with the House of Hanover, especially in 1715. We hear nothing more of George (or James either) for some time, but he seems to have been sent to Dalkeith grammar school,[2] as were his younger brothers after him. He was boarded at the school and seldom brought back to Penicuik because it had unsettled him.[3]

While George was away at school his mother would write to him, for example, to remind him that he should pray, read the scriptures and keep the Sabbath (NAS: GD1/1432/1.45, 17/5/1729). In the following year, however, he was sent away to Lowther School[4] near Penrith in Cumberland, just before his fifteenth birthday (BJC, p. 138). The Baron had heard about the good reputation of the school and the progressive outlook of its master, Mr Wilkinson; the bad memories of his own school days (Chapter 4) no doubt helped persuade him, but it was ever in the Baron's mind that his sons should learn 'the English language'. However, he did not go so far as sending him to Eton, as he had done with his first son, John, for while it had been a great success he had concluded: 'there was this bad consequence from an English Education, that Scotsmen bred in that way wou'd always have a stronger inclination for England than for their own Country' (BJC, p. 99). Penrith was in England, but only just, and it would have been hard to find any school in England that was closer to home. George was therefore packed off to Lowther with this encouragement from his father: 'You have nothing else to depend on but your being a scholar and behaving well' (BJC, p. 139n1).

The same theme was expanded on in a letter to George the following February. The Baron hoped that his son was beginning to love learning, was acquiring English language, and following the good example of his teacher. He reminded him that he should be good towards his schoolmates since they could well be of help to him later on in life, and that if he were to be seen as being a mere trifler and a bad scholar, he would never be able to live it down.[5] After visiting Dumcrieff that August, the Baron proceeded to Penrith to pay his son a visit at Lowther. He stayed three days and 'found all going very well with him' and likewise he found that Mr Wilkinson fully lived up to his expectations (BJC, p. 139). While George received plenty of letters, he seems to have been a reluctant correspondent; his mother longed to hear from him and told him so in May 1732 when she wrote telling him of his new baby brother, Matthew, and of Mr Wilkinson's good opinion of him.[6]

A year later, however, both his mother and his sister Betty were still chiding George for not writing. Betty in particular, who was some ten years his elder, complained that she never heard from him, and when she sees him next she fears they will be strangers. Apart from once more reminding him of his failure to write, his mother asked him what sort of career he had in mind.[7] Interestingly, his cousin Dorothea, then aged twelve, also wrote to him.[8]

The year 1734 finds George nearly nineteen and still at Lowther. In the

August of that year, the usual time for his annual peregrinations, the Baron had George brought up to join the family at Dumcrieff, and thereafter they all set off for Carlisle so that his wife Janet and daughter Anne could see a little of England. Having stayed only a few days, they returned to Drumcrieff (BJC, p. 143), but the Baron omits to say whether George came back to Penicuik with them then or went on back to Lowther. It would seem that the approach of his nineteenth birthday would have been an appropriate time for him to proceed to university. Lord Eldin (Clerk, 1788, p. 51) informs us that he duly went to the University of Edinburgh, but there is no mention of him graduating there (Laing, 1858). Even James Clerk Maxwell, George's great-grandson, did not do so; they both went on to complete their education elsewhere. For George, it was a case of following in the footsteps of his father, his uncle William and his brother James, who all went to Leiden.

An Early Marriage

The decision to send George to Leiden, or at least the timing of it, was affected by a singular event, for on 17 July 1735, while George was still only nineteen, he married his cousin Dorothea,[9] who was then just a month short of her fifteenth birthday (see Chapter 5). Since her father William's death in 1723 the Baron had been one of Dorothea's tutors, and following her mother's death some five years later, she had been brought into his care and presumably would have lived with him and his family at Penicuik. She and George had therefore ample opportunity to become acquainted. In spite of her young age, the marriage was legal, for the legal age of marriage for girls was then just twelve (Lorimer, 1862), and amongst the upper classes marriages between such young couples had often taken place as a way of binding the families together with the real business of married life beginning sometime later when the young couple were judged mature. Nevertheless the marriage was *irregular*. According to the Baron, the marriage took place 'privately' (BJC, pp. 144–145), which probably means that they went before a minister without proclamation of banns or any other of the usual formalities. There were indeed ministers who might do this sort of thing,[10] and perhaps for a suitable consideration. We know for a fact that the wedding was irregular because some years later George paid a 'fine' to the kirk session of St Mungo's in Penicuik: '1740 Jany 20 Given in by George Clerk for his Irregular marriage with Dorothea Clerk'.[11] Presumably in return his marriage to Dorothea then became regular.

The Baron later said of the liaison:

I had no hand or concern in the Match, but I hope it will prove a happy Marriage to both. I never recommended her to George, since I was her Tutor, but she had this advantage, that her Mother, before she died, frequently recommended George to her. (BJC, pp. 144–145)

She was just seven years old at the time of her mother's death, and George was then only twelve. One wonders if his remarks here were simply to gloss over how the marriage would have looked to others. Is the Baron therefore trying to distance himself from the notion that it may have been his own wish, for the young heiress would have made an excellent 'prospect' indeed for one of his own sons? In his *Life of James Clerk Maxwell*, Lewis Campbell clearly felt obliged to indicate a hint of incredulity at the idea of it being Agnes' idea.[12] Agnes could indeed have said such words to her daughter in the playful sort of way that one does with children, but hardly more. If she seriously had any such match in mind, it would have been something that she would have put to the Baron himself, for without his approval it would be no more than a fond wish.

As Dorothea's tutor and latterly curator, the Baron had had many years to ponder over Dorothea's future, for although she had a decent legacy from her stepfather Major Le Blanc, her inheritance through the entail of Middlebie was heavily burdened with a string of debts (Chapter 7), a messy problem that the Baron would have to untangle. First, we can be fairly sure that the Baron did his best to see that Dorothea was brought up in his own Protestant religion. Not only would that have been natural for him to have done, it also meant that there would be no problem with the troublesome 1701 Act. Furthermore, her marriage, just short of her fifteenth birthday, to one of his own sons would be likely to dampen any mention that her mother had been Roman Catholic. This would help to avoid her having to take the oath prescribed under the Act, which she would have been required to do, by *the age of fifteen at the latest*, if she were not by then accepted as being manifestly Protestant. Failure in this matter would have resulted in Dorothea being debarred from her inheritance. But she could not marry his eldest surviving son, James, for he would inherit Penicuik, the requirements of which were mutually incompatible with those of the Middlebie entail. Moreover, he also needed to provide a decent future for his next son, George. It would have to be George and Dorothea.

That settled, his next task would be to unburden the Middlebie estate from debt, to which end it was necessary to find a means of getting out of the entail. Furthermore, the times being what they were, he would have had to worry about what would happen if Dorothea died without issue. It would be a pity if Middlebie ended up being lost to the Clerk family and went instead to the

Areskines who, as the heirs of Grizzel Grierson, were the next in line to succeed (see Chapter 6 and note 31 of Chapter 4).

Work on George and Dorothea's Future

After the marriage, the Baron got to work arranging things for the young couple. First of all, he got Dorothea to make a will in George's favour,[13] which was followed up sometime later by a disposition and assignation to him of all the non-entailed lands, goods and monies that might belong to her on her decease.[14] We may guess that the intent of this was to secure matters for his son as far as was possible under the constraints of the entail.

The accumulated debts on Middlebie were a different matter, some of them dating as far back as 1720, they came to a total of £1,759 1s 6½d.[15] Against this, the property entailed by John Maxwell in 1722 comprised the following:[16]

In the County of Dumfries
   The Twenty Merk Land of Middlebie
   The Merk Land of Kirktoun of Kilmahoe
In the Burgh of Dumfries
   The south-most tenement of a land near the Friars Vennel, with yard
   A tenement in Bells Wynd, near the port of the Friars Vennel, with large yard
   A yard in Friars Yard
   A house at the head of the White Sands, with yard
   Five acres of land at the back of the Castle Gardens
   A yard adjacent to the above land, with barn at the foot of the Upper Green Sands
   A quarter's salmon fishing on the Nith, west and south of the town, from Powson's to the Powfall Burn
In the Stewartry of Kirkcudbright
   Nether Corsock
   Over and Nether Blackhills
   Little Mochrum

Figure 8.1 (opposite) The eight parcels of land making up the estate at Middlebie proper
Synthesised from the eight individual sketches, of which Plate 12 is one example, in DGA: GGD56/19, Plans of Middlebie Estate, eighteenth century.

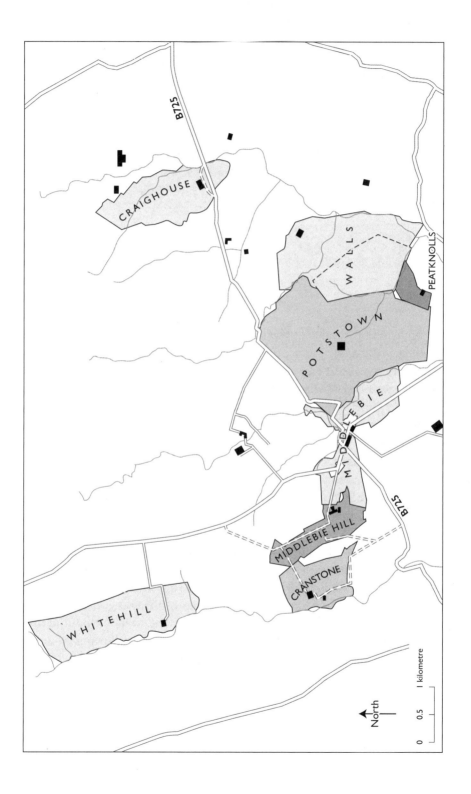

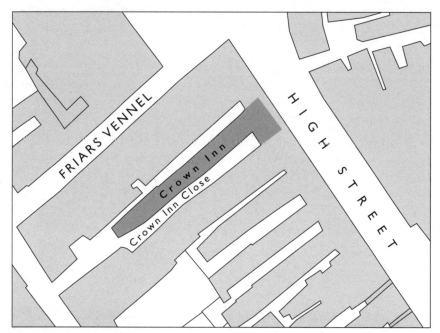

Figure 8.2 Location of the former Crown Inn, Dumfries, on a modern town plan
This was one of the tenements in Dumfries which came to Dorothea Clerk Maxwell through the Middlebie entail of 1722. A consequence of this entail was that, as the Crown Inn, it was inherited by her great-grandson, James Clerk Maxwell, in 1856.

The locations of the major parts of the estate and Middlebie itself can be found from Figures 2.1 and 8.1, while the original plan for Middlebie town is reproduced in Plate 13. Figure 8.2 shows the location of one of the main properties in Dumfries, which in James Clerk Maxwell's time was the Crown Inn. Who would have thought that James Clerk Maxwell owned a 'pub'?

The Twenty Merk Land of Middlebie would sell for £2,225 4s 4⅔d, calculated as twenty-three times the annual rental. The debt of £1,759 1s 6½d could then be paid off leaving a surplus that could be put to the purchase of some other small property nearer to the remainder of the estate in and around Dumfries. But in order to do this, a private Bill had to be introduced into House of Lords to obtain the requisite Act of Parliament allowing the entail to be broken: the Baron certainly had the legal skills and the necessary connections to ensure that it would succeed.

In the meantime, however, following the marriage of George and Dorothea the Baron sent George to Leiden to continue his studies, while Dorothea remained in the care of the Baron and his wife at Penicuik, his reasons being:

> As my s[ai]d son seemed very intent to study the Law in Leyden, and his Wife and he being too young to live together, I sent him to Holand in January 1736, where he had the advantage of staying with his eldest Brother James. (BJC, pp. 144–145)

Despite them 'being too young to live together' it seems as though the marriage was consummated, for a child was born in the following year.

While George actually departed in January 1736, the intention of sending him away was already raised by the previous October, for his uncle Hugh had found him a travelling companion.[17] But from George's standpoint, it was probably just a bit too soon, for not only did he have a marriage ball to give, he would also be reluctant to be parted so quickly from Dorothea. Moreover, by then they may also have been aware that their first child was on its way, which we will examine in more detail later on in this chapter.

George eventually set off for Leiden in the New Year and while he was ever the reluctant writer when at Lowther, he may have found some consolation in exchanging letters with Dorothea. However, only his general correspondence seems to have been preserved; for example, letters concerning his classes at Leiden, an intended tour of Germany with his brother James, the sighting of a comet, and failed attempts to get his father a Great Dane.[18]

In the meantime, Dorothea spent some time back at Dumfries, having visited an uncle, Robert Maxwell, there in the autumn of 1736.[19] Just who he was is not absolutely clear. Agnes Maxwell had a younger brother Robert who, if he was still alive in 1722, was not mentioned in the Middlebie entail, but her father, 'the Entailer', did have a half-brother, Robert Maxwell of Shalloch,[20] who died in the winter of 1737 and was therefore alive when the letter was written.[21]

On finishing his studies at Leiden in the summer of 1737, George accompanied his elder brother James to Germany, where he visited Hamburg, before returning home.[22] James, on the other hand, lingered on and extended his stay, just as his father had done before him.

The Rise of Dumcrieff and Middlebie

When George returned from Leiden in 1737, he had reached his majority and was, as we know, already a married man. While the Baron had at first diverted him from setting up house with his young bride, he was now in for a rather pleasant surprise. The Baron, having finished building a new estate and country house at Dumcrieff only four years before, now presented it to George for his very own country seat. He had spent five and a half long years building it, and no doubt a considerable amount of money, but he had its use for only five summers. Nevertheless, he would have been thinking all the while about his future provision for George and Dorothea, and the fact that he had given it away to George was of no great consequence in the grand scheme of things, and perhaps as it would turn out even a convenience. Since Penicuik would come to his eldest son James by right, he was already provided for, and he could also have the use of Mavisbank nearby when it came his turn to manage his father's coal works at Loanhead (Chapter 4).

Moffat lies in Annandale just over twenty miles from Dumfries, from where Middlebie and Nether Corsock lie about twelve miles to the east and fifteen miles to the west respectively. It would therefore be fairly convenient for access to Dorothea's properties in these three places. Furthermore, when the Baron wanted somewhere to stay at Moffat for the summer season, he and his family could stay as guests of George and Dorothea, and could travel there without having to bring their entire retinue along with them from Penicuik. And so, in a way, he gained as much as he lost by his generous gift! The entire process was complete when George obtained the Crown Charter for his new estate in 1738, and was officially able to style himself 'George Clerk of Dumcrieff'.[23] But by a different route, by the demands of the Middlebie entail, he was also George Clerk Maxwell.

Once installed as the new Laird and Lady of Dumcrieff, George and Dorothea did not simply settle down and live the lives of country gentlefolk, for George had to start a career. First of all, the Baron's plans for sorting out their financial position had to be put into effect, beginning with George and Dorothea signing off on the Baron's management of their affairs to date.[24] Dorothea no longer required the Baron to act as her curator since she was now the wife of George, who had reached his majority the year after they married. Just coming up to the tenth anniversary of Agnes Maxwell's death, the Baron had Dorothea served as heir of entail to John Maxwell of Middlebie.[25] This was followed by the private Bill which he was able to have introduced into the House of Lords 8 March 1738 (House of Lords, 1738, p. 222ff.). These key steps all took place within the space of a month. This must have been quite a feat, but the Baron clearly had both the necessary acumen and standing to achieve it.

The tenor of the Bill was that all the other heirs of entail had now been exhausted apart from two, first in line being Dorothea Clerk (and any future children) and, second, her distant relation James Areskine [Erskine], then a minor, the younger son of Charles Areskine, his Majesty's Advocate for Scotland (see note 31 to Chapter 4). The Bill presented the Baron's scheme of selling off the 'Twenty Merk Land' of Middlebie, the proceeds to be used to pay off the debts and to allow *Dorothea* to purchase in their place 'other Lands or Tenements, fituated as near as conveniently to the faid George Clerk's other freehold Eftate of Drumcreif' (House of Lords, 1738, p. 222ff.). Apart from the substitution of these new lands for the Twenty Merk Land, the original entail was to stand exactly as it was, and to make sure it was all done according to the Act, trustees were appointed to oversee the ensuing transactions. They seem to have been an eclectic bunch, but most likely they were well-chosen friends and associates of the Baron, with some input from Charles Areskine.

The Bill's progress through both Houses of Parliament was fairly rapid by today's standards and was given the Royal Assent on 20 May 1738 (House of Commons, 1738). The plan to clear the debts on Middlebie was then put into action. In addition, since George and Dorothea's financial state of affairs had now been made reasonably clear and secure for the foreseeable future, the terms for a marriage contract could be agreed;[26] and it is here, apparently, that the name Clerk Maxwell emerged for the first time. The lands of Middlebie were duly sold, only to be bought back by the Baron himself and made over to George. Why, if he had the money to do this, had he simply not handed over to George and Dorothea just half of that amount, which would have cleared off the debt without the bother and expense of the legal fees and Act of Parliament?

The Baron was, as usual, simply being astute. No doubt he would have mentally written off the issue of the debt many years before. Dorothea had brought a good deal into the Clerk family by way of land and legacies, but unfortunately it was she, not George, who had *title* to the entailed property. While it was she who owned the estate, in practical terms he could treat it as though it were his own 'by right of his wife'. The sting in the tail was that if he needed to do anything formal, he would require Dorothea's agreement and signature. But as a married woman, the protocol to be observed was that she would include the words, 'on the advice and consent of her husband, George Clerk Maxwell'. Of course, his signature would have been accompanied by suitable matching words. It may have been irksome for George, and probably just so for Dorothea too, but to some extent it created a balance of power between husband and wife – neither could act without the agreement of the other. From the point of view of an eighteenth-century gentleman, however, he would be in much better standing if the property were his alone.

One of the intentional effects of an entail is that the property in question cannot be signed over to a second party, George for example, nor could this be done in the Act of Parliament, because that would have been more than strictly necessary for the purpose of clearing the debt. By disentailing and then repurchasing the Middlebie lands, the debt was paid and the property could become George's. Possibly George and his father felt that it would prove a constant source of embarrassment if the Clerk Maxwells of Middlebie did not own the land whose name they bore. Although he did not obtain his official charter over Middlebie until much later,[27] he could now justifiably call himself either George Clerk Maxwell of Middlebie or George Clerk of Dumcrieff. In fact, he seems to have used whichever of them he felt best suited the occasion, and this is reflected both in his correspondence and elsewhere. Whatever may have been the Baron's motivation, first with Dumcrieff and now Middlebie, George had done very well out of him.

Dorothea's succession was not technically complete until sasines for her properties had been recorded in her favour. The sasines on Nether Corsock, the two Blackhills and Little Mochrum were registered in September 1738;[28] others are recorded in the *Index to the Register of Sasines for Dumfries etc* [...] 1733–1780 (SRO, 1931) from which it appears that, in due course, George also held property there.

Career and Family Life

Thanks to the Baron's good planning, his acumen and generosity, by the end of the year 1738 George and Dorothea were comfortably set up to make their own way in life. They were well-off indeed, for not only did they have a splendid country seat with parkland, plantations and a farm at Dumcrieff, they also had the entailed Middlebie estate as a further source of income. George had done little to achieve such good fortune other than through circumstance and simply being his father's son. This is reflected in the words of Lewis Campbell, in his commentary on the notes on the Clerk family written for him by James Clerk Maxwell's cousin, Miss Isabella Clerk;[29] they amply capture the general tenor of George Clerk Maxwell's progress through life: 'This George Clerk Maxwell probably suffered a little from the world being made too easy for him in early life [...] In some respects he resembled John Clerk Maxwell, but certainly not in the quality of phlegmatic caution' (C&G, p. 18). However, it is not that he was some sort of wastrel incapable of lifting a finger to do anything on his own behalf. Far from it, he was an imaginative and industrious man, ready to turn a hand to anything that interested him; but therein lay the problem: 'His imagination seems to have been dangerously fired by the "little knowledge" of contemporary science which he may have picked up when at Leyden' (C&G, p. 18).

As to what others may have thought about George, Campbell says:

in all relations of life, he seems to have won golden opinions [...] The friendship of Allan Ramsay [senior] and the affectionate confidence of the 'good Duke and Duchess of Queensberry', sufficiently indicate the charm which there must have been about this man. (C&G, p. 19)

Although George had been keen to study law at Leiden, he had soon come to the realisation that he was not cut out for it. Clerk (1788) gives us a good idea of where his aptitudes really lay:

a skilful engineer and draughtsman, as appears from various roads, bridges, and other public works[30] [...] executed under his direction, or on plans which he delineated. Nor were his talents in designing confined to this more mechanical species of drawing.

It therefore seems that he was naturally of a technical bent and so it is unsurprising that the minutiae of legal work would have left him uninspired. He wanted something rather more 'hands on',[31] and he began with some farming. This is evident in a letter he received while at Dumcrieff from his brother Patrick, who was then at Penicuik, enclosing an improved recipe (a pint of honey to a pint of tar) for making an effective paste with which to smear his sheep, presumably to kill off ticks and parasites in much the same way as dipping does today.[32] Such things he dutifully preserved in his personal notebook, wherein there is another recipe for a sheep salve that he recorded almost exactly thirty-six years to the day thereafter.[33] It seems it was something he took quite seriously, for he wrote about it to the Trustees for Manufactures and they ordered the publication of two of his letters 'Observations on the Method of Growing Wool in Scotland' and 'Proposals for Improving the Quality of Our Wool' (Clerk Maxwell, 1756).

By the following year of 1739, however, George had set his sights on a fresh line of interest when he embarked on a more philanthropic yet seemingly practical venture that aimed to encourage and improve local manufacture by the setting up of a spinning school in Dumfries.[34] His petition to the Trustees for Manufactures in Scotland met with success.[35] At the same time, he also set up a linen factory there. While the spinning school's purpose was mainly to teach girls to spin local wool, their skills would also have been employable in spinning flax to make linen yarn for the factory, of which, unfortunately, no details have been discovered. George and Dorothea based themselves in Dumfries for this purpose, and while we do not know for certain where they lived, we can at least be certain that they did live there by August 1739 because the Baron recorded it, 'Drumfrise, where my son George and his wife had taken up their residence' (BJC, p. 154), and in addition about this same time George also received letters addressed to him as George Clerk 'at Dumfries' and 'at his house in Dumfries'.[36] It will be noted that his surname was then still largely known as being Clerk rather than Clerk Maxwell, an issue that we will return to in due course.

In August 1740, in continuation of their longstanding custom of vacationing at Moffat, the Baron and his wife stayed at Dumcrieff, with George and Dorothea, for a month. They visited the local spas and the menfolk, at least, enjoyed the shooting and fishing. He made a note of his stay there that gives us some idea of what the house was like inside.[37] It had been finished nine years before, and for the last three years it had been in George's possession. Given its relative newness, it should have been in good condition, and there had been ample time to get it suitably furnished. The biggest room in the house, which was on the upper floor and about seventeen feet square was 'ill provided with furniture' and could have done with some sort of sofa bed. The ceiling was corniced, but the walls were bare and needed to be decorated with some printed linen wall hangings. As to the similarly sized dining room on the ground floor below, it was 'not in order'; it needed to be decorated with painted wallpaper and some decent framed prints. Money may have been tight for George at this time, for during his stay his father paid for all of his family's living expenses, including the mutton they had eaten and the grazing for his horses. Ever the kind and thoughtful father, the Baron also gave George one of his horses, and intended giving him the feu on a property in Moffat that would bring him a reasonable amount of money.

If the interior of the house was not to the Baron's entire satisfaction, evidently the house as a whole was not to George's, for two years later he had it extended and remodelled to such an extent that the central part had to be pulled down.[38] It would be reasonable to assume that he was now financially better off and able to afford such an undertaking. According to Window Tax records, sometime between 1748 and 1772 the house may have been enlarged once again, for the number of windows increased from twenty-one to thirty-five (Prevost, 1968, p. 204). However, since many householders took measures to avoid paying the tax, such as blocking off all but essential windows, the increase could equally well have been as a result of unblocking some that already existed. As to the progressive laying down of plantations, which had been one his father's great passions, George was not quite as diligent. He did, however, lay down one fairly large plantation in 1774 at Aikrig, which borders the old Moffat to Carlisle road. It covered nearly twenty-five acres, but only the central part lying mostly to the north of the road, survives today (Prevost, 1968, p. 204).

As already mentioned, at this time George's main residence was in

Dumfries where he had his linen factory, but in 1748 there was an abrupt change for he was forced to close the factory down. Prevost (1968, p. 205) gives no details, but all the signs seem to point to it having taken place in that year for it was then that he took up a post with the Forfeited Estates, in consequence of which he and Dorothea moved to Edinburgh.[39] The post-1745 Board of Commissioners had not yet been set up and the Forfeited Estates had been placed under the control of the Barons of the Exchequer, of whom, needless to say, his father was one. At the same time he borrowed, together with his father, £320 from Archibald Tod, a businessman from West Lothian, who had been one of the trustees under the 1738 'Act for Dorothea'.[40]

Children

While the available evidence relating to the births of George and Dorothea's children is somewhat fragmentary, a fair reconstruction is given in Appendix 4. They were, in chronological order:

John, born in 1736 (with an outside chance it was late 1735!)
Janet, baptised 7 July 1738, at Penicuik
Agnes, born September 1739
Joan, baptised 16 March 1741 at Dumfries, died November the same year
George, born November 1742
William, born 14 June 1744, died about May 1746 following inoculation against smallpox
James, baptised 20 September 1745 at Dumfries
Dorothea, baptised 27 February 1747 at Dumfries
William, Robert and Johanna, all probably born at Edinburgh

Janet, known as Jenny, married William Anderson WS, Clerk to the Signet, in 1775. Widowed in 1785, she was still alive in 1791.

Agnes married John Craigie, a Glasgow merchant who was the second son of the Lord President of the Court of Session, Robert Craigie of Glendoick. Their daughter, Barbara Craigie, married Colonel Lewis Hay, and in turn their daughter, Agnes Clerk Hay, forged a link with the Irving family by marrying John Irving, half-brother of Janet Irving, James Clerk Maxwell's paternal grandmother (Chapters 9 and 12). A son, Captain George Craigie of the 40th Regiment of Foot, was killed in the final years of the American Civil War at the battle of Groton Heights, New London, Connecticut, in September 1781, of which more later (*Ladies Magazine*, 1781, p. 613; Clerk, 1788).

John served in the Royal Navy but retired sometime after 1777, when he married Mary Dacre (also known as Mary Dacre Appleby and later Rosemary Dacre Appleby) of Kirklinton in Cumbria (1745–1834). She was the granddaughter of Baronet Fleming of Westmorland, the Bishop of Carlisle. John succeeded as 5th Baronet on his father's death and died on 24 February 1798 (Foster, 1884).[41] As he had no children, his young nephew George Clerk, the eldest son of James Clerk HEICS and Janet Irving, was retoured as his heir. An account of what little is known of this John Clerk and his wife Mary Dacre is given in Chapter 9.

There is little information on George other than that he was admitted as an advocate on 21 December 1767, lived at the same address as his parents in James' Court in Edinburgh's Old Town Edinburgh and died unmarried on 5 October 1776 (Foster, 1884; EPOD, 1773, 1774, 1775; Grant, 1944, p. 36 [wherein it wrongly gives 'Dunbarney' for 'Dumcrieff']).

The third son was James, who also became a seaman but in the merchant fleet rather than the Royal Navy; he eventually joined the HEICS[42] as a lieutenant and possibly reached the rank of captain. He then settled in Edinburgh, where he married Janet Irving in 1786 (Chapter 9). They were to be in due course the paternal grandparents of James Clerk Maxwell. James Clerk never lived to inherit the Middlebie estate from his mother, Lady Dorothea, who, by dying on 28 December 1793, outlived him by just two weeks (Mackay, 1989).

Dorothea married David Craigie of Dunbarney WS (d. 1796) who was the brother of John Craigie of Glendoick, her sister Agnes' husband, in 1779 (Grant, 1922). They had a son who was named after his grandfather as George Clerk Craigie;[43] he became an advocate, married and had children.

William, the fourth surviving son, was so named in memory of his dead brother, which was not an uncommon practice at the time. Robert was the sixth born and last son. Both he and William went into the army and became lieutenants in the 1st and 56th regiments of foot, respectively. Sadly, both died in service. Robert died towards the end of 1781 at Gibraltar, where the British garrison had been held under siege by the French and Spanish since the winter of 1779. If he was killed in action, then this would possibly have happened during the sortie that took place on 27 November, when a party of the besieged British troops left the safety of their garrison by night in an attempt to forestall an imminent all-out assault by the enemy. The sortie was successful in delaying the assault for many months.[44] William died just a few months later during the siege of Brimstone Hill, at Basseterre on St Kitts, which began on 19 January and ended with surrender to the French on 13 February (Clerk, 1788, pp. 55–56; Foster, 1884).

Johanna, the fourth daughter mentioned by Foster (1884), was doubtless a later daughter named in memory of the earlier Joan who died in 1742. According to Lord Eldin (Clerk, 1788), 'about 1782' an unmarried daughter of Sir George Clerk, 4th Baronet, died of grief over the loss of her nephew and brothers in the war. As Janet, Agnes and Dorothea had all married, the unmarried daughter John Clerk was referring to must indeed be Johanna. The records do show that a Miss Johanna Clark died in November 1781 from 'decay',[45] which in the language of the time meant that she had simply wasted away. William did not die until after Johanna, who in fact died almost exactly at the same time as Robert. Such a tragic coincidence must have left a great impression on the minds of her friends and family.

Lord Treasurer's Remembrancer to the Exchequer

In 1743, George Clerk was appointed Lord Treasurer's Remembrancer to the Exchequer in Scotland. This obscure administrative office was quite different from the present day combined office now called the Queen's and Lord Treasurer's Remembrancer which, despite the incomprehensible title, deals with such things as ownerless goods, treasure trove, heirless estates and the like. It seems that the function of the original Lord Treasurer's Remembrancer was to audit the accounts of the sheriffs – presumably their personal accounts for sitting in court and perhaps extending to the expenses of the trials that they conducted. At any rate, in return for a broad range of benefits and expenses the incumbent was not expected to know too much about it, nor was he expected to attend his office 'but infrequently'. That was how the system worked.

Such positions were obviously highly sought after, and for a man of appropriate standing and connections, the only reason for not seeking such an office would be seeking one that was even better. In July 1742, George was but yet only twenty-six years old and still accustomed to his father doing things for him; for example, by the age of twenty-three he had been given one estate and had the debts and legal entanglements of another resolved. Now his father was starting to pull some more strings to help him find a regular source of income. According to Scott (1981, pp. 162–163), William Allanson, the incumbent Lord Treasurer's Remembrancer, had not attended court for over twenty years. The Baron had therefore approached him in the hope of persuading him to make way for George. It was not quite as straightforward as that, however, because Allanson's nephew, Wyvill Boteler, was already acting as his deputy during the interim. The Baron's angle was that Allanson's nephew and George could share this post worth £200 a year. Allanson was not to be moved, but the Baron continued to press the matter and, to sweeten the request, he eventually offered Allanson a pension of £100 a year for life to step aside.[46] The offer was at first refused but when it was eventually accepted in the year following, 1743, George and Wyvill Boteler were appointed jointly to the office of Lord Treasurer's Remembrancer (Shaw, 1903). In a letter to the Duke of Queensberry some twenty years later, George revealed that he was only getting £86 10s out of it;[47] possibly the final negotiation ended up somewhat different from the anticipated 50/50 split, or George was having to pay back his father.

The Jacobite Rebellion of 1745

According to Lord Eldin (Clerk, 1788), it had been one of George's fancies to see action as a soldier, and so when came the Jacobite rebellion of 1745 he joined the Royal Hunters, a private brigade of gentlemen volunteers who, equipped and provided for at their own expense, had come out in support of the government army led by the Duke of Cumberland (Prevost, 1963). Also known as the Yorkshire Hunters, they were commanded first by Major-General Oglethorpe, and from 1746 by General Hawley (Ferguson, 1889). Having joined General Wade at Newcastle, they were deployed in gathering information on the progress of the rebel forces. They followed Wade on his journey south to join forces with Cumberland[48] and thence followed the retreating rebels northwards in December 1745 as they headed back from Preston to Carlisle. While Lewis Campbell says that the Hunters assisted in retaking Carlisle (C&G, p. 192), Lord Eldin is more cautious, saying only that George was: 'on different occasions, employed by the Duke of Cumberland (who knew him well) and, in particular, to conduct the forces to the proper ground for opening the siege of Carlisle'. This rings true, because George had first-hand knowledge of the area around Penrith from his schooldays. In a footnote, Ferguson gives some fragments of the pursuit:

> a column under General Bland [. . .] [including] some Yorkshire Hunters [...] was endeavouring to get in front of the Highland artillery by a lane through the Lowther enclosures. The Duke with the main body was three miles behind.

We can imagine George being to the fore and, remembering the lane, suggesting it as a way to gain ground on the rebels. Notwithstanding George's efforts on behalf of king and country, we find a slightly less daring impression of the Hunters from one of the Baron's letters to his son during the campaign; it was sent from Durham where he was taking refuge with his wife and eldest daughter:

Tuesday 29th October 1745

> General [Oglethorpe] was very friendly and told me that you was amongst the Hussars and that he wou'd take particular care of you. He regretted that his regiment was not more numerouse but said that you all wou'd be safe, tho' you should never come within gun shot of the ennemy, and said he never designed to expose any of you. He told me you had marched to Morpeth to intercept deserters [...] I expect James every day. (Prevost, 1963, pp. 237–238)

The Baron's concerns for his son's safety are understandable given that another son, Patrick, had died at the battle of Cartagena in 1741. Also on his mind was Patrick's twin brother Henry, who was then somewhere in the East Indies, and by now his concerns for him were growing by the day. It is clear that regiments such as General Oglethorpe's allowed cautious young gentlemen to uphold family honour by serving their king and country without unduly exposing themselves to any real danger, and so the Baron had been much happier that George was serving his king in this way rather than enlisting in the army itself. Nevertheless, it was by no means 'a picnic'. Not only was the late December weather very severe, the Hunters got close enough to the rear-guard of the rebels to get involved in a couple of dangerous situations. One such had taken place when they were marching on Lancaster. Having been surprised by some Highlanders who were hanging back at the rear of the rebel army, one of the Hunters was killed and another taken prisoner.

On 18 December, the Hunters were involved in a skirmish at Clifton, a village about halfway between Lowther and Penrith, an area George would have known intimately. Indeed so, for he sent his father a detailed sketch of the area, showing the dispositions of Cumberland's troops and the rebel forces, and the locations where the engagements took place.[49] During the retaking of Carlisle at the end of December, the Hunters were stationed two miles to the north at Kingsmoor, but there is no mention of them having taken any further part in the campaign. They were dismissed from service on 30 December, upon which George, having acquitted himself well, returned to his wife and children at Dumfries. Dorothea had been forced to billet rebels, but they had caused little trouble and were now long gone and heading for Glasgow. From then on it was simply a matter of Cumberland's army pursuing the rebels further and further north until a final battle became inevitable.

It is intriguing that so much is known about George's exploits with the Hunters. In part we owe it to the surviving correspondence between him and his father, as reported by Prevost (1963). George, ever sparing with his replies, was now writing regularly, but the really surprising thing is that the correspondence kept flowing even in the thick of a military pursuit. Despite the times and conditions, the weekly post seems to have been able to find the Hunters wherever they went!

The final battle of the rebellion took place to the south-east of Inverness at Culloden Moor in the April of 1746. The Jacobites were done for, and although Bonnie Prince Charlie escaped, many of his loyal supporters, both high born and low, either perished on the field of battle or were ruthlessly hunted down by Cumberland and his men. Little mercy was shown, and so many were captured and condemned to die that it was decided that a bloodbath on so large a scale was unconscionable; lots were drawn for those who were to hang, with the remainder being transported as slaves to the West Indies and Australia, and this was seen as no act of mercy. One of those executed was Sir John Wedderburn of Blackness, who went to his death on 28 November 1746. His son James, having fled to Jamaica, returned to Scotland and settled at Inveresk near Edinburgh, marrying into the Blackburn family. In due course his son, James Wedderburn, became Solicitor General for Scotland and married George and Dorothea's grand-daughter, Isabella Clerk, of whom we have already heard in Chapter 2; we will hear more of the Blackburns and Wedderburns in Chapter 9.

Commissioner for the Forfeited Estates

After the 1745 rebellion and the ensuing reprisals, most of the Jacobite lords and lairds who had taken part had either fled to the Continent or had suffered death, either in the field of battle or on the scaffold. Their property was, by default, escheat to the Crown. In Scotland, the final number of confiscated estates amounted to forty-one, and covered a vast area of the Highlands (Smith, 1975), so much so that it would have been possible to travel from Kippen in the south-west to Cromarty in the north-east without ever stepping outside it. When an estate or title was forfeit, the king would bestow it upon someone favoured, in reward for their good services, as in 1581 when the 8th Lord Maxwell briefly gained the Earldom of Morton (Chapter 6). But here the scale of things was so immense that this was not possible; for example, the clans living on these estates would be hostile and unmanageable, few trustworthy people could be found there, while any that could be found elsewhere would be unwilling to take on a far-flung and usually debt-ridden estate in 'the Land of the Mountain and the Flood'.[50] Furthermore, absentee landlords would simply create the sort of conditions that would allow further rebellion to foment.

In those remote areas of Scotland which are generically known as 'the Highlands', most of the inhabitants were spread out thinly over rough land from which only a subsistence could be wrought, and many found that stealing cattle offered better prospects. The men, being proud and warlike, were disinclined to do ordinary work. Scores were settled by combat, and feuds between clans endured for generations. Allegiance to one's laird and clan was supreme, and the Highland laird not only possessed his land, he possessed his people. The following quotation from Prebble (1968) gives a flavour, unpalatable though it may be, of how they lived:

> [southern visitors] were usually disgusted by the houses in which these heroic figures lived, comparing them to cow-byres, to dung-hills [...] Windows where they existed were glassless [...] Peat smoke thickened the air [...] [yet] Each house was an expression of the people's unity and interdependence.

For the government, the problem was plain enough: the Highlands needed the imposition of 'civilisation'. Nevertheless, it took them a long time to get round to doing anything about it. In the interim, the Barons of the Exchequer had to deal with the consequences of the forfeitures as best they could, for example, they organised factors to look after the day-to-day running of things and the collecting of rents, but the social circumstances remained unaltered, and if anything disaffection worsened. Rather than face the possibility of a third rebellion, in 1752 the government eventually set up a Commission for the Forfeited Estates[51] with the remit of implementing a document entitled *Hints Towards a Settlement of the Forfeited Estates in the Highlands of Scotland* (Smith, 1975, Appendix C, pp. 387–390). Although its ideals were liberal and seemingly benevolent, its aims were essentially practical: to bring about a great transformation in the Highlands through the imposition of sweeping changes; the old order had long been at fault and it had to go.

Amongst other things, it recommended: the widespread building of roads, bridges and harbours; suppression of cattle thieving; suppression of tartan and Highland dress except in regimental uniforms; tight control over leases to make sure every tenant would be law abiding; creating villages to provide decent housing and work; and the formation of strategically placed towns where travellers would converge and soldiers could be stationed. Ministers of religion were to be settled in communities and act along with the factors as Justices of the Peace. The general idea was that the income from the estates would be ploughed back in as a means of financing the necessary improvements. It was to be radical social re-engineering, but by and by it would bring about some semblance of normal civilisation.

A survey of the forty-one forfeited estates having been carried out, thirteen were selected as a manageable number to be permanently annexed to the Crown; it was an action gauged to break up the Highlands permanently, for areas with strong Jacobite sympathies would be isolated from each other. The annexed estates would be the easiest to supervise and control, and so they could be prevented from following any clan figurehead that might lead them astray. When a Board of Commissioners surfaced at last in 1755, it was headed by the reformer Archibald Campbell, 3rd Duke of Argyll, otherwise known as Lord Ilay, first governor of the Royal Bank of Scotland (Murdoch, 2004).

In 1764, when the Commission had been going for nine years and Lord Ilay was already dead, George Clerk Maxwell was appointed as one of the Commissioners (Smith, 1975, Appendix D, p. 394). It was just the sort of appointment that would have appealed to him. One of the things he became involved with was the building of the three-span bridge at the head of Loch Tay. This bridge, built by John Baxter[52] in 1774, formed a key component in a route cutting through the Highland glens from Crianlarich in the west to Ballinluig, on the road from Perth to Inverness. George Clerk Maxwell and Lieutenant-General Adolphus Oughton (1719–1780), deputy commander-in-chief of the army in 'North Britain', were keen to authorise the project but extra funding had to be requested from the Crown. Clerk and Oughton were involved again in the inspection of the foundation-work. It certainly seems to have been satisfactory, for after 240 years it is still the crossing over the Tay at Kenmore. At least fifty-eight bridges throughout the southern Highlands were funded by the Commissioners (Smith, 1975, p. 304).

Harbour improvements were another concern of the commissioners. George Clerk Maxwell was involved in the work carried out at North and South Queensferry, between which was the principal ferry crossing on the Firth of Forth. John Smeaton estimated £980 for repair work and the erection of new piers to accommodate the ferry. The Board awarded an initial £400 towards the cost and paid out a further £100 based on George Clerk Maxwell's expression of confidence, in 1776, that good progress was being made (Smith, 1975, p. 330). George also reported on Peterhead harbour proposals.[53]

Before the arrival of canals and railways, getting fuel to the places where it was most needed was always problematic. The provision of decent roads and bridges helped, but in the latter half of the eighteenth century only canals could carry bulk freight at a reasonable cost. As the nascent canal fever in England had not yet spread across the border to Scotland, the Commissioners initially acted by funding surveys seeking out viable canal developments. Some of the ideas they considered would be thought risible today. One such was for a canal from Perth to Coupar Angus. George Clerk Maxwell was satisfied with the engineer James Watt's conclusion that the project was not worthwhile, but not so with his cost accounting, saying: 'great share of genius [...] he is particularly fortunate in arranging his thoughts but [I] am sorry to observe that he is not so good at stating his account' (Smith, 1975, p. 335).

On the other hand, a canal from Loch Fyne through to the Sound of Jura would provide the Clyde fishing fleet much better access to the herring grounds along the Atlantic coast by cutting about 85 nautical miles off the journey around the Mull of Kintyre. Over six hundred years before, Magnus Barelegs, King of Norway, had claimed the peninsula of Kintyre from Malcolm III, King of the Scots, by having his men carry his longboat over the narrow neck of land that separates the head of West Loch Tarbert from the fishing village of Tarbert on Loch Fyne. On the face of it, this would appear to be an ideal location for such a canal. Watt was once again engaged to do the necessary surveys (Watt, 1771–73); George Clerk Maxwell duly went through Watt's report and made his recommendations to the Commissioners.[54] Watt recommended a seven foot deep canal along the longer route from Arisaig to Crinan, and George agreed that, at an estimated £35,000, it would prove the most cost-effective solution and his assessment was accepted by the Commissioners. It was indeed the route that was eventually taken.

By applying themselves to the well-intentioned policies in the guidance document, the Commissioners had started in earnest a process of transforming both the Highlands and its traditional way of life. But even so, the roads, bridges, churches and new settlements affected only a proportion of the population, and so the rest remained wedded to the old way of subsistence amongst the hills. While the Commissioners were working in one direction, many Highland lords and lairds began working in another. Rather than move the indigenous population to the new villages where they could find some sort of work, a look at their account books told them that they would be better off just replacing their tenants with sheep. And so began the saddest and sorriest episode in Highland history, 'the Clearances', described so well by Prebble (1982). Nothing that the Commissioners could ever have dreamt of could have been more effective in breaking up the old way of life. From time immemorial the clansfolk had been the lairds' children, with whom the lairds could do as they pleased, and now it pleased the lairds to forsake them.

Today, much of the Highlands still lies beyond the reach of public roads, and its vast empty regions of mountain and moorland are home mainly to the sheep, the grouse and the deer. Given the immense task the Commissioners faced, it could hardly have come out any differently. An idea of what they did achieve is given in the Appendices of Smith's thesis (1975) and in Telford (1838). In addition to twenty three harbour works and fifty-eight bridges (all of which had to be connected up by roads) and the subsidies for canal building all mentioned by Smith, there were numerous disbursements for improvements on a smaller scale. Even after the Commission was stood down in 1784, the spirit of improvement continued, and notably forty-three so-called 'Parliamentary Churches' were built at the hands of Thomas Telford in the early nineteenth century.

The Forth and Clyde Canal

While the Commissioners for the Forfeited Estates were surveying the Highlands, one of the major works going on outside their purlieu was the Forth and Clyde Canal. George Clerk Maxwell, however, was very much involved.

In September 1754 George received an interesting letter from one John Smeaton (1724–1792), then in Edinburgh, telling him that he had obtained the backing of both the Duke of Queensberry and the Earl of Hopetoun for his plan to drain the Lochar Moss in the Maxwell heartlands.[55] He submitted his report to the Duke shortly after writing to George, but nothing came of it (Skempton, 2002, pp. 622 and 625; Smeaton, 1754 [publ. 1812]). In the same letter, however, Smeaton mentioned that he had also just delivered his plan for the development of Leith Harbour in response to a Parliamentary Act for the project passed in the previous year. Unfortunately for Smeaton, the Act had neglected to provide the wherewithal to carry out the necessary work, and nothing came of that either. But it was the start of something, for John Smeaton was later involved with Lord Hopetoun, George Clerk Maxwell and the trustees of the Board of Manufactures, to construct a canal from Falkirk to Glasgow with the object of greatly facilitating commerce between the ports on the Clyde and the Forth. Edinburgh, in particular, would then be able to benefit from the riches that were flowing into Glasgow from the new world; sugar, tobacco and cotton. Smeaton delivered his report in 1767 (Smeaton, 1767) and an Act enabling the project was passed in March 1768. Now, two

Commissioners for the Forfeited Estates chose to invest in the canal; one was Lord Elliock (1712–1793), a judge, MP and landowner, and the other was George Clerk Maxwell. Although the route of the canal was south of the 'Highland line', there were some who considered the project to be of such great importance that they pressed for money from the Forfeited Estates to be used to subsidise it. It is to the credit of these two Commissioners that this did not happen, and instead the Act required that the initial capital should be raised from 1,500 shares, to be subscribed for by interested parties at £100 each. George Clerk Maxwell went in for five shares, as did his friend the geologist Dr James Hutton (1726–1797) (O'Connor & Robertson, 2004a) who had also been involved in promoting the grand design and advising on its route. Both men served on the project's executive committee from 1767 to 1774, and were amongst those directly involved in the on-site management of works (Daiches et al., 1986, p. 120).

Although Smeaton was able to complete half the work by the end of 1770, he withdrew because of ongoing difficulties with landowners and because the funding was running out. He had another crack at it in 1775, but the money ran out again when the canal was just six miles short of its Clyde terminus (Priestly, 1831). The war with the American colonies that began in the following year and dragged on until 1783 had a double impact; firstly, the cost of the war was a huge drain on the national purse, and secondly, trade with the American colonies and elsewhere was effectively brought to an end. Imports into Glasgow were badly hit so that the income from the completed section of the canal was less than half of what the investors had been hoping for. The project was left in limbo until things recovered after the war. It was at this point, in 1784, that £50,000 was eventually loaned from the Forfeited Estates account[56] to finish the job under a new engineer, Robert Whitworth, who may have previously worked under Smeaton on the Calder and Hebble waterway (Skempton, 2002). Even so, the canal was not finished until 1790 (Groome, 1885). But the money came too late for George, who had by then been dead for several months, and since it had been granted under the 1784 act for winding up the Commission, it could no longer have represented a moral dilemma for the one shareholder to have been on the Commission until the end, Lord Elliock.

George ended up losing out, for by 1775 not only were the canal shares worth only 40 per cent of their original value, there had been extra calls for cash along the way.[57] Amongst other woes regarding his finances, fifteen years of protracted tribulations with the project must have contributed to the deterioration of his health in his final years – something we should bear in mind.

Commissioner for Customs

Realising that he could well benefit from a boost to his regular income as joint Remembrancer to the Exchequer, in 1762 George Clerk asked for the help of his friend the Duke of Queensberry in finding an administrative post for him, preferably as Postmaster in Scotland. The word 'friend' in this context needs to be treated with caution, for their relationship was essentially that of patron and henchman. Just as had been the case between their fathers at the time of the Union of the Parliaments (Chapter 4), friendship under these circumstances could only go so far. However, the Duke turned out to be favourable to the suggestion and sounded out the Prime Minister, Lord Bute,[58] who seemed well disposed to the idea but introduced a note of caution, saying that ways and means would have to be found. The Duke therefore advocated to George that he should make some proposals for improving the revenue so that he might better his chances.[59] George duly complied with the Duke's request and sent him his observations and ideas on the subject, which Lord Bute was well pleased with.[60] Negotiation about ways and means, however, was still ongoing by the end of the year.[61]

At the start of the following year, the Duke was at his estate at Amesbury.[62] Writing to George, he reassured him that the position was his, and went on to say that when he got back to London he would speak to Baron Mure,[63] a close friend of Lord Bute, in an effort to expedite matters.[64] Despite the Duke's fairly positive indications, the outcome some two months later was that George was instead offered a position as a Commissioner for the Board of Customs. According to the Duke, Lord Bute thought this post would actually work out better.[65] It certainly carried a useful annual salary of £500, as much as his father had been getting as a Baron of the Exchequer. George wrote back to the Duke expressing his thanks and saying that he would gratefully accept the offer.[66] He received the Duke's and Duchess's congratulations a few days later,[67] but having hardly had time to celebrate his good fortune, he received a further letter from the Duke which betrayed the Duke's true position; he could be charming and helpful, but in the end George was merely his minion. Having got George the post, he was now requesting that he should forgo the salary of £82 10s from his existing post as Remembrancer to the Exchequer. He brushed aside any protest by insinuating that he was asking this out of courtesy, for otherwise someone else would come along and demand it from him. It would understandably have been a slap in the face for George to be treated in this manner, but he stood his ground and told the Duke that he could not afford to comply, stating that he had to put out twice that much on fees for his new post. Unfortunately, we do not know whether the Duke

Plate 1 John Clerk Maxwell aged sixty-three

In a letter of 5 March 1854 John Clerk Maxwell wrote to his son: 'Aunt Jane stirred me up to sit for my picture, as she said you wished for it and were entitled to ask for it, "qua" wrangler. I have had four sittings to Sir John Watson Gordon, and it is now far advanced; I think it is very like ... It is Kit-cat size, to be a companion to Dyce's picture of your mother and self, which Aunt Jane says she is to leave to you' (C&G, p. 207). He was therefore sixty-three at the time, not sixty-six as stated earlier in the book (C&G, facing p. 11). (By Courtesy of Sir Robert Clerk of Penicuik)

Plate 2 Edinburgh Academy c. 1830, from Shepherd's *Modern Athens*
The school, founded by Lord Cockburn, opened in 1824 about seventeen years before James Clerk Maxwell's first attendance there. Note the oval cupola atop the roof, which James would have seen every day, perhaps inspiring his proposition on 'Ovals'.

Plate 3 James Clerk Maxwell, dressed much as he would have been on his first day at Edinburgh Academy
The tunic, and the square-toed shoes fastened with buckles rather than laces, the frilly collar, are all as mentioned by Campbell (C&G). (By courtesy of the Cavendish Laboratory)

Plate 4 James aged 12
This may well be the picture he referred to as having been done by 'Mrs Tis', possibly his Aunt Isabella, who sometimes sketched and painted with Jemima. Whoever the artist was, they were close enough to the family to comment disparagingly on John Clerk Maxwell's grammar. (By courtesy of the James Clerk Maxwell Foundation)

Plate 5 George Street looking east, from Shepherd's *Modern Athens*
Before they flitted to 11 Heriot Row, Robert Hodshon Cay and his family lived at 2 George Street, the far house on the right side of the street, just before St Andrew Square. The gable-end of the house faced onto the street (follow upwards from the right-hand lobe of the hat of the lady on the extreme right). Many such buildings have since been replaced, and so the character of the street is now quite different. St Andrew's Church (now St Andrew and St George's) was where John Clerk Maxwell would take his son James to the morning service when visiting Edinburgh.

Plate 6 (*left*) **James Clerk Maxwell at about 24 years of age**
Here we can see the mutton chops that were a precursor to James growing a full beard. His hair is already receding a little, but the little tuft sticking up at the front seems to be a throwback to the ringlets seen falling down his brow in Dyce's portrait of him as a child in the arms of his mother. In the full picture, he holds one of his colour wheels. (By courtesy of the Master and Fellows of Trinity College, University of Cambridge)

Plate 7 (*above*) **The multi-pointed lightning rod from Maxwell's house at Glenlair**
The rod was not fitted by Maxwell, for he argued the case against having one. It seems to have been erected after Andrew Wedderburn Maxwell enlarged the house. Fittingly, it now makes a good amateur radio aerial. (By kind permission of Dr James Rautio)

Plate 8 (*left*) **James Clerk Maxwell's statue at the east end of George Street, Edinburgh**
The statue by Stoddart was erected in 2008 through public subscription under the aegis of the Royal Society of Edinburgh. Maxwell is seated with a colour wheel in hand and his faithful companion, Toby, ensconced below. The plinth has two bronze side panels, one depicting the connection between Maxwell and Newton through the science of colour, and the other depicting his connection to Einstein by way of his idea of field theory. Ironically, the column and statue of Viscount Melville towers massively behind Maxwell's statue.

Plate 9 Sir John Clerk 2nd Bt. of Penicuik by his cousin William Aikman
He is seen here in his robes as a Baron of the Exchequer. (By courtesy of Sir Robert Clerk of Penicuik. Photograph: Gary Doak)

Plate 10 William Clerk, the Baron's brother, c. 1700
There is a very similar portrait of the Baron by John Medina, painted about the same time. The portrait of William, however, is catalogued as being by John Aikman, which, given the similarities, would seem to be a mistake. In addition, although William Aikman had a son John, he would have been far too young to do such a work. (By courtesy of Sir Robert Clerk of Penicuik)

Plate 11 Agnes Maxwell, wife of William Clerk
(By courtesy of Sir Robert Clerk of Penicuik)

Plate 12 Dorothea Clerk Maxwell
Although the subject is believed to be Dorothea, this cannot be verified with absolute certainty. The facial similarities with Agnes Maxwell (Plate 11), however, do tend to support these portraits as being of mother and daughter. (By courtesy of Sir Robert Clerk of Penicuik)

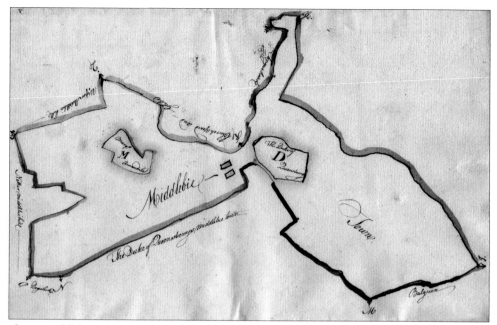

Plate 13 Middlebie Town as in the original sketches
There are eight such pen and ink sketches depicting the individual parcels of land that made up the Middlebie estate as shown in Figure 8.1. The Middlebie Town property, shown here, is very irregular in shape and almost split in two by being squeezed in by Church land on the north and the Duke of Queensberry's property on the south. In addition, the Marquis of Annandale owned an enclosed area within the western part the property, while the Duke of Queensberry had a similar piece out of the eastern part. The 'town' itself is just a few buildings. We must assume that this, on its own or with some of the contiguous pieces of land, was the originally entailed Twenty Merk Land of Middlebie. The Merkland of Kirkton of Kilmahoe was entirely separate and lay a few miles north of Dumfries (see Figure 2.1). (DGA: GGD56/19, eighteenth century, by courtesy of Dumfries and Galloway Libraries, Information and Archives)

Plate 14 Wanlockhead in 1775, by John Clerk of Eldin
(By courtesy of Dumfries and Galloway Libraries, Information and Archives (local studies collection: CO003976))

Plate 15 Princes Street when the Clerks and Clerk Maxwells lived there
The houses were then uniform and plain-fronted, in obvious contrast to the present mish-mash of heavily modified frontages and modern buildings. The extension of the earthen bank across the drained Nor' Loch, seen in the centre of the picture, indicates the commencement of the Mound emerging to the west of the gap between the buildings at Hanover Street. From the etching by Elphinstone in Arnot (1788). There is an earlier version of this print in which the buildings on the far left are unfinished.

Plate 16 Alexander Irving of Newton
In this detail from 'Twelve Advocates that Plead Without Wigs', John Kay (1838), Alexander Irving appears top centre, while his contemporary Sir Walter Scott is on the bottom left. (Reproduced by Permission of the National Library of Scotland)

Plate 17 Newton House, built by Alexander Irving of Newton
Although this plate in Irving & Murray (1864) is not captioned as being Newton House, it fits with available information as to its location, and with other views of the area, not to mention the fact that it was property of the first named author, George Vere Irving. (Reproduced by Permission of the National Library of Scotland)

Plate 18 The tablet commemorating the Franklin Expedition on Lt John Irving's memorial in Edinburgh's Dean Cemetery
The scene depicts the sailors heavily clad against the freezing cold on a rocky shore with their icebound ships in the background. The sailors behind the sled are lined up as on parade while one of the two figures on the far left holds a spade, suggesting that a funeral is taking place.

Plate 19 Milnes Court

(a) (*right*) Milnes Court from the head of the West Bow around 1900. Dr Thomas Irving's flat was two floors above the entryway. A rear exit door, on the same level as the flat, provided access to the court, a half flight of stairs below, as seen in the detail in (b) (*below*). It is here that they must have had the wooden stairway which has since been replaced in stone. The building, though internally renovated, still stands.
See RCAHMS (2013), Wallace (1987, pp. 26–27).
(© RCAHMS. Licensor www.rcahms.gov.uk)

Plate 20 Chessel's Buildings in Chessel's Court off the Canongate
This engraving by Storer (1820) was published when 'Mrs Dr Irving' was living at Chessel's Court. From 1810, the large building at the rear of the court had been used as an asylum for the deaf and dumb. The buildings on the left are now gone but the former asylum still stands. The court entryways are adjacent to 234–238 the Canongate. (Reproduced by Permission of the National Library of Scotland)

Plate 21 Anderson's pills, made by Dr Thomas Irving in the eighteenth century and sold widely
Clockwise from top left: (1) the label as it appeared in Wootton's *Chronicles of Pharmacy* (1910); (2) the pills and their packaging; (3) label bearing the initials of Katherine Anderson; (4) Raimes Blanchards & Co outer box label. (By courtesy of Lindsay & Gilmour Pharmacies)

Plate 22 St John's Church on Lothian Rd and Princes St, from Shepherd's *Modern Athens*
The Edinburgh Cays had a close connection with this Episcopalian church where James' Aunt Jane would take him for afternoon service, notwithstanding his attendance at St Andrew's in George Street, a Presbyterian church, in the morning.

Plate 23 Early calotype of John Cay, seated, with his sons Robert and Edward (Edinburgh Calotype Club, nineteenth century, vol. 1 p. 14. Reproduced by permission of the National Library of Scotland)

Plate 24 William Dyce's charcoal sketch of Puck, 1825
Possibly the first of several he did in this vein, the one of James Clerk Maxwell below being just one of them. (By courtesy of Aberdeen Art Gallery & Museums Collections)

Plate 25 William Dyce's charcoal sketch of James Clerk Maxwell as a child, c. 1833
Compare with Dyce's orignal concept, Plate 24. (By courtesy of the the James Clerk Maxwell Foundation)

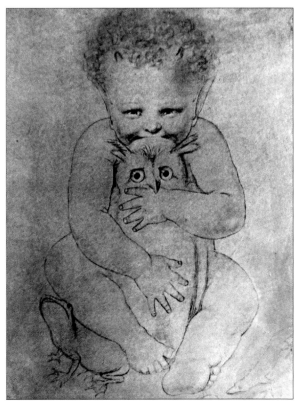

Plate 26 Early calotype of Robert Dundas Cay
From the Albums of the Edinburgh Calotype Club of which his elder brother John was a member (Edinburgh Calotype Club, nineteenth century, vol. 1, p. 55. Reproduced by permission of the National Library of Scotland)

Plate 27 Isabella Dyce
Painted by her brother William Dyce about the time of her marriage to Robert Dundas Cay. (By courtesy of Aberdeen Art Gallery & Museums Collections)

simply dropped his demand or whether he left George no choice in the matter.

In spite of not getting his preferred appointment, George took to his new post of Commissioner of Customs with his characteristic enthusiasm and a determination to make something of it. He started by addressing the problem of smuggling from the Isle of Man, which lies in the Irish Sea some fifteen miles south of Burrow Head on the Mull of Galloway. The nub of the problem was that it was not part of the United Kingdom; back then it was the sovereign property of the Duke of Atholl rather than the British Crown, and as such it was beyond the reach of the excisemen. The level of smuggling from Man into Galloway was such that it seriously curtailed revenue on imports such as rum and brandy. All the smugglers had to do was to get incoming ships to unload their goods on Man and then wait for a suitable opportunity to land the contraband on some remote area of the mainland coast, for which purpose they could use small local craft that would pass unnoticed by day or night. The smuggling was particularly rife along the south-west coast of Scotland which, from the Solway Firth to the Mull of Galloway, forms the nearest landfall north from the Isle of Man.

In 1764 the Prime Minister, George Grenville, tasked the Board of Customs in Scotland to come up with a plan to suppress this illicit trade, and the Board in turn asked George, perhaps because he knew the area around Dumfries, to make a survey of the problem and report back to them. This George duly undertook, but he soon came to the conclusion that not only was the coastline too vast and remote to be effectively patrolled, the local inhabitants were fully complicit in the smuggling. He therefore decided that it would be very difficult to achieve any measure of success by simply trying to suppress it:

> every farmer's servant who could purchase half a cask of spirits, was engaged [in the smuggling] for his share [. . .] the whole inhabitants on the south-west coast [. . .] had followed scarcely any other employment than this pernicious traffic. (Clerk, 1788)

The problem needed a radical solution, and he offered one: the government should buy out the sovereignty of Man from the Duke of Atholl. It was a sound idea but an expensive one, and Grenville had a different plan, for he was ready to send out additional naval ships to patrol the seas between Man and the Scottish coast. Although William Craik of Arbigland[68] was not actually a Commissioner of Customs, he was a local inspector that lived near the Solway coast. This position was something of a sinecure that gave him status, income and a sizeable share of the proceeds of all seized goods. He was a friend of George and no doubt one of the first people that George would have consulted about the problem. When the proposal to buy the Isle of Man was made, however, both men went to London to put the case to the government. It seems that George brought Craik to London as a reliable expert witness (*Farmer's Magazine*, 1811).[69] After several months of discussion, however, Grenville was at last persuaded that naval patrols would eventually cost more than the purchase of the Isle of Man. Although Man became part of George III's realm by July 1765, and the Duke of Atholl became £70,000 richer, it retains its own peculiar status as a British Crown dependency (Cahoon, 2014). As to the effect its purchase had on the smuggling, Lord Eldin (Clerk, 1788) tells us the eventual outcome:

> the whole inhabitants on the south-west coast [. . .] now earn their subsistence by a more honest application of their industry. The face of the country, which formerly never could raise a sufficient quantity of grain to support its own inhabitants, is totally changed.

A great success perhaps, but we cannot let it pass without mentioning some additional information about William Craik that seems not to have come to George Clerk Maxwell's notice or, if he did know about it, he was keeping his cards very close to his chest. Kirkbean Heritage Society has in its possession an old letter from an excise officer at Dumfries who, in a report to Edinburgh, stated: 'Sloop in Arbigland Bay [. . .] would not go so far as to say that the Laird was involved [. . .] but many of his servants and horses were' (Blackett, 2010).

The article goes on to assert that Craik was making a fortune out of playing the game on both sides. If George had suspected him, then it may have been the very thing that convinced him that neutralising the Isle of Man was the only way forward; if he did not suspect, then his survey of the problem had failed to get to grips with the real issues involved. Given that the solution he opted for was the right one, there is a fair chance that he did indeed suspect Craik and was playing his own double game in taking him to London to see Grenville!

• • •

Robert Burns was an exciseman in Dumfries from 1788 to 1796; the smuggling, of course, was still going on. In March 1792 Burns was personally involved in the capture and impounding of the Plymouth schooner *Rosamond*, which had run aground in the Solway River at Sarkfoot Point, immediately south of Gretna. Sarkfoot is quite far up the Solway, a firth notorious for tidal sands that the smugglers had badly misjudged. But the incident demonstrates George Clerk Maxwell's contribution to the suppression of smuggling in the region, for there was no longer any point in the smugglers dropping off their wares on the Isle of Man. The master, Alexander Patty, was forced to take his chances by bringing a sizeable ship far up the Solway laden with cargo, and he had paid dearly for it. While he lost both his ship and his cargo, he at least managed to escape the gallows. After putting up a stiff resistance involving several exchanges of cannon fire, he and his crew were able to get over the side of the vessel and make it along the treacherous sands onto the English coastline, where they soon disappeared (*Burns Encyclopaedia*, 2014).

• • •

In his time at the Board of Customs, George worked closely with two particularly interesting people. The first was an ex-soldier, Basil Cochrane (1701–1788)[70] who had also been appointed a Commissioner in the same year, 1763. Cochrane was the great-uncle of the journalist James Boswell, but more interestingly, in his military capacity he had been governor of the Isle of Man from 1751 to 1761, which makes it a bit harder to believe that the purchase of Man had been *entirely* George Clerk's idea. The other interesting party was none other than Adam Smith (1723–1790), author of the *Wealth of Nations*. Appointed as a Commissioner in 1778, Smith was also a friend of James Hutton and, along with Joseph Black, George Clerk Maxwell and his younger brother John Clerk of Eldin, they were members of the Oyster Club that met regularly to dine and converse in an unobtrusive Grassmarket tavern.[71]

• • •

In the *Correspondence of Adam Smith* (1987, pp. 405–412) there appears an account of a remarkable incident that took place off the Scottish coast in September 1779, during the War of Independence with the American colonies. The account comes to us via Smith because the Commissioners at the Customs House in Edinburgh[72] had been particularly tasked to keep an eye out for any such occurrence, and four 'French' warships had been seen out at sea somewhere between Dunbar and Eyemouth. Commissioners Adam Smith, George Clerk Maxwell and Basil Cochrane were in attendance, and they duly alerted the shipping in the Firth of Forth to the danger. On 16 September they sent a customs cutter out to reconnoitre, and it seems that her captain got within 'a pistol-shot' of what he took to be a fifty-gun French warship off the Isle of May;[73] in reality it was the 42-gun colonial Continental Navy warship *Bonhomme Richard*[74] under the command of Captain John Paul Jones (1747–1792). Jones was a native of Kirkcudbright and, coincidentally, the son of a gardener on William Craik's estate at Arbigland. Born plain John Paul, he became a merchant sea captain. He later fled to Virginia after killing one of his crew, and changed his name to cover his tracks. When the War of Independence was declared, Jones did not hesitate, and in a few short years he was in command of a colonial frigate and back in British home waters looking for a fight.

The captain of the cutter briefed Clerk Maxwell and Smith on the situation and although they gave orders for three revenue cutters to be placed in readiness, Jones did not sail up the Forth. Instead he engaged the British Baltic merchant fleet a week later and claimed his first great naval victory, thereby earning himself the title of 'Father of the United States Navy' (Potts, 2011).

Mineral Spas and Mining

George Clerk Maxwell was ever interested in what the land could be made to produce, and its natural resources were no exception. In his sketch of George, Lewis Campbell portrays him thus:

> We find him, while laird of Dumcrieff [. . .] practically interested in the discovery of a new 'Spaw' [spa] and humoured in this by his friend Allan Ramsay, the poet:– by and by he is deeply engaged in prospecting about the Lead Hills, and receiving humorous letters on the subject from his friend Dr. James Hutton. (C&G, p. 18)

From an early age, George would have taken part in his family's annual peregrinations to take the goat's whey and the spa waters for the benefits of their health (Chapter 4; BJC, p. 138). He would therefore have been well acquainted with spas both as a source of public benefit and as a means of making money. A letter from the Duke of Queensberry[75] provides some detail concerning George's involvement with the recently discovered 'spaw' on the slopes of Hart Fell, about three and a half miles north of Moffat.[76] Being on the Duke's land and near Dumcrieff, the Duke would have been certain to get George busy on it. Indeed so, for it was over a mile from the nearest track, not to mention a considerable uphill climb. In 1753, George visited the spa and offered to help with making a better access road. The entrance to the well also needed shoring up with some stonework to prevent it from becoming blocked. While the Duke was prepared to spend £30 on the work, in laying out such a sum he would have had the health of his coffers in mind rather than the health of the populace, for such healing waters could be bottled and sold.

Each spa was associated with a specific range of benefits as indicated by its characteristic chemical composition. There was consequently a great deal of interest in seeing what a newly discovered spa could offer. The new spa in question may have been awkward to get to, but there would always be people eager to try it in the hope that it might be just the one for them. As it turned out, the Hartfell spa was of a complementary composition to the local Moffat waters, for it contained salts of iron rather than calcium, and it was still rather than effervescent. It had been discovered in 1748 by a local eccentric, John Williamson,[77] who is said to have been involved in some mine workings nearby (Groome, 1885, vol. 3, p. 248). Although Williamson would have needed the water for the mine works, it appears that he did not start to take it regularly until 1751, when he noticed that it relieved a persistent stomach problem (Horsburgh, 1754).

Although the water at the Moffat Well could be taken away in bottles for later consumption, it was effervescent and did not keep so for long. The new spa had the commercial advantage that, being still, the bottled water could be kept more or less indefinitely, meaning that it could be distributed to a much wider market. Seeing such an opportunity, the Duke suggested building a well-house in the following year. The undertaking of such projects made people like George useful to the Duke, and in return, it would be hoped that he would help them, as George knew only too well.

The Duke was as good as his word, for in Horsburgh (1754, p. 1) we find the footnote: 'his Grace the Duke of QUEENSBERRY [...] was, for the public benefit, generously pleafed to give money to make a convenient road ; and to build a fhade or arch over the fpring to preferve it from dirt and rubbifh'. The bottled product was even exported to the West Indies (Turnbull, 1871), from which trade the Duke may have benefited from a percentage, as he did with his mines. Turnbull also mentions that George was the one behind the erection of a memorial to Williamson over his grave in Moffat churchyard:

>In Memory of Jno. Williamson, who died 1769,
>Protector of the Animal Creation,
>The Discoverer of Hartfell Spa, 1748:
>His life was spent in relieving the distressed.
>Erected by his friends, 1775
>(Turnbull, 1871, p. 101)

Indeed, George had been Williamson's friend and fellow traveller in prospecting for spas and minerals. In fact, he was 'as great an enthusiast in mineralogy as Williamson, and what is more, as little successful in his theories and operations' (Ramsay, 1888, p. 335).

The bulk of John Ramsay's account appears to be first-hand, and his visits to the area are contemporaneous; he would therefore have been in a very good position to take note of what was being said about George Clerk Maxwell's reputation in that sphere of activity, that is to say prospecting and mining. Gray (BJC, p. 96n1) said of George: 'he also set on foot various mining schemes for lead and copper, through some of which he suffered great loss'. It is perhaps because of this purported lack of success that there is little said about what he actually did achieve; it is important that we should explore this, for George did run into a severe financial crisis during the last few years of his life. How then did it come about?

George's interest in mining came naturally; firstly he was interested in just about anything that was either agricultural or industrial, and secondly, mining was in his blood. His brother James mined coal at Loanhead, as did their father, the Baron, and their grandfather, the 1st Baronet (Chapter 4) before them. In addition, in 1739 the Baron went out of his way to see how coal was hewn from under the sea at Whitehaven, and he was called on by his patron the Duke of Queensberry to inspect his mines at Wanlockhead (BJC, pp. 153, 165, 175). As far back as 1718, he and his father had also had their own designs on mining for minerals rather than coal. He says that they went to Leadhills, 'to view the Lead works belonging to the Earl of Hop[etoun], for we had then a design of purchessing the lands of Glendorch in the nighbourhood' (BJC, p. 97).

These lead mines at Leadhills and neighbouring Wanlockhead cover an area of many square miles just below the tops of the Lowther hills between Elvanfoot and Sanquhar. Plate 14, from an engraving by George's younger brother, John Clerk of Eldin, gives a realistic impression of what they were like. At an altitude of some 1,300 feet, their environs are truly inhospitable when the weather is anything other than fair and but for the mines the separate villages of Wanlockhead and Leadhills would not, and could not, have existed. Mining had been going on in the area for some hundreds of years by the time the Duke of Queensberry let out his mines to Ronald Crawford's company in 1755. Ronald Crawford of Restalrig was an Edinburgh merchant who had the capital for this sort of operation. The system was that the landowner owned the mineral rights and he could either do the mining on his own behalf or lease them for a 'tack' (take) of the product, for example, in the case of the Duke one bar of lead in every six. But mining was not an easy operation, particularly in these locations and with no mechanisation other than waterwheels. In August 1758, for example, lack of air at the Cove vein was a problem: 'The small progress made was owing to the miners not having air enough to enable them to work' (Museum of Lead Mining, 2014).

In August 1762, this same vein had a problem with flooding, which led to George Clerk Maxwell again being consulted by the Duke. Any decrease in output would have directly affected the Duke's tack, and so he wanted to know what could be done about it, and George was the man who would know. Months later, in December, the problem apparently was just the opposite, a lack of water. The mine manager proposed cutting a drift (horizontal passage) to collect the water from local burns, but the Duke once again wanted George's views.[78] However much or little it was that George knew about mining, the Duke certainly had confidence in him.

At other times, and with other veins, they met with more success. Production fluctuated because old veins would run out and new ones had to be found. For the year ending January 1762, production was seriously down, having progressively slumped to 40 per cent of what it had been just three years before (Harvey & Downs-Rose, 1976).

A Mine of His Own

There is no mention of George Clerk Maxwell being directly involved in the Crawford mining partnership, and his connection with Wanlockhead itself seems only to go so far as his occasional intercessions on behalf of the Duke of Queensberry. A tantalising hint from Lewis Campbell, however, implies that he was actually concerned in some other mining venture in the area: 'by and by he is deeply engaged in prospecting about the Lead Hills [...] After a while he has commenced active operations, and is found making fresh proposals to the Duke' (C&G, p. 18). The traces of what transpired are to be found in correspondence and agreements between George and his business partners[79] and from the hints these give in turn.

In 1755, Cuthbert Readshaw, a merchant from Richmond in Surrey who dealt in lead, had turned his mind to improving his profits by producing the raw material for himself. However, a venture of that sort always needed a good deal of capital to fund it, so that some sort of partnership was required in order to put it together. He and an associate William Hynd corresponded with George Clerk Maxwell and William Carruthers, a mining master from Dumfries, about the possibility of a joint venture to prospect for lead and copper in Dumfriesshire,[80] where George had probably done a good deal of prospecting already.

As already mentioned, mining had for some time been one of the Clerk family's business interests. Some forty years after his father and grandfather had visited Glendorch, about three miles south of Crawfordjohn, it fell to George and his associates finally to take up the idea of mining there. According to Irving & Murray (1864, p. 63n2) there was indeed a seam of high quality lead ore in the area: 'About the end of the eighteenth century "an astonishing and unprecedented width of 18 feet of pure galena [lead sulphide]" was discovered in the Glendorch mines.' The nearest surviving trace of mining in that area is actually in neighbouring Glendouran.[81] We can guess that the ore had originally been found in Glendorch itself, but once it was played out fresh seams would have been looked for nearby. It was therefore a natural place for George to consider mining, but it is interesting that it was only in the year that his father died that he acted on it. Whereas George was inclined to leap into things, his father and grandfather had been 'canny', that is to say, they achieved their good fortune through patience and shrewdness, qualities that were not reflected in George. His father would have exerted what calming influences he could on George, but now George would not only have been free to act as he pleased, he would very likely have found himself with the wherewithal to do it.[82]

In the following year, George and his partners reached an agreement with the owner of the land around Glendouran, William Maxwell 5th Baronet of Calderwood (d. 1789), for 50 per cent of all mines and minerals there.[83] From the scale of what remains to be seen at Glendouran, any operations there were limited and it appears they were too early to be the ones who made the big find mentioned by Irving and Murray.

It seems that the partnership continued to go through a formative period while a more suitable site was looked for. James Hutton may have been consulted, for he mentions in a letter to George that he had investigated a site around Leadhills and found some signs of metal ore.[84] On 1 January 1758, however, George and his partners took out a thirty-one year lease on the Earl of Hopetoun's property at South Shortcleugh.[85] Since this was over three miles to the south-east of their original mine, they were clearly starting afresh.

Only days later, further agreements were entered into. While one of the partners, Hynd, dropped out, Joshua and Caleb Readshaw came into the discussions along with Samuel Smith of Washton, a smelter. Nevertheless, the scheme to mine at South Shortcleugh was overtaken by events in 1763 when a soldier working on the new military road from Dumfries to Port Patrick happened to discover a quantity of lead ore at Blackcraig (Statistical Accounts, 1791–99, pp. 54–55), about two miles south-east of Newton Stewart[86] on the land of Patrick Heron of Heron and Kirroughtrie, a banker and later MP for the Stewartry of Kirkcudbright, and the rights were his.[87] Although it was in a different area altogether from the mines in the Lowthers, in fact some sixty miles to the south-west, there were several attractions to mining there: it was not an inaccessible, inhospitable area; it was near both to the sea and the new military road, so that shipping lead or even the ore would be straightforward; and there were as yet no competing mines in the immediate vicinity.

The new mining venture began slowly to get under way. Discussions on the structure of the partnership were still being held in 1764, and when the company articles of the Craigtown Mining Company were finally signed in 1766, Smith had dropped out to be replaced by Philip Jackson, another London merchant. Although Patrick Heron did not come in as an actual partner in the company, he would get the customary tack on any lead produced. However, as 'joint-adventurer', he would have had some additional form of interest for in it, for example, he could have put up some capital. The partners being now fully focused on Blackcraig, the lease over South Shortcleugh was duly abandoned in 1769.[88]

A 1768 drawing[89] shows the site plan. The first mine was probably adjacent to the original find on the old military road where it passes Blackcraig farm on the present map. The group of buildings at Craigton, or Craigtown, is clearly the source from which the partnership took the name of their company. Although some of the houses there are now of a much later construction, the company were granted permission to build accommodation, offices and works buildings in and around the settlement. While a few of these older houses are still to be seen where the old military road forks uphill away from the single street, the name Craigton has disappeared from modern maps and only the names of individual houses are left as reminders.

Operations at the mine must have started by 1770 because a suit for damages was filed by neighbouring prospectors, Thomas Paton and Richard Richardson, who were businessmen from the north-west of England. Alerted by news of veins of lead ore, these gentlemen had taken a lease on the adjacent property of Patrick Dunbar of Machermore,[90] which ran between Patrick Heron's property and the River Cree and was where Blackcraig itself actually lay. Calling their venture the Blackcraig Company, these wily opportunists clearly thought that since the veins were near the boundary line, there would be a good chance of them extending over onto their side. Not only did their hunch turn out to be right, George and his partners began to suspect so as early as 1768, a stark warning for them about the highly risky nature of their business.[91] At least it turned out that the ore on the Blackcraig Company's side appeared not to amount to much.

Meanwhile, the Craigtown mining company got into production. Later in the same year, George and his brother Sir James borrowed £1,500 from the Royal Bank of Scotland, an enormous sum at the time.[92] Though not a partner, Sir James had been in the background of George's project, presumably providing advice and financial backing, for there are accounts of the dealings between themselves and with the Bank of Scotland during this period.[93] At this stage, with the mine now in production, James clearly had sufficient confidence to put his name alongside George's on a bond of this magnatude for the purpose of funding the substantial investment that was required to get the mine producing in earnest.

By 1778 at the latest, William Mure had been put in charge of the Craigtown mines[94] and he regularly reported to George on how things were going; there was a smelter on site (but evidently not a very effective one); an ore crusher; a water wheel to power the ore crusher; water to power the wheel and to wash the ore; and lades to bring in the water from the hill. With forty or so employees, annual production reached about fifty tons of lead, representing about £1,000 of output, by comparison only a tenth of Wanlockhead's (Donnachie, 1971, p. 122). At first the lead cost more to produce than it sold for, but even in a year when a profit of £150 was achieved through cost cutting, the dividend of each partner would hardly amount to much. It was certainly not the kind of return that would have been needed to pay off the capital costs, not to mention the ongoing investment that would be required to boost production to a sustainable level. It is hard to see how the investors would have ever got their money back. Nevertheless, a shot mill was built at Creetown in 1780 so that they could get some added value on the lead produced, but even if that doubled their profits, it had also added to the total investment of capital. The fact that war with the American colonies and its European allies ended shortly afterwards could hardly have helped matters, for the demand for lead shot would have plummeted as a consequence.

By 1791, output had fallen to very low levels, only thirty tons a year (Statistical Accounts, 1791–99, p. 55) and at some point production must have ceased with huge losses all round. In the meantime, however, the Blackcraig mining company had continued, quite literally, to 'dig around' with some success and they had eventually been rewarded with 450 tons or so of lead ore per year, far more than the Craigtown company had achieved. Even so, their luck was short-lived, and the ore began to run out (Donnachie, 1971, p. 124).[95] The consequences of all this for George Clerk Maxwell were to be grave indeed.

Other Activities and Interests

In the eighteenth century, travel in the Highlands involved many privations and was not to be undertaken lightly. Nevertheless, in May 1739, George had accompanied his father on a tour of the north of Scotland via Perth and Dunkeld to Inverness, where they had visited their relative, Duncan Forbes of Culloden (see Chapter 3). They returned by the coastal route by way of Elgin, Aberdeen and Dundee (BJC, p. 152). In 1764 he was to repeat the experience in the company of his friend Dr James Hutton. They followed a similar route to George's earlier journey, but from Inverness they went on through Easter Ross to Caithness (Playfair, 1803, p. 45), perhaps because the area around Helmsdale is rich in geological interest, notably the Helmsdale fault. George was interested in Hutton's geological work and may even have done some field sketches for him, but having been appointed as a Commissioner for the Forfeited Estates in the previous year, he would have been taking in all that he saw by way of opportunities for improvement.

Hutton was also the close friend both of George's younger brother, John Clerk of Eldin, and later his son John, who became Lord Eldin, and between them they accompanied him on three other excursions to Glen Tilt, to the south-west of Scotland, and to the Isle of Arran. The Clerks recorded many of the places they visited in etchings and drawings, especially John Clerk of Eldin (Bertram, 2012a and 2012b).

It seems, however, that George made at least two separate journeys to the Highlands around June of both 1766 and 1767, most likely on Forfeited Estates business. One of the places he visited at that time was Rannoch, about thirty miles west of Pitlochry (Millar, 1909, p. 242), on the forfeited homelands of Clan Robertson, where a barracks had already been constructed in 1746. The record of this event is a petition that he received from one Donald Robertson of Rannoch, who had a workshop there with three looms and was endeavouring to obtain three more, but the trouble was he could not afford them. In his petition, he reminded George that, during his visit of 'June last', he had offered to have the additional looms funded by the Commissioners.

• • •

In 1760, George Clerk Maxwell was appointed to yet another office when he was named as one of the trustees for Fisheries, Manufactures and Improvements (Clerk, 1788), a role which fitted well with his own involvement under his father's tenure of this office, such as the spinning school mentioned above that he had set up at Dumfries some twenty years before and was still going in 1751. However, a few years later, at the time he was appointed as a Commissioner for the Forfeited Estates, the two roles tended to overlap since the Commissioners were trying to encourage the Board of Manufactures' efforts to be applied within their own sphere of activity. As to George's efforts to promote manufacturing closer to home, he now proposed to the Duke of Queensberry that weavers in various parts of Dumfriesshire should be offered subsidies to spin wool and to knit stockings.[96] In the same year there was also mention by the Duke of being in favour of proposals, in which Lord Elliock was also involved, for setting up a paper mill,[97] but sadly little more information than this appears to be available.

Having been appointed a trustee for the Board of Manufactures and Fisheries, George then became involved with the Edinburgh Society for the Encouragement of Arts and Sciences and Agriculture[98] under whose aegis a drawing school, the Trustees Drawing Academy, was set up in 1757. There was an existing drawing school, the School of St Luke, which operated during the winter from a room in the Old College. The Society offered attractive cash prizes to those pupils who produced the best drawings featuring some practical subject matter, such as a landscape design, a carpet design, or some floral patterns for prints; what is more, the offer applied equally to girls as well as to boys. In 1760, the year George joined the board of trustees, it appointed a permanent drawing master with a decent salary, and it thereafter became known as the Trustees Drawing Academy of Edinburgh, the precursor of Edinburgh College of Art. The third drawing master there was Alexander Runciman who, on his return from Italy in 1772, was painting scenes from Ossian on the walls and ceilings of the hall of the new Penicuik House, for George's brother, Sir James Clerk, the 3rd Baronet (Runciman, *c.* 1772; Laing, 1869).

• • •

George Clerk Maxwell is recorded as being a founder member of the RSE, that is to say he was already a fellow of the Philosophical Society of Edinburgh of which both his father, Baron Sir John Clerk of Penicuik, and his father's cousin, Dr John Clerk, had been co-vice-presidents (see Chapter 4). He submitted an article on agricultural improvement which was read in 1761 and later published in their *Essays and Observations Physical and Literary* (Clerk, 1771). Shallow ploughing, where only the top three inches of soil was turned over, was practised in Norfolk and he was making the suggestion that it might do well in Scotland. Lord Kames, however, disagreed and his viewpoint on the subject was published in the immediately following article. This was not, however, the only article of George's to be published by the Philosophical Society, for in Clerk (1756) he had already reported on bones, supposedly of an ancient elk, that had been found near Dumfries.

George's article on shallow ploughing, however, also reveals something of himself as a practical man, one for trying things out rather than learning from books:

> Some are convinced from what they hear or read; nothing will make impreffion upon others but occular demonftration; and, for my part, I am of the laft clafs, an unbeliever that could not fee the weight of arguments advanced in fupport of the practice, till I had an opportunity of [...] obferving. (Clerk, 1771, p. 56)

This chimes with his own preference to take up a practical profession rather than to become a lawyer. The initial pages of the same article also inform us that some time before his paper was read in February 1761, he had had the opportunity of travelling to England to observe for himself the farming methods in Norfolk and Lincolnshire, and possibly even Derbyshire. Since we already know that it was in 1764 that he went to London with William Craik to discuss his proposal for purchasing the Isle of Man with the prime minister, this was clearly an earlier and entirely separate expedition.

George was also a founder member of the Society of Antiquaries of Scotland[99] that was formed shortly before the RSE gained its charter. He was a keen antiquary and assisted the military surveyor General William Roy (1726–1790) in his search for Roman remains in Annandale (Prevost, 1968). In 1764 Roy had discovered a temporary camp at Cleghorn near Lanark, whereupon 'he requested his friend, Mr Commissioner Clerk', to search for similar camps in Annandale. George subsequently found the remains of two possible camps, one of which was at Torwood Moor, half a mile west of Lockerbie. While the discovery of this site is attributed to Roy,[100] the camp's original finder was obviously George Clerk Maxwell (Macdonald, 1920, p. 90).

• • •

Given the great number of things that George Clerk Maxwell was involved in, it is perhaps surprising that he had the time to be involved in anything more. Nevertheless, on five occasions between 1763 and 1780 he was the commissioner representing the Burgh of Sanquhar[101] at the Convention of Royal Burghs (Brown & Anstruther, 1894, Appendices B and C). This ancient convention was held annually in Edinburgh for the purpose of advising parliament on local regulations and bye-laws, particularly those affecting trade. The manner in which George's name is recorded gives clear evidence of its variability. It appears first as 'George Clark Maxwell of Drumcrieff', then 'George Clerk of Middlebie', and finally as 'George Clerk of Drumcrieff'. In 1768, it seems that the role was instead given to his son George, who is described as 'George Clark, younger of Drumcrieff', who appears to have also represented the burgh at the General Assembly in 1769–70.

The Fall of Dumcrieff and Middlebie

In 1774 George Clerk Maxwell had laid down a large plantation of trees at Aikrig on the south-western edge of his estate at Dumcrieff, and consequently we may well think that shortage of cash was not one of his problems at the time.[102] By 17 January 1777, however, he was being asked by Dumfries Kirk Session to pay off his debts to them, with a further request on 21 July 1778.[103] Such debts may have been relatively trifling, but they certainly seem to indicate that he was financially embarrassed at the time.

In the language of his era, George was an 'adventurer', what we would now refer to as an entrepreneur. He had taken a number of chances over the years, and that he was impetuous is suggested by his 'irregular' marriage to his cousin Dorothea in 1735. His first significant enterprise had been the linen factory that he opened in Dumfries, c. 1739, but then had to shut down in 1748. There had been some farming in the background and later an involvement in road building which, admittedly, may have been contributing to the public good. Then there had been his investment in the Forth and Clyde Canal, which cost him £500 plus further cash calls. Nevertheless, on the plus side he had secured three government posts and some other minor appointments which paid him well and, as was common at the time, probably brought him a number of additional perks and opportunities. Not only was he managing to undertake these things with great energy, he somehow also found time to go casting around Dumfriesshire for mining opportunities, and when he finally got the Craigtown Mining Company together, the effort of prospecting was simply replaced by the problem of getting it to yield enough lead.

But if he had travelled this road in hopes of succeeding with mining, his arrival was quite a disappointment. As we have seen, instead of profiting from his ventures, it had probably bled him dry by consuming his capital without ever yielding anything like a proportionate return. He had borrowed the necessary capital with the help of his brother James, by then Baronet, but whereas for James his Penicuik estate and his income from the coal mines probably gave him some extra slack with his creditors, for George it was a different story:

> Sir George Clerk, by Disposition of [April] 1782, registered in the Books of Council and Session 15th January 1783, in which he is designed George Clerk Maxwell, Esq. of Dumcrieff, one of the Commissioners of His Majesty's Customs, conveyed with consent of Mrs. Dorothea Clerk Maxwell, his spouse, inter alia, these lands[104] to Alexander Farquharson, accountant in Edinburgh, as trustee for his creditors. (BJC, pp. 249–250nN)

That is to say, he had had to surrender to an administrator all the properties that his father had given him as his start in life. Of course, we have found no direct evidence that it was principally the mine that occasioned this financial misfortune, but its failure to repay anything but a fraction of its total capital cost seems to make it fairly certain.

George's difficulties would have been exacerbated if his brother James' financial position had also been severely stretched, for example, as a result of all the expenditure he had laid out on building and lavishly appointing the new Penicuik House. James was very ill by late 1782, a fact which must have impressed on their creditors some urgency in calling their bonds in, and upon George and James the necessity, above all else, of saving Penicuik. It was therefore decided between them that the best thing to do would be to set aside the debt that George owed to James and concentrate on paying off the external creditors (Mackay, 1989, p. 2).

In financial terms, the situation was not fatal, just very bleak; George had no prospect of eventually paying off his debts out of his income alone, for the lack of return on his speculations was the very root of the problem. Whatever lesser measures were taken to claw back money for the creditors we do not know, but by his action in April 1782, Dumcrieff and the lands at Middlebie proper were put into Farquharson's hands as security for his creditors, to be sacrificed, if necessary, to save Penicuik. It would have been difficult to do anything similar with the rest of the Middlebie estate, that is to say Dorothea's collection of properties in and around Dumfries, because the greater part of them were still under the entail of 1722.

The calamity took place amidst all the other family tragedies that beset him and Dorothea in that year; the disposition was formalised in the courts at Dumfries the following month,[105] and thereafter in the Court of Session in Edinburgh in January 1783 (as mentioned above). The pains of an illness that he was now developing must have been as nothing compared with the pangs of regret that he must also have been feeling at this time. How diligently his father had conceived and executed his plans for Dumcrieff, only for George to lose it by risky speculations that were the antithesis of his father's measured way of proceeding. How skilful his father had been in freeing the Middlebie estate of its burden of debt and buying the property back again to put in George's hands, only for it to be lost for good. How great the loss of face, to lose the two names, Dumcrieff and Middlebie, that had been synonymous with him for most of his married life. How low the standing of a man once so esteemed and now found to be sadly wanting. Despite all his good deeds, he would seem to be only a faint shadow of what his father was before him.

On the death of Sir James in February 1783, George became the new Baronet and his debts to his brother were consequently eliminated. Even so, in stepping into his brother's shoes, he still did not have the wherewithal to save Dumcrieff and Middlebie, and so Farquharson proceeded with the usual humiliating form of disposal by public roup.[106] By the sale of Dumcrieff he hoped to raise £5,600, and it went at the first attempt for £5,300, not far from the valuation: 'Farquharson, as trustee, with consent of Sir George, disponed [Dumcrieff] on 22d May 1783 to Lieutenant-Col. Wm. Johnston, of the Royal Artillery'[107] (BJC, pp. 249–250nN). The parcels of land at Middlebie had been expected to raise £7,200, but they did not sell at the first attempt. They were offered separately on two occasions, but each time with no success. Finally, the price was dropped and they eventually went in 1787 for £6,300 to David Ewart, a writer in Edinburgh.[108]

By the time Dumcrieff was sold, George's own health was failing. According to his nephew John Clerk:

> the most acute bodily pains gradually wasted his constitution. He bore all his distresses with unshaken fortitude to the last, and attended to the duties of the publick Boards of which he was a member, with the same assiduity and perseverance as ever. (Clerk, 1788)

In the last few years he had lost a brother, two sons, a daughter and a grandson; his mining venture had done poorly; and the total consequence of all that he had worked for was that he had been nearly bankrupted and had suffered the humiliation of not only losing Dumcrieff and Middlebie as properties, but from his name. It could have given him little comfort to succeed as 4th Baronet of Penicuik. Nor did he live to put these dreadful years behind him, for he died on 29 January 1784, less than a year after succeeding his brother. It would be left to the next new baronet, his son John Clerk, and his widow Dorothea, to cope with the aftermath.

Soon after George's death, Dorothea moved out of James Court in the Old Town to Princes Street, the New Town's splendid main street overlooking the sunken gardens newly being created in the basin of the now drained Nor' Loch (Plate 15).[109] We cannot exactly say whether this was consequent upon her husband's death or if the move had already been planned in the wake of George inheriting the baronetcy from his brother in 1783. Having been ill for some time he may have wished to retire to the more salubrious surroundings of the New Town. Dorothea's house number was not given until the next issue of the Post Office directory, in 1784, when it was number 30 Princes Street, but from 1788 it was given as number 52; as frequently turns out to be the case, this could simply have been a numbering issue rather than an actual move of house.

George and Dorothea's second surviving son, James, having come home from sea, lived with his widowed mother at Princes Street and continued to do so upon his marriage to Janet Irving in 1786. Unfortunately, Dorothea suffered a stroke in 1788 from which she did not properly recover, leaving James to manage her day-to-day affairs.[110] James died on or about 14 December 1793, and she followed him a fortnight later (Mackay, 1989), on the 28th of that month.[111]

CHAPTER 9

The Successors to George and Dorothea

George and Dorothea had five sons in all: John, George, James, William and Robert. John and James were the only two of the brothers to survive their father. John, being the eldest, was his successor to the baronetcy of Penicuik, in consequence of which James became the heir apparent to Middlebie, his mother's estate.

Sir John Clerk, 5th Baronet and Mary Dacre

John Clerk, 5th Baronet of Penicuik, was the eldest son of George Clerk of Dumcrieff, later known as George Clerk Maxwell, and his wife Dorothea, heir of entail to John Maxwell of Middlebie, and was probably born in 1736 (see Appendix 4). In January 1784, he succeeded his father who had been baronet for only a year. The new baronet was then about forty-one years old and a retired naval captain. Unfortunately, we know very little of him and what he achieved in his lifetime. His wife, Mary Dacre, on the other hand, was comparatively well known, and we shall hear rather more of her later.

John Clerk was probably born at the old Penicuik House, Newbiggin, which was the home of his parents George and Dorothea, who were then too young to have a home of their own. In Chapter 8 it was surmised that he attended Penicuik school when his parents were living at Dumfries. When we first hear of him being at sea, he was already about twenty-one years of age.[1] Since it was more usual to sign up for a naval career as a boy of about twelve or so, it seems likely that he had been afforded the opportunity of continuing his education, perhaps at university, with the prospect of taking up a profession but, like his father had done in his time, he had decided that he wanted some other sort of career. His father's brothers Henry and Adam had also been to sea (Chapter 4) and a cousin, Charles Inglis, was then a lieutenant aboard the 74-gun HMS *Monarch* and was destined to become a rear-admiral (National Galleries of Scotland, 2014).

While service in the army and navy did provide opportunities for intrepid young men to distinguish themselves, and hopefully thereby gain promotion and a share of booty or prize money, it was not the ideal preparation for an eldest son whose prospect would be to take over the running of a family estate. At one time, George himself had a hankering to serve in the army, and did get a little taste of it with the Royal Hunters during the '45 rebellion. In the end, two of his sons, John and James, went to sea; another two, William and Robert, went into the army; while the remaining son, George, was the exception who stayed at home, studied for a profession and qualified as an advocate.

During 1768–70, however, John Clerk seems to have had the opportunity of travelling in France and Italy, and when in Rome he had his portrait painted by his third cousin, Ann Forbes.[2] In 1775–76 he was back at sea aboard HMS *Enterprize*,[3] a newly commissioned 28-gun frigate that served throughout the American Revolutionary War as cruiser and convoy escort (Wikipedia, 2014). Meanwhile John's uncle, Sir James Clerk, 3rd Baronet of Penicuik, was pulling strings with the intention of getting him promoted, and by November 1776, he had gained the assurance of the old Clerk family ally, Charles Duke of Queensberry, that he would do all he could.[4] Nevertheless, by October of the year following, John Clerk had left his ship with the intention of finding a shore job somewhere in the admiralty (NAS: GD18/5511). Presumably this was with the intention of settling down, for within weeks, he had pipped William Scott, later Baron Stowell, to the post by marrying Mary Dacre of Kirklinton[5] in Cumbria on 23 December 1777. Mary had received William Scott's proposal just days before her marriage to John Clerk, but she held fast to her original acceptance, saying in reply to Scott that she would indeed have married him, but he was too late (Wilson, 1891, p. 159)!

In the following year, however, despite his marriage and earlier wish for a shore position, John was back at sea, only to return again complaining of a recurrent malady.[6] Several men in the Clerk family suffered from gout and 'the gravel' (stones in the bladder), including his father, uncle and grandfather, and the Baron had concluded that it was hereditary.[7] John's ill health is confirmed in letters to George Clerk Maxwell, from his friend Lord Eliock who was in London.[8]

In early 1780, John Clerk of Eldin wrote to Henry Dundas requesting a transfer for his nephew, then a lieutenant on HMS *Alfred* at Spithead.[9] Dundas replied that he had approached Lord Sandwich, then First Sea Lord, and within weeks John had heard some favourable news and was hoping to be transferred to no less a ship than HMS *Victory*.[10] Whether this actually took place, and whether he was ever formally promoted to the rank of captain, we do not know. If he had been on the *Victory*, it is likely that he would have been present at the second battle of Ushant (1781) and possibly the relief of Gibraltar (1782).

In 1782, he was given a charter for part of his uncle Sir James' estate around Penicuik, including Mountlothian and Ravensneuk.[11] Sir James must have been by that date quite ill, and he died within six months, whereupon John's father, George, who was himself an ailing man, became the 4th Baronet, with John as the heir apparent to Penicuik. It will be recalled that George held his baronetcy for just one short year, and so on his death on 29 January 1784, Captain John Clerk became the 5th Baronet of Penicuik and his wife became Lady Mary. A well-known backlit double portrait of the middle-aged couple was painted by Sir Henry Raeburn in 1791 (Colour Plate 10).[12]

Sir John had become, like his father George and his uncle John Clerk of Eldin, a fellow of the RSE on its formation in November 1783, by right of having already been a fellow of the Philosophical Society of Edinburgh. In addition, he was a director of the Highland Society of Edinburgh[13] from 1785 (Waterston & Shearer, 2006). According to Wilson (1891, p. 100), Sir John was well regarded, and a good landlord, but other than that there is little known about his life as 5th Baronet.

Before his marriage, Captain John lodged with a Mrs Shaw at James Court in Edinburgh, near his parents who lived in the very same 'land' (EPOD, 1774 and 1775). Since he was in the Royal Navy, we take it that this is where he stayed when he was on leave, and so he simply kept it as a forwarding address while he was at sea. The first Post Office directory to appear after his marriage in 1777 to Mary Dacre, however, was for 1784,[14] the year in which he succeeded to the baronetcy, when we find his address given as Parliament Square, while for her we find the entry 'Clark Lady, Crichton Street', which is just off the east side of George Square (Figure 1.1). We find her there listed alongside a 'Clark Miss, Crichton Street', who we may infer was her husband's Aunt Barbara, or Babie, one of the Baron's unmarried daughters. This Miss Clerk was also listed at that address in each subsequent directory until the year 1788, and since she would have then been only about sixty-three years of age this is not at all improbable. In the year 1784, 'Clark Lady (Dowager)' is also mentioned, living at Dickson's Close, but because the qualification Dowager is explicit, it is clear that this must refer to Sir James the 3rd Baronet's widow (Chapter 10). From 1786, however, the new baronet, Sir John Clerk, is listed as being at 37 Princes Street,[15] whereupon Lady Clerk's independent entry vanishes.

As to the question of when it was that John finally left the Navy, the prospect of him soon falling heir to the baronetcy would have become evident soon after his uncle James' death in 1783 when his own father's health was also beginning to fail. By the autumn of 1783, the war with the American colonists was effectively at an end and it may then have been possible for John to think of returning to dry land for good. Nevertheless the business of retiring from naval service may have been protracted, delaying his permanent return. His 1784 directory address being different from Lady Clerk's seems to concur. Parliament Square also suggests another forwarding address that would have been more appropriate for an office than a residence. His cousin, the advocate John Clerk, may have had a room there, but the directory is definite 'Clark, Sir John, of Pennycuick'. It is therefore likely that his permanent return was not until sometime later that year.

• • •

Mary, or Molly, Dacre was born on 2 or 3 November 1745,[16] during the second Jacobite uprising just at the time when the Jacobite army was setting out for Carlisle. Mary's father was Joseph Dacre of Kirklinton Hall,[17] who was descended from the Dacres of Lannercost and was a distant kinsman of Lord Dacre of Naworth Castle (Family Tree 5). Although Joseph was born in 1711 as Joseph Appleby-Dacre, in 1743 he dropped the Appleby, presumably as a result of inheriting the manor of Kirklinton in Cumberland, not far from Carlisle (Nicolson & Burn, 1777); consequently, his youngest daughter Mary's surname was simply Dacre. Joseph had married Catherine, daughter of Sir George Fleming, the Bishop of Carlisle,[18] in 1736. A well-connected and comparatively well-to-do couple, they already had six children by the time Mary came along (Prevost, 1970, facing p. 176).

In view of the possibility of the Jacobite army advancing on Carlisle rather than Newcastle, Joseph had previously taken the precaution of sending his family out of harm's way to Rose Castle,[19] which was then the residence of his father-in-law the Bishop. It was just as well, for Joseph was then colonel of horse in the King's army, and he was one of the soldiers responsible for guarding Carlisle Castle when it surrendered to the rebels on 15 November 1745. While the surrender had been ignominious, a Mr Dacre, in all likelihood Joseph, dared to speak out while the rebels enacted a ceremony of surrender in which the city magistrates were compelled, on their knees and in their ceremonial robes, to present the keys of the city to Prince Charles Edward Stuart, better known as 'Bonnie Prince Charlie'. In the teeth of this, Dacre shouted out a toast to the health of King George II, to which the Prince responded graciously, in the tradition of a Highland gentleman, in saying that Dacre should not be punished for expressing his loyalty (Jefferson, 1838, pp. 62–83).

During the siege of the Castle or shortly thereafter, the Jacobites came to Rose Castle looking for any King's militia in hiding and hoping, in the course of events, to plunder victuals for themselves and fodder for their horses. Mary Dacre would then have been just two weeks old. The occupants of Rose Castle, probably fearing the worst, had sent out a servant to plead with the leader of the rebel soldiers, one Captain Macdonald, that Mrs Dacre and her children should be left in peace, particularly in view of her only recently having been delivered of a child. It is said that the Captain then took the white cockade[20] from his bonnet and placed it in Mary's crib, saying that it would be honoured by any Jacobite as a token of immunity, and they would come to no harm (Prevost, 1977, p. 163; Jefferson, 1838, p. 70n).

Even after their main army headed south, the Jacobites continued to hold Carlisle until it was relieved on 30 December (Jefferson, 1838, p. 77), some six weeks later.[21] When Joseph Dacre got back to Rose Castle shortly thereafter, he would have been thankful to find that his family had come to no harm, but he was much concerned that he and his fellow officers would face a court martial for letting Carlisle Castle be taken. In the event, nothing was said and Joseph was appointed to the bench of judges who tried the rebels that had been captured and imprisoned there (Prevost, 1970, p. 180). One of those unfortunates was a Henry Clerk, a relative of the Baron,[22] who was sentenced to death for consorting with the rebels, notwithstanding testimony that he did not seem to have done so of his own free will. At least he died of fever before the sentence was carried out.[23] Mary Dacre was to keep her white cockade as a treasured memento, and she adopted the habit of wearing it every birthday, as a result of which she became known as the 'White Rose of Scotland'. It was thought that it had been lost sometime after her death, but it was rediscovered by the Clerk family about 1974 amongst some old family letters (Prevost, 1977).

In later years, when all the turmoil and bloodshed of the rebellion had died down, Mary Dacre became friendly with the Macdonalds of Kinlochmoidart, to whose family the gallant Captain Macdonald is believed to have belonged.[24] About 1794 the young Donald Macdonald 7th of Kinlochmoidart presented her with a dress made in his own clan tartan. The wearing of tartan had been proscribed after the rebellion (Chapter 8) and this had only been repealed about the time when the Commission for the Forfeited Estates was abolished in 1784. Even in 1794, many people would still have had clear memories of the invasion of Edinburgh by a tartan army nearly fifty years before. An Edinburgh gentlewoman in a tartan dress would have been an oddity indeed, if not an outrage.

It was not until nearly thirty years later, when Mary's friend, Sir Walter

Scott,[25] orchestrated King George IV's visit to Edinburgh, that tartan was fully rehabilitated. King George himself appeared at the royal ball given at Holyrood Palace in the outfit of a Highland chieftain, and Scott had let it be known that, unless in uniform, all the gentlemen attending should wear Highland dress. The irony was that Scott had persuaded the 4th generation Hanoverian monarch that he was a Stuart, as indeed he was by direct descendancy from Elizabeth Stuart, daughter of James VI of Scotland, albeit five generations previous. He came to visit Scotland in Royal Stuart tartan, the same tartan that his distant kinsman, Charles Edward Stuart, had worn to raise it in rebellion. On this occasion, however, Mary did not see fit to wear her tartan dress, but wore a dress of white satin (Prevost, 1970, pp. 169–170).

Let us turn now to the question of how Captain John Clerk, who was frequently away at sea on long voyages, came to find and marry Mary Dacre in very short order in late 1777. It turns out that Joseph Dacre may have had the opportunity of knowing the Clerks of Penicuik, for he attended Edinburgh University sometime around 1731 to 1734, when he was in his early twenties. Not only may James Clerk, the future 3rd Baronet, have been there about the same time (Chapter 10), his brother George may have also been there about 1734 (Chapter 8). Joseph had an aunt and uncle in Edinburgh with whom he lodged;[26] and his aunt was Dorothy Gilpin (1703–1797), daughter of William Gilpin, antiquary friend of the Baron (Chapter 4) (Gilpin, 1879, pp. 10–40, 44–48). Even on its own, the Gilpin connection might have merited a social introduction between the young Joseph Dacre and the Clerk family.

Further to this, as a teenager George Clerk Maxwell, John Clerk's father, had been at the Lowther School near Penrith, He was there between the ages of fifteen and nineteen and may well have met with many of the local families, particularly if they had boys at the school. In addition, the Baron made a visit with his family in the late summer of 1734, probably with the idea of doing some sightseeing and bringing George back to Dumcrieff. They visited Carlisle Cathedral, where George Fleming, Mary Dacre's grandfather, had been installed as Bishop.

Finally, when George Clerk Maxwell was with the Royal Hunters in 1745, he ended up in the environs of Carlisle while the occupation of the Castle was relieved by the King's army. He therefore had every chance of meeting Joseph Dacre. While there is no actual evidence of any familiarity between the Clerks and the Dacre families that extended beyond the 1730s, it can at least be said they would not have been total strangers, but what particularly brought together Mary Dacre and John Clerk, who were from locales 100 miles apart, remains quite unknown.

Both Lady Clerk's paternal grandmother and her great-aunt, the sisters Susannah and Dorothy Gilpin, had had a hard time of it in their marriages, particularly Dorothy, for they were left to manage as best they could their husbands' affairs as well as their households (Gilpin, 1879, pp. 44–50). Lady Clerk seems to have learned the principles of household management by their good example, and since her husband was frequently away at sea, she would have had to put this into practice. And when he became the baronet, according to Wilson (1891, p. 158) he was: 'much indebted to the wise help and counsel of his wife, who was a woman of excellent abilities, and of great shrewdness and force of character'.

More or less all that we have of Sir John and Lady Mary Clerk as a couple is the Raeburn double portrait of 1791 and the story of an impromptu visit paid to them at Penicuik House by Sir John's young cousin, William Clerk, and his friends, amongst whom were Walter Scott and John Irving; we can place the story as being sometime about 1790 (Lockhart, 1837, vol. 1, p. 41). After Sir John's death in 1798, Lady Mary Clerk continued to live at 37 Princes Street, being listed there as 'Lady Clerk of Pennycook'. In 1807, however, she moved from number 37 to number 57. After 1811, the house was renumbered as 100, which is the house number with which she is more generally associated, for example in Grant (*c*. 1887, vol. 3, pp. 124–125).

Here at number 100 Princes Street she became something of a local celebrity, someone with her own characteristically quaint ways, such as the wearing of her Clan Macdonald tartan dress. By the year 1800, Walter Scott was living in South Castle Street, not far away from her home; since their meeting when the young Scott visited Penicuik House, they had become firm friends. On one occasion, after he became Sir Walter, he went into Constable's bookshop at the east end of Princes Street. Lady Clerk happened to be at the serving counter as he made his way to the bookshelves within. Although he had not noticed her (either intentionally or otherwise), she had detected him out of the corner of her eye, and called out 'Oh, Sir Walter! Are you really going to pass me [by]?' Given that she was, as usual, in one of her eccentric outfits, Sir Walter apologised, adding, 'I'm sure, my lady, by this time I might know your back as well as your face.' (Grant, *c*. 1887, vol. 3, pp. 124–125). Clearly, they knew each other very well indeed.

At one time, the area on the north side of the sloping ridge of the Castle hill was the only attractive bit of greenery that could be seen from the houses on Princes Street. There were then no enclosed public gardens, the drained Nor' Loch was unfinished business, and 'Geordie Boyd's mud brig' was the mere beginnings of the earthen 'Mound' that now connects the Old Town with the New. It had been proposed that 'a row of about 20 or 30 little detached brick cabinets' would be built east to west along the slope, each having 'a hole in the wall, for a window, looking towards Princes Street'. Lady Clerk was outraged at the intended visual insult and straightaway objected in the strongest terms. When her mere protests did not have the desired effect, she kicked up such a public fuss that there was eventually 'a sort of Princes Street rebellion' with so many of the residents backing her campaign that she at last succeeded in having the project abandoned (Cockburn, 1874, p. 322).

Another interesting anecdote about Lady Mary Clerk relates to the aftermath of the relief of Carlisle and the execution of the Jacobite leaders that followed. Sir Archibald Primrose of Dunipace suffered his demise at Carlisle on 15 November 1746 (Hallen, 1891). In later years, his great granddaughter, Susan Buchanan, had a falling out with Lady Clerk, presumably because she had been careless enough to recall the part played by Lady Clerk's father in the sentencing of the prisoners at Carlisle. Incensed, Lady Clerk told her in no uncertain terms that she should be grateful to the Dacre family for, on finding Sir Archibald's head rolling about the streets of Carlisle, her father had picked up and given it a decent burial![27]

In later life, Lady Clerk took the idea of the 'Rose of Scotland' to heart, no doubt with the encouragement of Sir Walter Scott who was ever recalling romantic stories and conjuring them up anew, for she began adding Rose to her given name. This, however, was informal, for she continued to sign her name simply as Mary, as, for example, she did in her will and codicils (Prevost, 1970, pp. 162, 170–171). That said, in a letter published in 1817 by *Blackwood's Magazine*, she gave her own account of how she came to be given the Highlander's white cockade, and she signed 'Rosemary Clerk' (Prevost, 1970, p. 163).

While the year of her death is frequently given as 1835, Lady Clerk actually died at home in Edinburgh on 1 November 1834, just days short of her 89th birthday. She left an estate worth over £8,000 (Prevost, 1970, p. 163), which is perhaps an indication of just how well she had managed her affairs during nearly forty years of widowhood.

Captain James Clerk HEICS and Janet Irving

Of the life of James Clerk HEICS we have only a few scraps. He appears to have been born at Dumfries shortly before 20 September 1745, the seventh child of George and Dorothea Clerk Maxwell (Appendix 4). Like his siblings, James was not given the surname Clerk Maxwell, a construct peculiar to the circumstances of the entail by which his mother held the lands of Middlebie.

His elder brothers were John (subject of the preceding section) who was born in 1736; George, born in November 1742; and William, born in June 1744 (Appendix 4). Given that William died at about two years old, James is usually referred to as the third son. On 19 August 1745, the second Jacobite rebellion was launched when Bonnie Prince Charlie raised his standard at Glenfinnan to the west of Fort William. The start of yet another Jacobite uprising was certainly not the appropriate occasion for a loyal citizen to christen his son James, the name of Bonnie Prince Charlie's father and 'king across the water' for whom it had been proclaimed. The infant's grandfather, the Baron, had refrained from calling his own son James during the previous rebellion of 1715, calling him instead George, after the incumbent Hanoverian king. But George and Dorothea already had a son called George, and another called William, the same name as the Duke of Cumberland; it therefore seems that they felt they had already done quite enough to demonstrate their loyalty. The name James had probably been in mind before the rebellion broke out, and it seems the problem was solved simply enough by leaving their new son's given name blank on the baptismal certificate!

The next we hear of James is in the year 1775, when his father made out a bond of provision for him (see note 1), but in no spirit of generosity, for in doing so he felt compelled to point out to his son that he had already given him substantial sums of money. In the following year, we hear of James in a letter sent by the Duchess of Queensberry to his father (who was both henchman and friend of the Duke). Writing from London, she tells George that James 'has gone through a great deal of loss and trouble but appears a very deserving person', and while she had been hoping for another visit from him before his next voyage, he was being kept very busy by the captain of his ship. James was therefore a sailor. [29]

The next trace of him is nearly two years later, when in 1778 he is at Portsmouth on the *Hector*, having recently returned from India when there was a great storm on the return leg of the voyage.[30] At that time, twenty-two months was about the right duration for a voyage to India and back, the only question being, was he on a naval ship or a merchantman? A 74-gun ship, HMS *Hector*, was launched in 1774, and no East India Company ship of the same name has been found. In addition, the date of this record is too early to refer to James Clerk, son of John Clerk of Eldin, who was a naval seaman. It is therefore a reasonable conclusion that, at this stage of his career, James Clerk had been in the Royal Navy, and by the age of thirty he would probably have been a lieutenant.

Although he seems to have started his seafaring career in the Royal Navy, some ten months later, he was trying to find a berth as an officer on an East

India Company merchantman.[31] Being potentially very lucrative, such positions were difficult to obtain, but while he would get to share in the profits made from each voyage, a hefty down payment was required before he could get taken on. In telling his father about his plans and the consequent need for yet more funds, he also let him know that the Duke of Queensberry had been taken mortally ill, as a result of which James was regularly attending Queensberry House in London on his father's behalf.[32] A week later, Lord Elliock informed George Clerk Maxwell that he had seen James, but that his eldest brother John, who was then still in the Royal Navy, was unwell.[33]

James must have succeeded at some stage in getting taken on by the East India Company, for Campbell gives him as 'a naval captain in the H.E.I.C.S.' (C&G, p. 3). Four years pass, and we find James taxing his uncle, John Clerk of Eldin, with the idea of getting a group of friends together to invest as private shareholders in an East India Company trading voyage which he would lead as the ship's captain.[34] In the same letter he flatters his uncle with good reports, from Admiral Barrington and Lord Howe, on his book on naval tactics (Clerk of Eldin, 1827).

Now, we do not actually know if James' proposed voyage actually took place, but C&G (pp. 3–4) informs us that his ship was sunk in the Hugli (anglicised as Hooghly) River, which flows southwards through Kolkata (Calcutta) into the Bay of Bengal, and that he 'retired early'. The East India Company ship *Major* was lost in 1784 off Calpee,[35] at Diamond Harbour in the lower reaches of the Hugli River and only thirty miles south of Kolkata. Over the years, many ships sank in the Hugli, but the dates and location make *Major* the most likely fit. The story of James Clerk's survival of the sinking is that he had with him on the voyage a set of bagpipes, which he could play well, which he used as a flotation device to help him get ashore. He then struck up a tune with great vigour, 'whereby he not only cheered the survivors, but frightened the tigers away' (C&G, pp. 3–4)! The same bagpipes were in James Clerk Maxwell's possession many years after and were kept at Glenlair as a curio to show his visitors (C&G, p. 407); even if the original story had been embroidered in telling over the generations, it does seem to have had a basis in fact.

If James Clerk did indeed captain that last voyage of the *Major*, setting out in late 1782 or early 1783, he had succeeded in his aim of getting together the investors he needed to fund it. Had he managed to return safely to Britain with a ship laden with cargo, he would have made a fortune for all concerned. Instead, there would have been financial losses all round, in which case the sinking in the Hugli would have finished his seafaring career. Having managed to return home by 1785,[36] he settled down to life in Edinburgh. Seeing the difficulty of his situation, his mother made out her will in James' favour:[37]

leaving him my Sole Heir & Executor & this I have done Because he has met with many Cases of Difsappointments in the way of his Profefsion & consequently requires to be afsisted, More than any of my other Children.

On 14 October 1786,[38] James married Janet Irving, daughter of George Irving of Newton (see Chap. 12) and in November the following year their first child was born, a son named George after both of his grandfathers.[39] The marriage had not been the usual prearranged affair that had taken place following the negotiation and signing of a marriage contract, which meant that Janet Irving had most likely married James for love rather than for her future security. James himself would have been in no position to offer Janet any more than the possibility that he might some day inherit what his mother had willed him and, in all likelihood, the Middlebie estate; there was even an outside chance of the Penicuik baronetcy. For the moment, however, they had no home of their own and lived with Dorothea at 52 Princes Street. In 1788, things shifted a little in Janet's favour when her mother–in–law put things right and agreed to a post-nuptial marriage contract whereby Janet would get one third of the income from the Middlebie estate, which of course, was Dorothea's property in her own right.[40] Furthermore, in the same contract she formally made James heir to Middlebie on the grounds that his older brother, Sir John, was excluded by the provisions of the entail on the estate of Penicuik.[41] In the following December, however, Dorothea, now aged about sixty-eight, suffered a severe stroke that left her incapacitated. James now took on the role of her administrator and looked after her affairs.[42]

On 11 April 1789, a daughter Isabella was born to James and Janet (see Note 38). It would appear that she was named after her maternal grandmother, Isabella Colquhoun (see Chapter 12), and it was she who was to be James Clerk Maxwell's Aunt Isabella, Mrs Wedderburn of 'Old 31' in Heriot Row. Following Isabella, a second son, John, was born on 10 October 1790, and he of course was James Clerk Maxwell's father. James and Janet's last child, a son named James born 24 July 1792, did not survive.

The next morsel about James Clerk comes from an unexpected source, for in May 1793 he took out insurance with the Edinburgh Friendly Insurance Society for the farm of Upper Lasswade.[43] The name refers to a property at Over Lasswade,[44] on the north bank of the North Esk River, a little to the east of Mavisbank (Chapter 4). It was a Clerk property that in 1793 would have belonged to his older brother, Sir John. In later days, James' widow Janet still held the property, renting it from her son George at a nominal sum, for the purpose of subletting it as a source of income.[45] Interestingly, it was rented by Sir Walter Scott as a summer retreat between 1798 and 1803.[46] Scott, as we already know, was on good terms with the Clerk family as a whole.

The next event we have a record for is the death of James' mother, Dorothea, on 28 December 1793,[47] and it is known from various sources that he predeceased his mother in that year; in fact (Mackay, 1989) gives the interval as being two weeks. In his memoranda, Andrew Wedderburn Maxwell certainly gives us December 1793, but he clearly did not recall the exact date at the time of writing and left a space for the day, which he later pencilled in as '28'. Clearly on looking it up he had confused it with the date of Dorothea's death. Therefore, James Clerk Esq., sometime captain in the HEICS, died on or about 14 December 1793, of what cause we do not know. That his mother followed him two weeks later suggests that it could have been an infectious illness, or that his death somehow helped hasten hers.

After the death of James and his mother, all of the contents of 52 Princes Street that were neither *jus relictae* nor family heirlooms, were sold off, raising the princely sum of £1,200 [48] and opening up a complex inheritance problem, for a new heir to Dorothea's estate of Middlebie needed to be found, and both her accounts and James' needed to be brought up to date. Another essential action was the appointment of tutors to look after the interests of the three children, in particular their financial provision and education. John Clerk of Eldin, James' uncle, was appointed for George, while Janet's half-brother Alexander Irving (Chapter 12) was appointed for John and Isabella.[49] Meanwhile Middlebie would have been placed in the hands of trustees, who would have mostly been lawyers and the like, with some connection to the family.

At the time of Dorothea's death, Sir John Clerk and his wife Lady Mary had no children, nor did they have much prospect of any, for she was then forty-eight years of age. It would have therefore been accepted that the Penicuik and Middlebie estates would now fall to the heirs of James Clerk, namely his two surviving sons, George and John, by his wife Janet Irving. Their daughter Isabella had no immediate status as an heir[50] and so her tutor, her uncle Alexander Irving, petitioned the court so that she could have £60 a year out of her elder brother George's inheritance. In turn, George's tutor John Clerk of Eldin obliged by making the provision.[51] Of course, that was not the full extent of his concerns for his niece, for it would have been uppermost in his mind that Isabella should marry, and be married well. Since she was regarded as a beauty, as reflected in Colour Plate 11,[52] her prospects of a decent marriage were good; she would have an ample settlement out of her marriage contract to whichever well-off gentleman was going to be lucky enough to win her hand.

Janet Irving was no more than thirty-five years old at the time of her husband's death. She was not without means, for she had one third of the rent of Middlebie out of her marriage settlement and her maiden aunt, Sarah Irving, had left her everything.[53] Where they lived immediately after James and Dorothea's death we do not know, but it is a reasonable conjecture that an agreement would have been made between her children's tutors and Sir John Clerk to ensure there was ample provision for both Janet and her children, for after all the boys would become the inheritors of the two major strands of the Clerk dynasty. Sir John and Lady Clerk had no children, and the previous baronets' widows were now all dead.[54] Furthermore John Clerk of Eldin and his family were already well provided for, and so the Clerks had no other major family responsibilities to concern them.

It would have been quite reasonable that Sir John would have taken them under his protection, and they would then have lived at 37 Princes Street or at Penicuik House according to the season. In this light, selling up the surplus furniture and utensils at 52 Princes Street would seem to be the rational thing to have done, and by no means would Janet have been left to look out for herself. Sir John died in February 1798, whereafter everything changed. The provisions made for Janet and her family just three months thereafter seem to support the idea of them having been previously under his care.[55] Consistent with this, by 1800 an independent entry for Mrs Clerk can be found in the Edinburgh directory – at 47 George Square.[56] (EPOD, 1800–09). The directory entries for Lady Clerk of Penicuik (Mary Dacre) and Mrs Clerk 'of ditto' (Janet Irving) in 1800 were, 'Clark, Lady of Pennycuick, No 37. Prince's ftreet' and 'Clark, Mrs, No 47. George fquare'. We know that it was our Mrs Clerk, for in 1805 the same entry appeared under 'Clerk, Mrs of Pennycuik'.

George Square would have been convenient, because it was in a fashionable location near to where she herself had once lived, that is to say the vicinity of Bristo Street[57] and its continuation, Buccleuch Street, and it was also very close to Buccleuch Place, where her two half-brothers Alexander and John lived. As we have seen in the previous section, Janet's sister-in-law, Lady Mary Clerk, was an assertive woman, who was free with her opinions and strict with money. One may imagine that once Sir John had died and Janet could well afford to look after herself and her children out of the provisions that had been made for her and so there was little to recommend them continuing to live all together.

By 1808, while Lady Clerk's address in Princes Street had changed to number 57, Mrs Clerk was still at 47 George Square: 'Clerk, Lady of Pennycuick, 57. Prince street' and 'Clerk, Mrs of ditto, 47. George square'. It appears that, as the mother of the 6th Baronet, she now allowed the privilege of appending 'of Pennycuick' to her name. In the year 1809, however, things changed in a major way, for the directory entries now read: 'Clerk, Lady of

Pennycuik, 57. Prince street' and 'Clerk, Mrs of ditto, Heriot row'. It is therefore clear that Lewis Campbell was wrong in saying of Mrs Clerk and her son John: 'About 1820, in order to be near Isabella, Mrs. Wedderburn, they 'flitted' to a house in the New Town, No. 14 India Street, which was built by special contract for them' (C&G, p. 4). On the contrary, it is clear that Mrs Clerk and her brood had been in the New Town, at what is now 31 Heriot Row,[58] since 1809, and as we shall see, she was the original occupant of that house. We can be sure of this because even here she is listed on the line directly below the entry of her sister-in law as 'Clerk, Mrs of ditto' and, as will be discussed in the following section, the Wedderburns did not live there until more than ten years later, shortly before James Wedderburn's premature death.[59]

Number 31 (Colour Plate 4) was, and still is, one of the grandest houses in that street, ranking alongside some of the best in the New Town. Heriot Row itself was begun at the start of what was effectively a second New Town plan (Chapter 12), in which building commenced to the north of the original boundary that ran along Queen Street (Figures 1.1 and 1.2).

The fact that Mrs Clerk had moved with her two sons and daughter Isabella from George Square to 31 Heriot Row in early 1809 is significant inasmuch as George, the young baronet, came into his majority in November of the previous year, whereupon he had unfettered access to his inheritance. He could, if he wished, buy a house befitting his station, and that is exactly what he did do just two months later. It is George's name that is recorded on the original sasine for the house.[60] That it was to be the house of a baronet perhaps explains why it was one of the grander houses.

On the eventual sale of the property in 1851, a further sasine summarised all the transactions that had taken place since the original purchase.[61] It is very interesting, for it first of all makes clear that the house had been tied up in George's marriage contract with Maria Anne Law[62] whom he married in 1810, the year after the original purchase.

As already suggested, it is more than likely that George's future had been long considered by his curators and the trustees of his estate, who would be keen to see that he was able to make the most of his inheritance and future prospects. They would have thought about his career and about a suitable marriage. A month past his twenty-first birthday, he was no longer a minor when he purchased the house, and so it is no more likely that George had simply taken it into his own head to do so all by himself. It would have taken some time in the consideration and planning, and so the process must have begun when he was still in his minority. Nor would he have decided all by himself that he was going to marry Maria Anne Law. Even if his future bride was the woman he had set his heart on marrying, the match needed negotiating and the agreement of all concerned, and the fact that the purchase was tied in with the marriage contract underlines the fact.

George and Maria married in August 1810 (Foster, 1884, p. 52), by which time he already owned 31 Heriot Row. Did he and his new wife stay there for any length of time, or did they simply take up Penicuik House as their residence? Certainly, he was at Oxford University until shortly before his marriage, and while he is not listed in the Edinburgh directories from 1809 onwards, in the less frequently issued county directories his seat (though not necessarily his residence) is listed as being Penicuik House. By 1811 he had begun a busy and fruitful political career by becoming the MP for Edinburghshire, which seat he held for many years. It may therefore have been that his focus was then on London rather than Edinburgh itself, and he no doubt would have had more use for a town house there than in Edinburgh. In her memoir, Jemima Wedderburn (Fairley, 1988, p. 123) mentions visiting her Uncle George at Park Street in London sometime in the 1840s, and in 1866, the year Maria died, the couple appear to have been living at Eaton Square, Pimlico (Lundy, 2015). An interesting sketch of Sir George is given in Wilson (1891, pp. 160–163) and his political career is extensively covered in Fisher (2009). Nevertheless, even in the 1830s Sir George was a frequent enough visitor to 31 Heriot Row, when it was his sister Mrs Wedderburn's house, for it to be taken as his own Edinburgh residence by a mob intent on stoning the windows of prominent New Town Tories (Fairley, 1988, p. 102). It is quite possible he was even in the house at the time!

In reality, therefore, it was Sir George's mother Mrs Clerk of Pennycuik, née Janet Irving, who was the principal occupant of 31 Heriot Row and lived there with her son John and daughter Isabella. On marrying James Wedderburn in 1813 (see the section following), Isabella left 'Old 31', as the house in Heriot Row came to be known (C&G, p. 52), to settle at her husband's residence at 126 George Street. In the year 1821, however, she returned to her former home with her husband and children. This had been facilitated by Janet Irving vacating the house and moving round the corner to 14 India Street, which she did under the name of plain 'Mrs Clerk'. Notwithstanding the fact that John Clerk Maxwell had purchased 14 India Street and hitherto had always lived with his mother, he now listed himself as though continuing at 31 Heriot Row (EPOD, 1821): 'Clerk, Mary Lady of Pennycuik, 100. Prince st.'; 'Clerk, Mrs, 14. India street'; and 'Maxwell, Clerk esq. advocate, 31. Heriot row'. Janet Irving lived at 14 India Street for at most a year. She died, aged about sixty-four, on 25 March 1822.[63] The fact that her name ceases to be listed in the directory in the following year concurs with this, whereas Lewis Campbell mistakenly states that the year was 1824 (C&G, p. 4).

Little more than six months after her mother's death, Isabella Wedderburn was widowed, leaving her with six children and another on the way. She did, however, have her brother John to support her, and he was a favourite uncle to her children: 'Uncle John was a great favourite of mine – he was always kind to me' (Fairley, 1988, p. 107).

> He was the confidential friend of his widowed sister Mrs. Wedderburn's children, who were in the habit of referring to him in all their difficulties in perfect confidence that he would help them, and regarded him more as an elder brother than anything else. This is abundantly confirmed by various entries in [his] Diary. (C&G, p. 10n1)

John Clerk Maxwell continued to keep his address as 31 Heriot Row, and not until after his marriage did he put his name in the directory under his house at 14 India Street.

Isabella Clerk and James Wedderburn

In Campbell and Garnett's biography of James Clerk Maxwell, his aunt, Isabella Clerk, is the 'Mrs Wedderburn' who lived at 'Old 31' in Edinburgh's Heriot Row. Born on 11 April 1789,[64] Isabella was the only sister of Sir George Clerk and John Clerk Maxwell (see Chap. 9). According to her youngest daughter, Jemima, she had: 'naturally a very cheerful disposition but was subject to fits of depression when ill' (Fairley, 1988, p. 97). In addition, she was regarded as being something of a beauty, for she was often referred to as the 'Daisy of Pentland' (Fairley, 1988, p. 97), which her portrait by Raeburn (Colour Plate 11) seems to justify.

Isabella came by her new name through her marriage to James Wedderburn, Solicitor General for Scotland, in October of 1813 (Grant, 1944). James was one of the Wedderburns of Inveresk, a picturesque village that overlooks the mouth of the River North Esk on its exit into the Firth of Forth, and it will be illuminating to say something about the colourful Wedderburn family history.[65]

James Wedderburn (1782–1822) was the youngest son of Dr James Wedderburn (1730–1807) of Inveresk. The father of this Dr Wedderburn, Sir John Wedderburn, 5th Baronet of Blackness,[66] came out in favour of the Jacobite rebels in 1745. Having been captured at Culloden, he was taken to London to pay the price for backing the wrong side. At great risk to himself, Dr Wedderburn managed to make his way to London to try to help his father, whom he was even able to see, to attend his trial, and finally to witness his life being mercilessly extinguished by hanging, drawing and quartering on Kennington Common: 'The recollection of his father's fate never left him, and till the end of his life he would, when in London or its neighbourhood, make any detour to avoid the scenes connected with that event' (Wedderburn, 1898, p. 305). It was said Sir John had gambled his life in the rebellion only for the opportunity of restoring his dismal family fortunes, but in the end he lost it all by having his title and estates escheat by the Crown.

James Wedderburn, thus left with no father, no home and no prospects, managed to escape to Jamaica to join his brother John, who was making his living there as a doctor, part of which involved looking after the health of slaves on the sugar plantations. James, by assisting his brother in his practice, acquired the necessary skills to become, albeit without formal qualification, a decent doctor. In this profession he managed to get together the wherewithal to become a planter and, because of the great demand for sugar, he eventually made himself a fortune. To modern eyes, however, the negative side to this was that he was therefore a slave-master. In addition, he had five children to two different slave women, the eldest of which was named Robert Wedderburn (Chase, 2004–13). Having made his fortune, Dr Wedderburn returned to Britain in 1773, when memories of the '45 had sufficiently faded to make it safe enough to do so. He bought Inveresk Lodge, and by the following year he had married and started his life and his family anew.

Dr Wedderburn's wife was Isabella Blackburn (1756–1821), daughter of Andrew Blackburn, a wealthy Glasgow banker. Of their four sons, the youngest was an advocate, the very James Wedderburn that married Isabella Clerk, the sister of John Clerk Maxwell. But let us now turn our attention to the Blackburn connection. Andrew Blackburn (Glasgow University Library, 2004; Teevan, 2008–10) was the son of John Blackburn of Househill in Glasgow, and he was also the brother of Peter Blackburn (b. 1728), a Glasgow merchant who was the grandfather of the brothers Peter Blackburn of Killearn MP and Professor Hugh Blackburn, who married, respectively, Jean and Jemima Wedderburn, daughters of James Wedderburn the advocate and Isabella Clerk. The Blackburns and Wedderburns who feature in the life of James Clerk Maxwell therefore had a previous family alliance as a result of which Jemima and Hugh were second cousins once removed.

• • •

Dr James Wedderburn brought the two youngest of his Jamaican children by the slave Esther Trotter back with him to Inveresk. Although they were raised separately from his new Scottish family, he did provide for them. Less welcome, however, was the uninvited appearance of his son Robert Wedderburn, who had come to Britain in 1779 and tracked him down to Inveresk some years later (Wedderburn, 1991). He was, however, spurned by his father and sent away with no more than a crooked sixpence. Although he worked in London as a tailor, some time after he had married he fell on hard times and found it necessary to apply to his brother Andrew Wedderburn Colvile, the eldest of Dr James Wedderburns' Scottish-born sons, for assistance, and was again rebuffed. Such rejection may have led Robert, who was evidently educated, able and of robust character, to turn his mind from gaining the acceptance of his own kin to agitation against slavery.

In 1824 he eventually found an outlet for his anger towards his father and brother by publishing a letter stating his case in *Bell's Life in London*, which was met with a swift rebuttal by his brother. While Robert maintained that Dr Wedderburn had never denied that he was his father, Andrew's case was that Robert's mother 'could not tell who was the father' and consequently 'her master, in a foolish joke, named the child Wedderburn' as she had had been carrying the child while she was Dr Wedderburn's slave. What the outcome of this was we do not hear.

Robert died about 1835 having led a very colourful life. His writings (1991) and the litigation by Joseph Knight (Cairns, 2009), who had been a slave belonging to Dr Wedderburn's brother John, must have done much to bring home to the British public many of the horrors of the slave trade.[67] In the event, however, any potential scandal was averted when Robert's attempts to claim kinship with his father were thwarted in the courts by his own half-brother, who may have had a lot to lose if things had gone differently.

John Blackburn (1756–1840), the father of Peter and Hugh Blackburn, had been a slave owner in Jamaica, where he was from about 1772 to 1805 (UCL Department of History, 2014a). After his return to Scotland, however, he continued as an absentee owner until the repeal of slavery throughout the British Empire in 1833 and left a fortune of more than £100,000 on his death.[68]

A further relative of James Clerk Maxwell who was party to the employment of slaves was his great uncle William Bullock,[69] whose daughter Emily Bullock had married John Cay, Sheriff of Linlithgow. Having been Secretary to the Island of Jamaica, he was at one time an attorney for absentee plantation owners, and since he occasionally factored for local planters, he conceivably even did so for John Blackburn. In 1832, he registered himself as having eighty-five slaves, but in right of his wife Elizabeth Smart, whose property they actually were (UCL Department of History, 2014c). Also, in the *Inventory of Cay Family Papers* (Cay Family, 1910), there is an item listed as being

'Valuation of Slaves belonging to Mrs Bullock – 3rd Octr. 1832'. There is therefore no doubt that slave ownership was still very much current around the time of James Clerk Maxwell's birth. Many a well-to-do and respectable family had either based its fortune on the profits of slavery, or at least had connections with those that had done so.

• • •

Returning now to James Wedderburn, the husband of Isabella Clerk, he was born at Inveresk on 12 November 1782 (Grant, 1944). He qualified as an advocate in December 1803, shortly after his twenty-first birthday, and by 1806 he had his own house and legal chambers at 61 (later 126) George Street. Within a few years he was appointed an advocate-depute, meaning that he could represent the Crown in the High Court, which is where the serious criminal cases are tried, and in March the following year (1811), he was made the Sheriff of Peeblesshire.[70] This seems like a very significant promotion for someone who had not long turned twenty-eight, but that was much the same age by which his friend Sir Walter Scott had become Sheriff of Selkirkshire.[71] The promotion to sheriff, however, did not mean that he had to move away from Edinburgh, for he would have had a local sheriff-substitute on hand to do his bidding and to cover in his absence.

When James and Isabella married in 1813, they settled down to married life at 126 George Street, the clearest indication of which is to be found from the birth records of their children (Wedderburn, 1898, pp. 313–14). The five children born at their George Street home were:

James (1814–1863), who became a doctor and died unmarried;

Janet Isabella (1815–1853), who married James Hay Mackenzie in 1838.[72] Their son, Colin Mackenzie, was the cousin, once removed, of James Clerk Maxwell. In later life Colin was an advisor to James and was by his side when he died (Chapter 2);

George (1817–1865), a Writer to the Signet who never married and continued to live with his mother;

Jean (1818–1897), the sister who married Peter Blackburn MP;

John (1820–1879), a major general in the army who married but had no children.

By 1820, however, not only was James an important man, he and Isabella had these five children, and they would have needed the run of a big house befitting James' status, a place where he could do business by day and entertain the cream of society by night. And the solution to this problem was the purchase of his wife's former home, 31 Heriot Row, where his name first appeared in 1821. The Register of Sasines (see the previous section) makes it clear that it was at the end of this year when James Wedderburn actually bought the house from his brother-in-law, Sir George Clerk 6th Baronet of Penicuik; and since his entry in the Post Office Directory would have had to be made quite early in 1821, the transaction must have been formalised after he and Isabella had been living there for over six months. As was pointed out in the previous section, Lewis Campbell's version of events is somewhat different (C&G, p. 4) in that he erroneously believed that Isabella and James Wedderburn were already living at 31 Heriot Row and that Mrs Clerk and her son moved to India Street in 1820 in order to live nearby. In reality, they had long been living at 31 Heriot Row, as did Isabella up until her marriage. However, it seems that in early 1820 the notion was conceived that it would suit all concerned to let the Wedderburns take over 'Old 31', and it was this that led to the building of a new house for Mrs Clerk at 14 India Street, just around the corner. This plan was put into action when John Clerk entered into a contract with Wallace the builder (see Chapter 15), and Mrs Clerk, at least, moved out in the following year.

After James and Isabella's move to Heriot Row, they had two more children, the first of whom must have been on the way at the time of their 'flitting':

Andrew (1821–1896), who joined the HEICS and became a senior civil servant in the administration of Madras in India;

Jemima (1823–1909), a well-known bird water-colourist (Fairley, 1988), who married Professor Hugh Blackburn of Glasgow University in 1849.

Andrew and Jemima Wedderburn had their own particular connections with their younger cousin James. Firstly, it was Andrew who inherited Middlebie and Glenlair on James' death in 1879. They would have barely known each other, for Andrew entered the HEICS College at Haileybury in 1842 and was thereafter posted to India. He did return to Edinburgh upon his marriage to Joanna Keir in 1847 but, while he may have had further visits home, these would have been few, for the round trip would have taken the best part of a year. Although he was due to retire from office in 1878, he was asked to stay on another year to help deal with an ongoing famine, something which he had had to do on two previous occasions. It is therefore unlikely that he and James ever had the opportunity to reacquaint themselves, and certainly he is not mentioned in Lewis Campbell's account of James' final days (C&G, Chap. 13).

Although Jemima was eight years older than her cousin, she was James' frequent childhood companion both at 31 Heriot Row and during summers spent at Glenlair. Significantly, as a budding water-colourist, it was she who recorded for us some significant events in James' early life, beginning with a visit with his father to Edinburgh in 1841, through to his visit to the Blackburns' estate at Roshven in 1857. Jemima was born on 1 May 1823, six months after the untimely death of her father. It seems that James Wedderburn developed some sort of brain fever that began as a chill that he had caught while visiting his sister Lady Jean Douglas, widow of the Earl of Selkirk, at St Mary's Isle near Kirkcudbright (Fairley, 1988, p. 97). James was buried there on 7 November, just days short of his fortieth birthday. Not only had he been appointed Sheriff of Peeblesshire before he was thirty, he had been subsequently appointed Solicitor General for Scotland, the highest law office in the land after the Lord Advocate, at the age of just thirty-three. By 1819 he had been reaching for the post of Lord Advocate itself[73] but did not live to see it. After barely ten years of marriage, Isabella was left a widow with seven children to look after.

After James' untimely death, Isabella would have come into the settlement that she was due as a result of her marriage contract,[74] together with any further provisions that James had put in his will, written at the time of their marriage (Wedderburn, 1898, p. 313). The will put James' estate in the hands of trustees, who saw to it in a rather forward thinking way that the their eldest son James jnr did not inherit 31 Heriot Row, for he was, after all, only eight years of age at the time of his father's death. Instead they got a decree from the courts allowing them to dispone it to Isabella, which they did by 31 July 1824,[75] the rationale being that James jnr would in due course be likely to inherit it anyway.

When James Clerk Maxwell arrived in Edinburgh in November 1841, it was of course to the house of his aunt Isabella at 31 Heriot Row that he came. Although this was several months after the census of June 1841 had been taken, by the greatest of luck we find him there with his father on a visit at the time of the census. The record[76] shows the occupants of the house on the day of the census as being:

Person	Age	Relationship to head of household
Isabella Wedderburn	50	Head
George Wedderburn	20	Son
James Hay Mackenzie	30	Son-in-law
Isabella Mackenzie	25	Daughter [Janet Isabella]
Colin Mackenzie	1 month	Grandson
John Clerk Maxwell	50	Brother
James Clerk Maxwell	9	Nephew
Caroline Colvile	14	Great-niece
Alice Douglas Colvile	10	Great-niece[77]

Isabella's household also reported seven female servants and a governess at the time of the census, though not all of them were necessarily her own. Likewise, the butler 'Hornie' mentioned in (C&G, p. 47), and depicted in Jemima's watercolour of James Clerk Maxwell arriving at the house in the November of that year, was not present at the time, nor was Jemima herself, who was then aged 18 and therefore unlikely to be away from home on a permanent basis.

Isabella continued to live at 'Old 31' for many years to come, indeed until 1850, the year her nephew James Clerk Maxwell went off to Trinity College, Cambridge. By then her children were grown up and, apart from a son George who never married, all had left home. Perhaps she and George then found the house too big for their requirements, for they more or less retraced her mother's move round the corner to India Street back in 1821, only this time the house concerned was number 18, only two doors away. After nine years there, where they were often visited by James and his father, they moved on to 25 Ainslie Place (Colour Plate 3).

Isabella Wedderburn would by then have been seventy-two, and so it may have been more George's idea than hers to move back to a grander house, for he was by then a relatively well-off lawyer. But within a few years, George was diagnosed with a consumption from which he eventually died on 1 May 1865, and his mother Isabella followed him almost exactly six months later, on 2 November, at Killearn, the home of her daughter Jean and her son-in-law Peter Blackburn MP. It was there that she was buried (Wedderburn, 1898, p. 313).

CHAPTER 10

Sir James Clerk, 3rd Baronet of Penicuik

Having remarried in February of 1709 to Janet Inglis, the Baron's first son by his new wife was born on 2 December of the same year and was named James after Janet's father, Sir James Inglis of Cramond (BJC, p. 2). He was the Baron's second son, the first being the child's half-brother John from the Baron's first marriage, almost exactly eight years his elder. We learn this, and much of what little we know of James' childhood, from the Baron's memoirs. He tells us that just about James' fourth birthday: 'my third son Hary [...] was a very strong healfull [sic] Boy as ever I saw in my Life, whereas his elder Brother James was very tender from birth, and continued so till he was 4 or 5 years of Age' (BJC, pp. 85–86). Sadly, Hary (Henry) died of smallpox, and so it had very much surprised the Baron and his wife that the comparatively weaker James not only survived but was their only child not to catch the full disease, escaping with just some sores and boils. Once again the Baron mentions his son's frail disposition: 'We were in great anxiety about him, because of his weak constitution and bad habit of body' (BJC, p. 86). Despite his apparent frailty, not only did he survive, his boils and sores were cured when he took smallpox proper just over a year later.

Perhaps one of James' earliest introductions to the world of antiquities was when, at the age of fourteen, he was allowed to accompany his father and his antiquarian friend, Alexander Gordon,[1] on a trip to see Hadrian's Wall in the spring of 1724 (BJC, p.117). However, we hear nothing more of him from the Baron until seven years later, when he set out for Holland to follow in his father's footsteps as a student completing his studies and making the grand tour (BJC, pp. 138–139). Nevertheless, we do know that James had previously been sent to Dalkeith grammar school (note 2 of Chapter 8), for the Baron received a letter concerning James' future education from William Simpson, schoolmaster there.[2] It must have been a satisfactory institution, for his younger brothers George and John were sent to the same school in due course.

James may then have attended Edinburgh University, but if so, he did as many others did and left without graduating (Laing, 1858). In 1729, he enrolled in St Luke's Drawing School[3] where he would have made connections with other young artists of his day. Nevertheless, the intention seems to have been that he should become an advocate, for it was with that aim he set out for Holland in April 1731, travelling first with his uncle Robert to London, where he stayed for several months. He then headed for Leiden in the following October; once there, however, he did not find the study of law as much to his taste as his father had hoped. Having stuck it for only one year, his failure to obtain a satisfactory result seems to have occasioned an appeal made on his behalf to the Lords of Council and Session, requesting that he be admitted as an advocate.[4] The attempt must have been in vain because he does not appear in the Scottish Record Society's publication of those who were admitted (Grant, 1944).

In the years that followed, James travelled about Europe looking for paintings and music to study and to buy, much of which was sent back to Penicuik for his father, including works by Rubens, Rembrandt and Poussin. In January 1736 he was joined in Leiden by his brother George who was belatedly beginning his studies there (Chapter 8), and when George had completed his first year he begged leave of his father to make a tour of Germany with James before he returned in the summer of 1737.[5] But James could not be persuaded to return along with George and stayed on in Europe for another two years, soaking up as much European art and refinement as he could.

It was just as his father had done many years before. The Baron, however, was against it because of the likely expense (BJC, p. 146, marginal notes), for he knew only too well how much debt he had racked up during his own time abroad. He also felt his son, having reached the age of twenty-eight, was letting time slip by. James, on the other hand, was having the time of his life. Nevertheless, in 1738 and 1739 he also found time to study the process used by the Dutch for bleaching linen,[6] which may have been done with the idea of helping George in setting up his linen factory at Dumfries.

In October 1739 he at last returned, much to his parents' relief and delight, and his father was further pleased to observe that, despite his long sojourn in Europe and the extent to which he had embraced all that he had found there, his head had not been turned against his native land (BJC, p. 154). Within two years, however, James was off again, ostensibly to London. But it seems that he already had a different plan in mind, for once there he wrote home begging leave to go again to Europe, saying he wanted to observe the election of a new emperor at Frankfurt (BJC, p. 163). His father protested, knowing that he could not really do much to prevent James from going, and then at last consented. James stayed on in Europe, and having spent something close to a third of his life abroad, set off for home only when he caught wind of the impending Jacobite uprising in 1745.

James' first thought on his return was to enlist in the King's army, and so when he reached London he obtained some letters of recommendation to help get him a commission (BJC, p. 192). The Baron knew well that despite his two eldest sons' eagerness to fight for King and country, they had not been bred to military life. He had managed to channel George's enthusiasm towards taking up with a private regiment, the Royal Hunters (Chapter 8), which engaged in supporting roles, such as scouting and acting as guides. But when at last he met up with James at Durham, where he was taking refuge from the invading Highland army with his wife and eldest daughter (Chapter 4), he managed to put him off the idea completely and encouraged him instead to make his way back to Scotland to 'help the people in our country'. The Baron told George, then at Morpeth, by a letter of 3 November 1745, 'I suppose you have seen James. *I wish he wou'd go on to Berwick, for he is not a case to be one of your Hunters*' (Prevost, 1963, p. 238, my emphasis).

As to what James actually did when he got back to Scotland, we know only that he was present, though not a participant, at the battle of Falkirk Muir on 17 January 1746, when the King's army was shamefully and unexpectedly routed by the rebels; many of its soldiers fled the field without even engaging the enemy. Although the Baron tells us many details, he merely mentions in a marginal note, which he added sometime later,

> N.B. – There were thousands of onlookers who did great harm, for as they came not there to fight, they ran off amongst the first, and came directly to Edin. *However, my son James, who was there, continued till our army retired.* (BJC, p. 195, my emphasis)

These spectators had gone in expectation of seeing the Jacobites defeated. But, because he makes no mention of James in the main narrative, no mention of any regiment that he was with, nor any other role that he might have been undertaking in any sort of supporting capacity, there is no question that the Baron means that James was with the spectators rather than the army – which was as only as far as his father wished him to go. Unfortunately, Grant states misleadingly in his introduction (BJC, p. xix), 'but his second son, George, served in the royal army, and James, his eldest son, fought bravely at Falkirk'. If the reference to George here is only mildly inaccurate, the part about James seems quite adrift and is no doubt a misconstruing of the Baron's note.

The last battle of the rebellion, indeed the last pitched battle to take place in the United Kingdom, took place three months later at Culloden Moor, much to the relief of the Baron and his family. The Baron was now seventy years of age and experiencing further decline not only in his health but also his vitality; he therefore began to think about the day when he would no longer be there to take charge of affairs. In 1748 he had discussions with James about the succession, showing him the family accounts and papers so that he would be adequately prepared when the time came.[7] While the Baron was still far from retired from the management of his affairs and his projects on the estate, by 1750 he had come to the realisation that there were things he could no longer satisfactorily cope with:

> I found my self very ill used by some whom I trusted at Lonhead in the management of my Coal affaires, therefor I put them in the hands of my son James, who had more strength of Body and more leisure to look after them [. . .] Besides, as to the choise of my son for chief Manadger, there was a necessity to breed him up a little in the management of these matters. This experiment I found succeeded to my Wishes, for the profits of my coal began to be doubled. (BJC, p. 225)

It therefore seems that James was more than just a lover of the arts and a gentleman of leisure, and that he proved adept at the practical affairs of business. The last the Baron mentions of being with his family at Mavisbank (Chapter 4), a house he built to be close to his coal workings, is in December of that year. Although there is no explicit mention of it, it would seem natural that James would now have lived at Mavisbank so that he could be the one on hand to see that things ran smoothly, which they did.

In 1753, the Baron went further, in allowing James to add some rooms onto the west side of Newbiggin, and to build a library (BJC, p. 228). All the same, he mentions 'as there was no pressing occasion for these things, the work proceeded slowly', which is tantamount to saying that, in spite of his agreement and in spite of his son having greatly improved his coal revenues, he did not see any urgent need to spend money on them. On the Baron's death on 4 October 1755, he was succeeded by James, who thus became the 3rd Baronet of Penicuik.

After about a month as the new Baronet, and no doubt having taken the opportunity to review his new situation and his future options, Sir James decided to go and live in Edinburgh for a while.[8] His father had referred to living in Edinburgh, when it was convenient to do so, at his house there in Blackfriars Wynd (Chapter 4). There is also mention of Sir James having either lived there at about this time, or at Sempill's House off the north side of Castlehill.[9] Unlike his father, he did not have reason to be in Edinburgh because of any office that he had to attend to and so, for a man of his tastes, the reason must have been social, perhaps the opportunity of finding likeminded company, or perhaps even a wife. Had the Baron already sorted out a match for James during his lifetime, he would have been sure to mention something of such import in his memoirs. It therefore appears that even on entering his late forties James had not yet found a life partner. He did eventually marry Elizabeth Cleghorn (d. 1786?), daughter of Rev. John Cleghorn (Anderson, 1878), who until his death in 1744 had been minister at Wemyss in Fife,[10] but no record or other particulars of his marriage has been discovered.

In the years immediately following his father's death, James attended to family affairs, such as discharging his father's bequests, providing an annuity for his mother and a bond of provision for his brother John, and helping his brother Adam get a commission in the Navy.[11] At Penicuik he made few changes while his mother was still alive, save for the erection of a monument to Allan Ramsay, a longstanding family friend, who died in January 1758; an obelisk, no doubt to Sir James' own design, was erected at a high spot near Ravensneuk, on the south-east of the estate in 1759 (BJC p. 229n1; Wilson, 1891, pp. 154–155).

At the end of January 1760, just shortly after his own fiftieth birthday, Sir James' mother, Janet Inglis, died (Foster, 1884, p. 50). James now felt free to ring the changes at Penicuik (Penicuik House Project, 2014). He designed for himself a new main residence and stable block[12] and began their construction with the aid of the builder John Baxter snr, the stonemason who had constructed Mavisbank for the Baron (Chapter 4). At first Sir James' intention was to remodel Newbiggin, the old family house, but he soon gave up the idea and tore it down, perhaps to the regret of his brothers and sisters; they simply had to accept that the prerogative was his alone. Sir James' second cousin, Colonel Robert Clerk,[13] offered his opinion on Sir James' design, and drew up some notes criticising it and proposing alterations (Thom, 2014, p. 121). Not only did he do that, he also sent a copy to the architect Robert Adam, son of William Adam who had worked with the Baron, and was now also Sir James' brother-in-law.[14] Robert Adam apparently agreed with the Colonel, but Sir James had been assiduous in developing his aesthetic acumen and would not have his ideas and desires dismissed as being idiosyncratic. He would again have his own way.

John Baxter snr turned Sir James' original drawings for the house into detailed plans, and in 1762 work began in earnest, using 200,000 bricks ordered for the project[15] and materials recycled from the now demolished Newbiggin. To help finance the project, in July 1763 Mavisbank was sold to another cousin by the name of Robert Clerk.[16] Meanwhile, Sir James was helping to support John Baxter jnr (Skempton, 2002), the builder's son, and Alexander Runciman (1736–1785), then a fledgling artist, who were both studying in Italy. While John Baxter jnr was in Rome, Sir James wrote asking him to commission copies of three statues suitable for Penicuik. Baxter sent him sketches of four to choose from: the Medici Apollo, the Borghese Faun, the Apollo Belvedere and the Campidoglio (or Capitoline) Antinous.[17]

According to Jackson (1833, pp. 341–342), Runciman had been one of the 'apprentice painter-boys' working on the new house who

> had executed the paintings under the colonnade at Penicuik House, so much to Sir James Clerk's satisfaction, that he sent him to Rome, at his own expense, to complete his professional studies.

Sir James' patronage of Runciman is confirmed in letters, including those he received from Runciman himself when in Rome, concerning the direction of his studies;[18] after informing his patron that he had finished a particular picture, in one such letter Runciman asked for an advance so that he could do even more. However, he did make clear that he was not seeking 'pecuniary advantage' and, on the contrary, 'my ambition is to be a great painter rather than a rich one'.[19]

When Runciman returned from Rome in 1772, Sir James commissioned him to decorate the interior of the house, which had been structurally completed by 1769. His initial idea was to have him paint the ceiling panels of the main drawing room with themes from the Baths of Titus, but he changed his mind to have it done in themes inspired by *The Works of Ossian*, a 'translation' of ancient Gaelic mythology[20] that had been fabricated by the Scottish poet James McPherson (1736–1796) and published during 1761–65. Although the authenticity of the work was challenged, the 'discovery' of an ancient Gaelic mythology caused a sensation amongst the devotees of classical romantic tales and epic poetry, and it is clear that Sir James had been one of them. His great drawing room thence became known as 'Ossian's Hall'.

While work was progressing on the house, it was also progressing on the stable block, which is notable for the full-size replica of Arthur's O'on[21] that was erected as a dovecot forming the centrepiece of its rear elevation. James' father, the Baron, had much admired the original, an almost intact Roman temple, and declared, 'I wish I could have redeemed it at the expence of 1000 guineas' (BJC, p. xxvi). While the replica, built as per the drawings of the monument recorded by Gordon (1726), was Sir James' touching memorial to his late father, from an architectural standpoint it looks at odds with the tall spire that dominates the front elevation of the building. He was clearly more taken with the notion of recreating the O'on than sticking to classically proportioned shapes based on straight lines and circles.

In all, it seems that Sir James had the financial wherewithal to complete his grand design and to decorate and furnish it in equally grand style within more or less ten years. During the time of its building, Sir James also got involved with John Baxter snr in submitting designs for Edinburgh's North Bridge, which provided the first convenient link from the Old Town to the New by spanning the chasm across the head of the Nor' Loch between the High Street and the site of the new Register House.[22] He even sent to John Baxter jnr, who was then studying in Rome, a request for a bridge design in the style of a Roman viaduct. It was, however, William Mylne (Skempton, 2002) who won the contract to build the bridge to a design of James Craig, architect of the New Town. There was much ado in 1769 when part of this bridge collapsed during its construction and killed five people; the bridge was too narrow for modern use, and was consequently demolished in 1896 to make way for the present one.

After Penicuik House and its stable block, Sir James turned his attention to the village of Penicuik itself, which was then very small, and created a new design around a spacious central 'square', at the south-west corner of which was to be a new kirk:

> about the year 1770, Sir James Clerk [. . .] planned and laid out a portion of the village as it now stands, giving at the same time pecuniary assistance towards the erection of not a few of the buildings. He also induced a doctor to settle in it, building him a house to dwell in, and providing a large park to graze his horse in the summer. (Wilson, 1891, p. 10)

After giving the schoolmaster notice to quit his property to make way for the development, the foundation of the new St Mungo's was laid in August 1770.[23] The ruins of the simple rubble-built old kirk still stand in the kirkyard a little to the east, midway between the Clerk family mausoleum and the new kirk. The old and new represent a complete contrast in ideas of what a church should be; small though the new church is, its design is fit for a much grander purpose. Sir James clearly could not resist the classical Graeco-Roman design with portico (St Mungo's, 2013), just the thing that Robert Adam and Colonel Robert Clerk had criticised in the design of Penicuik House, and which Sir

James had so staunchly defended.

In 1772, Sir James Colquhoun of Luss, Bt (Collins, 1806) commenced the building of a house at Rossdhu for his new family seat on the west shore of Loch Lomond. He had asked Sir James' advice on the design, which has some features similar to those of Penicuik House, including the raised portico. Indeed, James is attributed as being the architect, with John Baxter jnr as the builder.[24] Elsewhere it is suggested that there were inputs from Robert Adam (*The Gazetteer for Scotland*, 2013b. 'Rossdhu'), but if so, either they did not concern the portico or Adam had now deferred to Sir James on the subject!

If the 3rd Baronet spent lavishly to satisfy his own aesthetic aspirations, he was, like his father, a philanthropist who assisted his protégés, tenants and employees. Not only had he given money for the rebuilding of Penicuik and its church, he acted with all due care by providing financial assistance for rehousing those that were displaced. In particular, his treatment of the old schoolmaster was compassionate, for he gave him a present of a new house and yard (Wilson, 1891, p. 57). Having inherited his father's coal mines, he did not do as many a laird would have done and simply reaped the financial benefits, he became involved in trying to help his miners, who as a class were then treated as little more than indentured slaves. He interceded when they got into trouble and, in 1772, he even went so far as to write to the Committee of Coal Masters in favour of the abolition of the iniquitous practice of bonding miners to their employment.[25]

Fine tastes in art and architecture and long sojourns abroad apparently did not spoil Sir James' love of Scottish ways and simple fare. Jackson (1833, p. 342) recounts an anecdote in which Sir James was visited at the newly finished Penicuik House by Henry Dundas, later Viscount Melville, whose statue now stands atop the lofty column in St Andrew Square, Edinburgh. Dundas was seeking Sir James' vote in the forthcoming parliamentary election, but as Sir James did not know the man well he decided to put him to the test by serving up nothing more than porridge for dinner. Dundas was unabashed by his frugal repast, and thereby gained Sir James' vote by demonstrating that he was a man who appreciated ordinary Scottish values; of course, the anecdote is meant to reflect that the same quality applied equally well to Sir James.

In 1781, the Society of Antiquaries of Scotland, which had been formed just the year before, made Sir James a fellow.[26] Clearly he had been forgiven for demolishing Newbiggin, perhaps on account of his recreation of Arthur's O'on?

As we have already seen, Sir James' younger brother George Clerk Maxwell was involved in lead mining at Craigton on the Solway coast. While James took no active part in George's mining company, he did become financially interested. George no doubt persuaded him that in time he would get a good return for his money and so they borrowed jointly from the Bank of Scotland to finance George in his enterprises (Chapter 8). Between Sir James spending a great deal of money on his building projects and lavish furnishings for his house, and George investing in such things as the Forth and Clyde Canal, prospecting, and the eventual lead mine at Craigton, they both had racked up a fair amount of debt between them. As we already know, by 1781 things had gone badly wrong with George's mining ventures, and the debts were substantial. In 1782, when Sir James was seventy-two years of age and probably quite ill, the two men agreed that in the interests of paying the external creditors, the debts between them would have to be put to one side and Dumcrieff and Middlebie would have to go. Did Sir James have any inkling that, even if this was a hard cross for George to bear, it would not be long before his brother would inherit Penicuik? George's own health was also deteriorating, and so they may have appreciated that it would all sort itself out soon enough, and the key thing was to preserve Penicuik for the heirs.

The possibility that George would soon enough come into the baronetcy of Penicuik came to fruition within a twelvemonth:

> Sir James Clerk Baronet of Penicuik died at Leith whether he had gone for the recovery of his health upon the 6th day of February 1783 at 4 o'clock in the morning and was interred in the burying ground belonging to the family in the church yard of Pennycuik on 10th day of the said month.[27]

George then became the 4th Baronet of Penicuik, as Sir George Clerk, notwithstanding the condition in the Middlebie entail that required him, as Dorothea's husband, to hold to the name Maxwell. The validity of his juggling with the names of Clerk and Clerk Maxwell had in any case never come to any legal test. Sir James' widow, Elizabeth Cleghorn, lived thereafter in Edinburgh at Dickson's Close until her death, apparently in March 1786. She appears as the Dowager Lady Clerk in only one edition of the Edinburgh directory (1784–86), which is consistent both with the time of her husband's death and the date believed to be of her own.

CHAPTER 11

From Weir to Irving

James Clerk Maxwell's paternal grandmother was Janet Irving, daughter of George Irving of Newton. In turn, this George Irving's maternal grandmother was Margaret Weir, brother of John Weir of Newton, so that the Clerk Maxwells are connected to the Weirs through the Irvings. However, the tale of how Newton passed from John Weir to the Irvings came to be a family legend with which James Clerk Maxwell would have been very familiar. While the story is not mentioned in Lewis Campbell's biography, Newton is indirectly touched on concerning the journey between Edinburgh and Nether Corsock: 'Coming by way of Beattock, it occupied two whole days, and some friendly entertainment, as at the Irvings of Newton [. . .] had to be secured on the way' (C&G, p. 26).

And indeed, we find Newton to be very nearly the halfway point on that journey. But it is with the story of the Weirs that we must begin the tale.

The Weirs of Blackwood and their Cadets

In olden Edinburgh there were a number of families by the name of Weir. In addition to John Weir of Newton (*c.* 1650–*c.* 1713), an early Weir, of grand infamy, was Major Thomas Weir (1599–1670), while a later one, Janet Irving's great-grandfather, Dr Thomas Weir (*c.* 1725– *c.* 1804) ran a very interesting patent medicine business that endured through several generations of Weirs and Irvings. These three Weirs belonged to cadet branches of the Weirs of Blackwood: the Weirs of Newton, Kirkton[1] and Stonebyres respectively, all of which originated from an area around Lesmahagow, a small town in the upper Clyde valley about twenty-five miles to the south-east of Glasgow. Around the year 1400, Patrick, Bishop of Kelso (*c.* 1392–1411) had granted part of the church lands in the vicinity of Blackwood to Rotaldus (or Rothald) Weir, one of his baillies in Lesmahagow, in return for services rendered, which mainly consisted of carrying out 'inquests' on his behalf, a duty that probably ensured, by one means or another, that the Bishop got paid his rightful dues (Greenshields, 1864, pp. 83–84). Blackwood itself is a village situated just three miles to the north of Lesmahagow, while Stonebyres lies only two and a half miles to Blackwood's east, and Kirkton is some five miles to its north-east and on the opposite bank of the River Clyde. The histories of the families therefore did not take entirely separate paths, for they lived close enough to become intermingled, very likely as a result of alternating cycles of feuding (involving murder) and reconciliation (involving marriage) that were typical of the times.

According to Irving and Murray (1864, vol. 1, p. 80), a cadet of the Weirs of Blackwood subsequently acquired property about twenty miles further to the south, in the parish of Crawford lying in the upper Clyde valley,[2] whereupon they became known as the Weirs of Newton, a small steading on the property that gave it its name. On John Clerk Maxwell's family tree,[3] this cadet is identified as having arisen from the Weirs of Stonebyres. As to the place, Newton is a very common name simply because it means what it says, 'new town'; there are consequently Newtons scattered here and there all over the country, both north and south of the border. Our particular Newton (Figure 11.1) was the 'new toun' on the east bank of the River Clyde facing its junction with the Elvan Water (Irving & Murray, 1864, vol. 1, p. 80). No town is to be seen there, nor was there ever one that we would now recognise by that name, but as already mentioned, in the ancient manner of speaking in Scotland, a 'toun' could simply mean a single country house. Nevertheless, not far away, and on both banks of the river, there are many signs of previous habitation where the earlier toun could have been,[4] but when Newton was actually a new town must have been a very long time ago indeed.

While an estate of a few thousand acres went along with Newton, it was just part of the vast, empty moorland that forms the Upper Clyde valley. The M8 motorway now passes through the estate and indeed within a few hundred yards of the site of Newton House, but virtually all that is now left there is the name and a fragment of the buildings. It had once been the property of one William Were (alternatively spelled as Vere or Weir) who in 1512 served on an 'assize' (Irving & Murray, 1864, vol. 1, p. 80; Pitcairn, 1833, p. 87):

Figure 11.1 (opposite) Newton, the seat of the Irvings of that Ilk
(a) Elvanfoot and Newton as they appear on Ross' 1773 Map. The Edinburgh–Carlisle road was then on the west bank of the river.
(b) Newton House and its policies as they appear on the 1846, 6" OS Map. The bridge, access road, drives, trees and offices remain, but the buildings now at the site of Newton House are not the original and the M74 motorway passes close by.

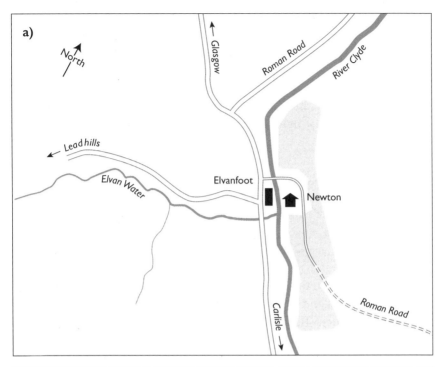

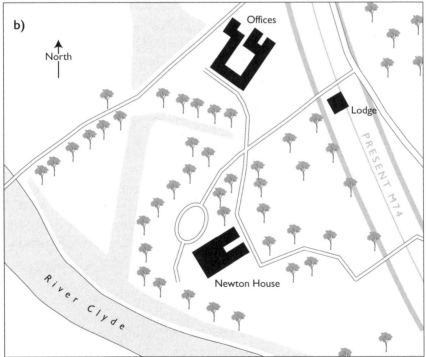

Slaughter of the Laird of Greifton.

May 31. –WILLIAM DEKESOUNE, fon of umq⸰ John Dekefoun, convicted of art and part of the Slaughter of George Myddilmeft of Greiftoun, committed by the faid William, John his father, and John his brother, upon forethought and culpable felony : *Item*, for being at the horn of the faid Slaughter.– BEHEADED.

ASSISA.

Thomas Lewis of Menner,	Gilbert Berde of Poffo,
Thomas Wylie of Bonytoune,	Ninian Paterfoun of Caverhill,
John Elphinftoun of Hendirftoun,	James Levingftoun of Girifwode,
John Caverhill of Foulage,	William Were of Newtoun,
John Carmichell, Cap⸰ of Crauforde.	

William Were lived in the turbulent times when war with the English seemed to be the status quo and any intervening peaces were short-lived. In particular, a war against Henry VIII that started in support of the French culminated in the infamous defeat for the Scots at the battle of Flodden,[5] which took place in the year after Were served on the assize. As already discussed in Chapter 6, cross-border reiving and retaliatory skirmishes were endemic, and both sides were equally culpable. The reiving flourished as far north of the border as Newton, for the Devil's Beef Tub, to the north of Moffat, is not far distant and, as the name suggests, it was a well-known repository for stolen cattle. On the Elvan side of the Clyde there is ample evidence of ancient buchts and bastles (Ward, 2012), the former being a sort of byre or sheep pen and the latter a fortified farmhouse, heavily built with slit windows to make it easier to defend in a raid.

In the seventeenth century, the reiving gave way to a period of more widely spread political and religious strife during which, in 1662, one John Weir of Newton was fined £360 Scots by the Scottish parliament for being a Covenanter sympathiser (Woodrow & Burns, 1824, p. 272), and as we shall see, the fact that he was soon in even greater trouble had a major bearing on our story, in which the Weirs of Blackwoods also have a minor connection.[6]

From Weir to Irving

Now, as has just been mentioned, John Weir of Newton had been accused of being involved with the Covenanter cause, and it is a matter of record that he was indicted for treason in one of the infamous trials that took place between the years 1682 and 1686. A dramatic story is recounted by George Vere Irving of Newton, of how his great-grandfather, George Irving, came to inherit the estate of Newton from this John Weir as a result of his trial (Irving & Murray, 1864, vol. 1, pp. 80–81; Irving, 1907, pp. 3–4). Vere Irving called the story a family tradition, but we will now examine it to see how close to the truth it might have been.

John Weir of Newton's sister, Margaret, was married to a Scottish court official by the name of James Irving. Irving was a common family name in places both to the north and south of Edinburgh. His family had descended from the staunchly royalist Irvings of Drum,[7] whose seat is about ten miles east of Aberdeen (Boyd, 1908). Since his appointment in 1683, James Irving had been macer to the Privy Council, which at that time was effectively the highest court in the land, for it had taken to intervening in the usual judicial procedures whenever it happened to be in the King's interest to do so. The mace was the symbol of the monarch's authority (as it still is in parliament today) and the macer was the senior court usher, amongst whose duties it was to carry the mace before the procession of Privy Councillors as they filed into Council. James Irving is also recorded as having been secretary to Alexander Stuart, 5th Earl of Moray,[8] who not only was a strong supporter of James VII but had taken part in suppressing the Covenanters. There is no way that James Irving could have held such a position had he not been trusted by the Earl as being, heart-and-soul, loyal to the Crown, and therefore an anti-Covenanter. How, therefore, he came to marry the sister of a prominent Covenanter we do not know. Margaret's testament,[8] however, informs us that she had been married before, to John Balmaine (died *c.* 1679), a baillie in the Canongate,[9] but of him there is little to report. It seems there were no children and so on his death she inherited his moveable estate. Now, it has been said that James Irving married Margaret Weir about 1684, in the April of which year he was granted the movable property of 'John Balman in Doning' (Irving, 1807, p. 4), who had been outlawed. Perhaps the similarity of the two names is just a coincidence, but if so it is a curious one.

Let us return now to George Vere Irving's family story of how the Irvings came into possession of Newton. During his trial before the Privy Council, Margaret Weir, now Mrs Irving, was said to have gained entry to the courtroom, whereupon she proceeded to plead vehemently for clemency on her brother's behalf. This she may have contrived to do with or without the connivance of her husband the macer for, as his wife, she would have been able to get into the courts armed with some suitable excuse. Being heavily pregnant at the time, that may not have been difficult, and for the same reason the judges may have tolerated her interruption. Her impassioned plea found its mark with them and a pardon was duly granted. A crucial part of the story, however, was that in gratitude for this providential escape from the gallows, John Weir declared 'be it lad or lass' he would leave his property, then comprising mainly the estate of Newton in the Upper Clyde valley, to his sister's child. Vere Irving goes on to say that the Lord Chancellor[10] himself was so affected by these entreaties that it was he who had relented, exclaiming 'Take away the woman, and make out the pardon!'

Vere Irving cites David Hume's account of Weir's trial (Hume, 1796, pp. 144–145) as supporting evidence for his story, but the Weir in that trial was actually a different one, namely William Weir of Blackwood[11] who was tried in 1682–83, as recounted and described in detail in Cobbett's *Complete Collection of State Trials* (1811, pp. 1022–54). By offering only 'Weir' as the name, Hume clearly left the door open to confusion. Nevertheless, our John Weir of Newton did stand trial for treason, along with twenty-one others, in a separate, later trial. We have an account of its lengthy progress from Fountainhall (1848, pp. 465, 529, 559, 568, 600, 709) and Brown (2007–13).[12] Weir, as we may surmise, was not amongst the accused that were eventually convicted; the principal charges against him were that he had been part of the rebel conspiracy, or at least had helped to conceal it, and more specifically he had sent a man to join the battle of Bothwell Bridge. To this, his defence was that he had sent his brother-in-law there, but not to join the battle; he had simply asked him to recover a horse stolen by the rebels. After an ordeal lasting more than two years, during part of which he had been held in Dumbarton Castle, he was finally acquitted on 18 February 1686 on the orders of the Privy Council, to which, of course, his brother-in-law James Irving was macer.

But it did not happen just as in the story, for the trials of the rebels went on from month to month, in fits and starts, sometimes in front of the criminal court, in the case of usual trial business, and sometimes in front of the Privy Council. Although Weir had managed successfully to petition the Privy Council on the grounds that the witnesses for the prosecution were tainted inasmuch as they had uttered 'threats and capital enmities' against him, such a procedure did not happen instantly even then. Neither is there any mention in the records of a woman having been present at any stage in the proceedings. Walter Scott's edition of Fountainhall's commentary, however, includes the suggestion that Weir may have got off by bribing the Lord Chancellor himself:

John Weir of Newtown, (who was an arrant Whigg rogue,) [i.e. a manifest rebel sympathiser] the dyet [was] deserted against him before the Justiciarie, for which, no doubt, he paid the Chancellor. (Fountainhall & Scott, 1822, pp. 163–164)

The Lord Chancellor at that time was James Drummond, Earl of Perth, a staunch royalist and prominent Roman Catholic who would have had no natural sympathy for a Covenanter, even an innocent one. If emotional pleading cut no ice with him, then perhaps some other form of inducement was indeed involved. One way or another, Weir did manage to 'persuade' the Privy Council that the prosecution witnesses 'had it in for him', and so they let him go:

18 Februarij 1686, poft meridiem.–At Privy Counfell, John Weir of Newton's petition is confidered ; and, in regard the Articles of Parliament had praecognofced his objections of minæa et inimicitiæ capitales, vented by the witneffes, viz. Hamilton of Gilkerfcleuch, Symonton, and Bailzie of Litlegill's brother, and found them proven, they fet him at liberty ; and ordained the Lords of Jufticiary (tho' some alleged this was to impofe upon the Justices, who were a fovereign Court,) to defert the dyet fimpliciter againft him. The Hy-Treasurer and his party oppofed this ; but loft it. (Fountainhall, 1848, p. 709)

The irony here was that the witnesses against Weir would have simply been saying exactly what the Privy Council would have wished them to. One cannot help but think that having a brother-in-law as macer to the Privy Council must have given John Weir some sort of advantage when it came to planting the seeds of his escape from the gallows. It is here that the promise of an inheritance for James Irving's son-to-be would most effectively have been put to work.

Despite Weir's release in February 1686, he was before the courts again in 1687, this time accused of extorting taxes on behalf of the rebels (Fountainhall, 1848, p. 813). Although he was found guilty, he was let off with a mere fine of 500 merks (about £25 sterling, or roughly £4,000 in present-day terms). But again Fountainhall seemed to think that this was more than just luck, for he alluded to Cato's observation about certain Roman censors having 'passed delinquents for money', and he pressed home the point by stating that such practices were 'as true of our times as of theirs'.

The persecution of the Covenanters waned soon thereafter and effectively ended in 1689 after the vengeful James VII took flight and sequestered himself and his court at St Germain in France. The English parliament had finally managed to oust him by appealing to William and Mary of Orange to come and take the Crown. The new monarchs were sympathetic to the Protestant cause and so, having come through the worst of it, John Weir managed to survive until about 1714, the year that George Irving inherited his estate of Newton.

The First Irving of Newton

Irrespective of whether there is any truth to the story of Margaret Weir's pleading on her brother's behalf, John Weir was certainly involved in the marriage contract, dated 29 October and 2 November 1711, between his sister's son, George Irving, and Sarah Weir, the daughter of Dr Thomas Weir and Bethia Blackwood (see Chapter 13). While he did subsequently make this nephew his heir, there is no way of knowing whether it was actually George who was being carried by his mother at about the end of 1685; the best we can say is that George would have been the eldest surviving male child. Margaret Weir died in 1695 and James Irving in 1698 (Paton, 1902) and so George was an orphan by the age of about twelve. He was his mother's next of kin (see note 8) and probably her only surviving child. As John Weir made George his heir, it would have been only natural for him to have become one of his tutors on the death of his macer father in 1698, and under the circumstances, he and his wife Jean[13] may even have raised George as one of their own, a situation that could well have led to him being named as his uncle's heir, regardless of whatever had taken place in front of the Privy Council.

The sasine giving George possession of Newton was recorded on 6 January 1714 (Irving, 1907, p. 4), and so we must assume that John Weir had died towards the end of 1713 or very early 1714. The record of his marriage to Sarah Weir (Paton, 1908, p. 279) states:

> [Irvine,] George, writer in the N. W. p.; Sarah Weir, d. of late Thomas W., apothecary, burgess, in S.E. p.
> 4 Nov. 1711, m.

What we know of George Irving, or Irvine as the spelling had become, is as follows. As already mentioned, his mother and father were Margaret Weir and James Irving. We may guess from the story of James Weir of Newton's trial that George would have been born about 1686. Since he was admitted as a notary public in July 1708 (Finlay, 2012) he must then have been above the age of twenty-one and was therefore born no later than July 1687, so the dates are indeed consistent. The same source informs us that, having had hopes of getting a position with the Privy Council dashed by the abolition of that body, he found employment with the Town Council in Edinburgh as a clerk making out burgess tickets. Later, he was set to work for the committee dealing with the parliamentary tax on ale, and by and by he showed himself to be sufficiently competent and reliable that he was made its secretary, even being sent to London on one occasion to negotiate on the town's behalf. By 1720, he was allowed the privilege of becoming a Writer to His Majesty's Signet, and his entry (Writers to the Signet, 1890) verifies that he was indeed by then George Irving of Newton:

> IRVING, GEORGE, of Newton. 8th February 1720.
> Apprentice to Alexander Home. – Son of James Irving, Usher of the Privy Council. Died 19th June 1742. Mar. 1711, Sarah, daughter of Thomas Weir, Surgeon in Edinburgh. Clerk to the City of Edinburgh.

This is also of help in the identification of Dr Thomas Weir, for the terms apothecary and surgeon were then mutually consistent, as the roles were frequently combined under 'apothecary surgeon'. For George Irving, promotion then seems to have followed promotion and eventually in September 1728 he was made joint Town Clerk, a position which he eventually held in his own right from 1736 until his death in 1742. Another gleaning from Finlay is that from 1712 he had a pew in Lady Yester's Church, which seems always to have been a Presbyterian charge (Hunter, 1864). If his father James, the macer, had followed the religious leanings of his master the Earl of Moray, of which there can be little doubt, George leaned in the opposite direction, towards his mother's family.

George Irving and Sarah Weir had three sons that we know of: Robert, George and Thomas. Robert was the first to inherit Newton, on the death of his father in 1742, but he died in 1748 and it then passed to his younger brother, George, who was served heir in July of that year.[14] Thomas, the youngest, became a doctor and served with the army in Lisburn in Ireland. He inherited a lucrative medical business from his grandfather, Dr Thomas Weir, the story of which features in Chapter 13. They also had a daughter, Sarah, who did not marry.[15]

An interesting footnote to the story of George Irving WS is prompted by the very late date of his mother's testament (see note 8), in 1718, for she died around the New Year of 1695. The answer lies in the astuteness of George himself, as the up-and-coming lawyer. He seems to have discovered a debt dating from 1658 that was owed by Archibald Earl of Argyll (1629–1685) to his mother's first husband, John Balmaine. As Balmaine's widow, Margaret Weir inherited both his debts and any monies that were owed to him, and since George, her son, was now her next of kin, sixty years later he was able to claim that the unpaid debt was now owed to him!

• • •

As we have seen, the Irvings and the Weirs were not only connected through the marriage of James Irving and Margaret Weir, they were connected again in the next generation through the marriage in 1711 of George Irving and Sarah Weir. We do not as yet know whether Sarah's father, Dr Thomas Weir of the Weirs of Blackwood, was related to George's mother and uncle, who belonged to the Weirs of Stonebyres, a sept of the Weirs of Blackwood. It seems plausible, but even so, any connection may have been distant.

As to the Irving side, it turns out that contemporaneously with James Irving the macer, there was a Dr Thomas Irving in Edinburgh, for between 1693 and 1696 he buried three children in Greyfriars kirkyard (Paton, 1902). Thomas was a name carried on down the Irving of Newton family. Were James and Thomas related?

It is sometimes the case that such things are purely coincidental, but coincidences do happen. Two cases came to court on the same day, on 21 July 1687. One was brought by Thomas Steill against Dr Thomas Weir, George Irving's father-in-law, of whom we shall hear later, while the other was the case brought by the Crown against John Weir, his uncle, for extortion on behalf of the rebels (see above). There was no connection whatsoever between these two cases, and so there could be no reason other than coincidence for them coming to court on exactly the same day.

CHAPTER 12

The Enlightenment of Edinburgh

It has already been mentioned that following the union of the Scottish and English parliaments in 1707 there began a remarkable period in which all branches of the sciences and the arts flourished, and it was all the more remarkable, for Scotland had been until then one of the smallest and most deprived countries in Europe (Daiches et al., 1986). As the economic situation improved during the eighteenth century, however, men of exceptional ability found outlets for their talents and thus emerged as towering figures who have in some way or another made a lasting impression on the world. John Amyatt, a Londoner who was appointed the King's Chemist from 1776 to 1782, had stayed long enough in Edinburgh to become a fellow of its former Philosophical Society, and although by 1783 he was no longer a resident, he was nevertheless admitted as a founding member of the RSE, along with many of the men that he was referring to in the following remark reported by his friend, William Smellie (1740–1795), editor of the first Encyclopaedia Britannica:

> Mr Amyat [...] one day surprised me with a curious remark. 'There is not a city in Europe, said he, that enjoys such a singular and such a noble privilege.' I asked, 'What is that privilege?' He replied, 'Here I stand at what is called the Cross of Edinburgh, and can, in a few minutes, take fifty men of genius and learning by the hand.' (Smellie, 1800)

Under the topic 'Scottish Enlightenment' on Wikipedia (2015), of the thirty-seven people suggested as being the most eminent of the Enlightenment, twenty-three flourished in the decade 1780–89, not far removed in time from when the remark was made. Given that these represent only the top stratum of such noteworthy people, largely concentrated in Edinburgh, Mr Amyatt could easily have demonstrated the underlying element of truth in his claim. Their number increased progressively from two in the decade 1710–19, to its peak of twenty-three in 1780–89 and thereafter decreased almost symmetrically to about two again in 1830–39. It is not that we had run out of such men of genius, it was simply that they stood out so much less amongst the growing number of well-educated, highly competent people across the world, while at the same time the bar for genius was consequently rising. If the Scottish heroes of the Enlightenment, notable examples of whom are given in Appendix 5, had been amongst the first to shine their light in the dark, by the time the nineteenth century was well under way there was light everywhere. By this one crude measure alone, as shown in Figure 12.1, we may chart a timespan of the age of enlightenment.

Edinburgh's new towns were not only a product of this age of enlightenment, they form a magnificent and lasting testament to it. The grandeur of the buildings, laid out to the overall design of James Craig (1739–1795) (Coghill, 2010, p. 143; Youngson, 1966, pp. 70–110, 288–289) gives some idea of the talent and ambition that abounded in this golden era. But before discussing the new towns, which were very much the scene for James Clerk Maxwell's immediate family and their circle of friends and relatives, we must begin with the Old Town.

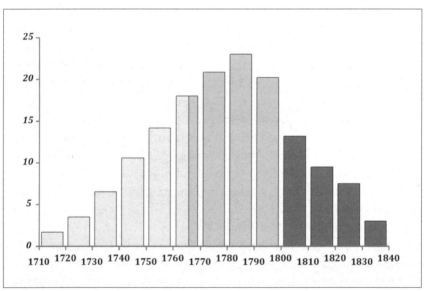

Figure 12.1 The number of leading figures of the Scottish Enlightenment active in each decade from 1710 to 1840
Out of a total of thirty-seven leading figures, twenty-three, about two thirds of them, were active during the decade 1780–90 when the population of Edinburgh itself was still less than 150,000. The first and second New Town eras are indicated by the two sets of shaded bars on the right.

The Old Town

James Clerk Maxwell grew up when Edinburgh's New Town was burgeoning. The Old Town, as shown in Figure 1.1, dated from mediaeval times and as we have seen, had once been the place to live for great and humble alike. Hemmed in by the great wall that had been thrown up as protection after the momentous defeat of the Scots at the battle of Flodden in 1513, the citizens were constrained to expand their town in an ever upward direction. Making use of the steep gradients that run down on each side from the High Street ridge that flows to the east of the castle, their stone tenements, or lands, were built to great heights, especially on the sides facing the valleys of the Cowgate and Nor' Loch (Topham, 1899, p. 2):

> The style of building here is much like the French: the houses, however, in general are higher, as some rise to twelve, and one in particular to thirteen storeys in height. But to the front of the street nine or ten storeys is the common run; it is the back part of the edifice which, by being built on the slope of an hill, sinks to that amazing depth, so as to form the above number.

Closely packed, and getting wider and more higgledy-piggledy as they rose, it is said that neighbours in the upper reaches of two adjacent lands could easily touch hands, or even lips, from their viewless windows (Chambers, 1980, p. 4), which notion is expressed very well in Figure 12.2.

Robert Louis Stevenson gives a romantic view of these lands of the Old Town at the time of his boyhood in *Edinburgh: Picturesque Notes* (1879, pp. 5–9). The tradesmen and shopkeepers occupied the floors nearest the ground, while the gentry occupied the airier floors just above. Thereafter, the spectrum of affluence rapidly decreased with height, with the uppermost levels being divided into many small apartments each packed with a large family of the poorest people. Notwithstanding, all was accessed by a narrow common stair, which in many of the oldest buildings would have been a wooden structure that rambled up the outside of the building. While fresh water was fetched from a number of common wells at street level, and had to be carried up to the giddy levels above, the problem of returning waste water of any kind, indeed waste of any kind, back to street level was much more easily dealt with. On the sounding of the town guard's tattoo at ten o'clock each night, and theoretically not before, the obedient citizens would duly tip their buckets out of the nearest window to the cry of 'Gardyloo', which was often greeted with a cry of 'Haud yer hand!!' from some panic-stricken reveller who had missed his way home.

BRILLIANT LIVES

Figure 12.2 'Dickson's Close'
Viewed on a map, the old High Street of Edinburgh and its multitude of steep and narrow wynds running downhill on either side appear like the bones of a herring. Many of these narrow passages remain, but there are few examples left of the type shown in this drawing. The tenement buildings on either side of the wynds, already barely separated at their base, were allowed to reach ever closer together as they went up. While this increased the available floor space, it compromised fresh air and light. The gentlemen sharing this book must have been good friends to have been able to put up with living so close together. (From Dunlop, et al., 1886)

The New Towns

From about 1750, in the Scottish Lowlands, security was less of a concern and prosperity was on the increase, at least for those who were well placed to grasp it. The town's gentry began to yearn for more refined surroundings, and many managed to escape the city by building villas to the south of the University, or College as it was then known (Chambers, 1980, p. 4):

> Edinburgh by-and-by felt much like a lady who, after long being content with a small and inconvenient house, is taught, by the money in her husband's pockets, that such a place is no longer to be put up with.

By the 1760s, however, a scheme was being discussed to build houses on a grander scale to the north of the town. Land was acquired and in 1767 work began on James Craig's formal plan (Youngson, 1966). Within ten years the 'New Town' was taking shape and such noteworthy folk as the philosopher David Hume decanted to the other side of the Nor' Loch (Chambers, 1980, p. 58). Each new house that was built realised just one small part of a grand master-plan, but they were no less grand in their own right. Typically they were built of quarried stone, comprised three floors and one or two basements, had large windows and were often ornamented with splendid columns and balustrades. A few, such as the house of Sir Lawrence Dundas, on the east side of St Andrew Square, were magnificent classically inspired edifices in the grand Georgian style. On the interior, many were decorated by

Robert Adam, or in his style; they represented an opulence that only a very few could aspire to by the standards of today. At that time, however, many of those who made it to the New Town were what we would consider as being moderately rather than very wealthy. Well-to-do merchants and master tradesmen built houses in the same streets as lawyers, physicians and lords and ladies. In Edinburgh at least, the snobbery associated with class distinction was something for the future, when people started equating the grandeur of each house to the social status of its inhabitants.

While Craig's New Town plan to the north of the Old Town was the officially adopted version promoted by Edinburgh's Lord Provost George Drummond, George Square to the south of the Old Town was part of a slightly earlier informal development which was begun in 1766. Much land had been freed up in the vicinity by the draining of another loch, actually more of a marshland, the old Burgh Loch which is now a public park referred to as 'the Meadows', and it had the distinct advantage of being quite level. Built by James Brown and named after his brother rather than the King, George Square was not part of any grand design similar to Craig's, and because Drummond's efforts were directed towards bringing Craig's plan to fruition, it did not spread to any great extent.

George Square is the largest of its kind anywhere in Edinburgh, but while its original houses are smaller and of a rougher external appearance to those in the Second New Town, they are actually similar to the majority of those that were erected about the same time in the First New Town. They would have been, almost literally, a breath of fresh air compared with the hemmed-in Old Town dwellings. From at least 1800 to 1808, 47 George Square was the home of Mrs Janet Clerk (Irving) and her three children, George, Isabella and John (Chapter 9). It was close by the University and also fairly convenient for the old Royal High School, which the boys would have attended, as did the the future Sir Walter Scott, who was somewhat older but who had also lived in George Square (see Figure 1.1, locations 5 and 6).

Within just thirty years of the commencement of the first New Town, a second New Town was called for, and about 1803 building began on Heriot Row. When Henry Mackenzie, author of *Man of Feeling*, was living in one of the first to be built there, he recalled the land as it had been before the new houses were started; he had shot snipe there in the fields of Wood's farm lying between Canonmills and Bearsford Park[1] (Grant, *c*. 1887, vol. 2). Another thirty years saw the area north of Queen Street (Figure 1.1) largely built up. Considering that many other developments to the east, south and west are not shown in the map, the overall rate of expansion of Edinburgh was truly enormous. Today, the New Town refers to the entire development of both the

First and Second New Towns, characterised by buildings of quality in the Georgian style harmonising with wide streets and numerous gardens conforming to a geometrical layout.

Many of James Clerk Maxwell's forebears and their kin lived in Edinburgh, and most of them who were alive in the New Town era made the move from the lofty old tenements (whether in the Old Town proper or just beyond) to the brand new terraced villas to the north and at George Square to the south. Using information collated from various sources, including the Edinburgh directories from 1773 on, Table 12.1 shows where they lived both before and after they moved. Figures 1.1 and 1.2 may be used to locate a number of the New Town addresses.

Table 12.1
Addresses of the Clerks in Edinburgh's Old and New Towns

Baron Sir John Clerk, 2nd Bt.	*Blackfriars Wynd*
Sir James Clerk, 3rd Bt.	*Blackfriars Wynd*; *Sempill's Close*
Lady Elizabeth Clerk (nee Cleghorn)	*Dickson's Close*
George Clerk Maxwell	*James' Court*
Dorothea Clerk Maxwell	*James' Court*; 52 Princes Street
Sir John Clerk, 5th Bt.	*James' Court* (c/o Mrs Shaw);
	37 Princes Street
Lady Mary Clerk (nee Dacre)	Crichton Street; 37 Princes Street;
	57 (100) Princes Street
John Clerk of Eldin	Shakespeare Square; Hanover Street;
	70 Princes Street; 16 Picardy Place
Capt. James Clerk, HEICS	*James' Court*; 52 Princes Street
Mrs Janet Clerk (nee Irving)	Buccleuch Street; 52 Princes Street;
	37 Princes Street; 47 George Square;
	31 Heriot Row; 14 India Street
Sir George Clerk, 6th Bt.	47 George Square; 31 Heriot Row; LONDON
Isabella Wedderburn (nee Clerk)	47 George Square; 31 Heriot Row;
	126 George Street; 31 Heriot Row;
	18 India Street; 25 Ainslie Place
Baron James Clerk of Rattray	George Square; 53 (92) Princes Street
John Clerk, Lord Eldin	70 Princes Street; 16 Picardy Place
William Clerk, of Eldin	70 Princes Street; 1 Rose Court; 4 Rose Court
John Clerk Maxwell	47 George Square; 31 Heriot Row;
	14 India Street; GLENLAIR
Frances Clerk Maxwell (Cay)	11 Heriot Row; 14 India Street; GLENLAIR

Old Town addresses are in italics while the underlined house numbers are pre-1811 designations. The numbers in brackets show the later designations, where available.

Colour Plate 1 Number 14 India Street, James Clerk Maxwell's birthplace
This category-A listed building has changed little since it was built in 1820. The false window on the top floor makes it look a little different from some of the other houses in the street.

Colour Plate 2 Frances Cay and her son James Clerk Maxwell by William Dyce RA
This painting was finished by Dyce at Glenlair in the late summer of 1835 when James was aged four.
(Photograph © Birmingham Museums Trust)

Colour Plate 3 (*above*) Number 6 Great Stuart Street, home of James' Aunt Jane, and 25 Ainslie Place, home of James' Aunt Isabella during her later years
6 Great Stuart Street is the door on the left with the astragal fanlight while 25 Ainslie Place is on the right just round the corner. The dummy windows to the left of the corner help maintain the elegant regularity of the frontage.

Colour Plate 4 (*left*) Number 31 Heriot Row, home of James' Aunt Isabella
Numbers 31 and 33 are the centrepieces of the west side of Heriot Row, being the only houses that have a third floor above ground level.

Colour Plate 5 Old College, University of Edinburgh
James Clerk Maxwell attended Professor Forbes' natural philosophy classes here. The lecture theatre and equipment store were located beside the three upper windows on the far left of the picture. In Maxwell's last year at the college, Forbes allowed him the use of the facilities there so that he could do his own experiments. A plaque commemorating Maxwell was erected in the quadrangle in November 2015.

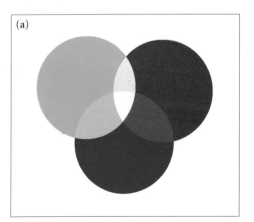
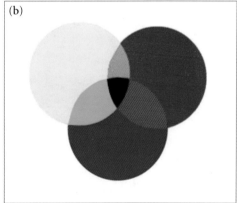

Colour Plate 6 James Clerk Maxwell's Theory of Three Primary Colours
(a) **Mixing primary colours with light.** All the colours together make white at the centre.
(b) **Mixing primary colours with pigments.** All the colours together make black at the centre.

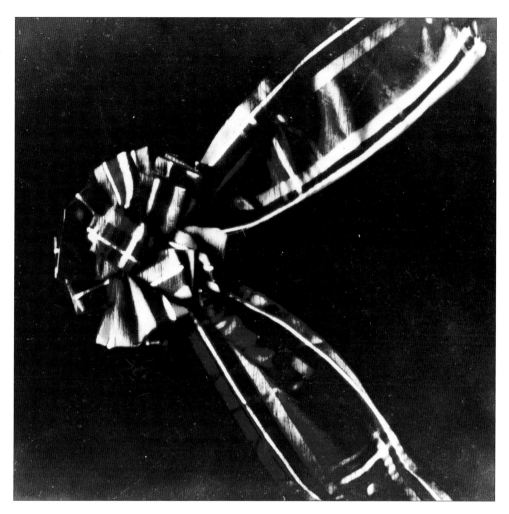

Colour Plate 7
(a) (*above*) The tartan ribbon colour image demonstrated by Maxwell in 1861
A reproduction of the first ever photographically produced colour image as demonstrated by James Clerk Maxwell in May 1861 at the Royal Institution in London. He had in fact put the idea forward much earlier, in 1855. (© National Media Museum / Science & Society Picture Library – All rights reserved)

(b) (*left*) Enlargement of part of the image as viewed on a typical LCD screen
An LCD screen is made up of a large array of tiny 'pixels' (<u>pic</u>ture <u>el</u>ements). In this case, it can be seen that each pixel is more or less square in shape and comprises three vertical bars, one red, one green and one blue, the brightness of which can be individually controlled. At normal magnifications the eye cannot resolve the individual coloured bars and so it just sees a local average of the colours they emit. The three separate red, green and blue images formed by the bars therefore correspond exactly to the concept invented by Maxwell 150 years ago!

Colour Plate 8 (*above*) **The wing James Clerk Maxwell added to Glenlair**
Glenlair House was destroyed by fire in 1929. This wing, which was built by Maxwell along the lines of a design that had been his father's, has recently been protected from the elements and the vestibule restored as a small museum. The original house, out of view on the right, was plain and functional because it was what his father could afford at the time.

Colour Plate 9 (*right*) **The Clerk Maxwell family grave at Parton**
James Clerk Maxwell, his mother and father and his wife lie within the ruined walls of the Old Kirk.

Colour Plate 10 Sir John and Lady Clerk of Penicuik, 1791, by Sir Henry Raeburn
(Photo courtesy of the National Gallery of Ireland)

Colour Plate 11 Isabella Wedderburn, John Clerk Maxwell's sister
Painting by Sir Henry Raeburn c. 1820. (Yale University Art Gallery (open access / public domain))

Colour Plate 12 Elizabeth Cay, née Liddell, James Clerk Maxwell's maternal grandmother
Not so much a self-portrait but her pastel copy of one of the two portraits of her by Archibald Skirving (1749–1819), from whom she received lessons. (By courtesy of the James Clerk Maxwell Foundation)

Colour Plate 13 Robert Hodshon Cay, Judge of the Admiralty Court in Scotland, by Sir Henry Raeburn
A copy of this painting, by his granddaughter Isabella Cay (p. 251), hangs in the exhibition room at 14 India Street. (Wikimedia Commons)

Colour Plate 14 *Glenlaer* by William Dyce RA
This is not one of the original watercolours that Dyce painted in 1832, but a later one from his second visit to Nether Corsock in 1835. The only loch on the Glenlair estate was Loch Falbae. If the view was taken from Upper Glenlair looking north, this loch would then have been in the middle distance, with high hills on its left flank and in the far distance, all as in Dyce's image. (© Victoria and Albert Museum, London)

From the Irvings of Newton to the Irvings of the New Town

Of the first George Irving of Newton we now know at least a little: his parentage, how he acquired Newton, his marriage and children, his qualification as a Writer to the Signet, and his career with Edinburgh City Council (Chapter 11). George died in June 1742, whereupon he was succeeded by his eldest son Robert, who then died in 1748 leaving it to his brother George, the second George Irving of Newton, about whom we have been able to find very little information. We may assume, however, that as the second son he was at the earliest born in 1713; he died in early November 1782,[2] having married twice. Referring to Family Tree 6, his first wife was Isabella Colquhoun,[3] who was the daughter of James Colquhoun,[4] Lord Provost of Edinburgh from 1738 to 1739, and his wife Janet Inglis.[5] George and Isabella were married on or about 22 March 1757 and a year later had a child, Janet Irving, James Clerk Maxwell's paternal grandmother (Chapter 9). Within the space of just a few years Isabella had died, whereupon George remarried Mary, or Molly, Chancellor, daughter of Alexander Chancellor of Shieldhill near Biggar, on 22 September 1763. Curiously, his younger brother Thomas, the surgeon, married Molly's sister, Jean or Jane Chancellor, of which more in the following chapter. Of Janet, the only information we have of her follows her marriage with James Clerk HEICS in 1786, the story of which was taken up in Chapter 9 above.

The children of George's marriage to Molly Chancellor were:

Alexander (1766–1832), who was an advocate (1788), mining manager at the Scots Mining Company at Leadhills, joint Professor of Civil Law at Edinburgh University (1800) and later a judge, Lord Newton (1826).[6] He married his cousin Bethinia in 1814. Since her parents were Dr Thomas Irving and Jean Chancellor, respectively the brother of his father and the sister of his mother, there was a high degree of consanguinity in this marriage.

John (1770–1850)[7] was a Writer to the Signet and an early friend of Sir Walter Scott. Amongst other things he was a director of the Bank of Scotland, and became wealthy. In 1804 he married Agnes Clerk Hay whose grandmother was Agnes Clerk, daughter of George and Dorothea Clerk Maxwell, and as we will see, some of their children's achievements were noteworthy.

Thomas (1774–1852) left Edinburgh to go into the Civil Service and eventually settled in London. His first wife Margaret Colhoun was connected through her mother Rebecca Napier to the Napiers of Merchiston. It is not known whether her father, John

Colhoun, had any connection with Isabella Colquhoun (frequently also spelled Colhoun), his father's first wife.

There were also three sisters, of whom we know but little. Elizabeth we know of only because of her death, in April 1808 (*Scots Magazine*, 1808, p. 399) which Sir Walter Scott said affected John Irving very much.[8] The two other sisters that we know of are Mary and Jeannie (Fairley, 1988, p. 111) who were likely to have been the two elderly ladies depicted in Jemima's watercolour (Wedderburn, *c.* 1841) of James Clerk Maxwell's arrival at Newton House in November 1841 on his way to Edinburgh (Chapter 2).

Janet was therefore the half-sister of Alexander, John, and the other children.

One might suppose that George Irving never lived much on his estate at Newton, for although there was a house there, the countryside for miles around was desolate moorland surrounded by hills and transected by water. Newton itself was not the entirety of his estate, for according to Irving (1907), it also comprised lands at 'Partraith, Over Fingland, Shortcleugh and others'.[9] Now, it was on South Shortcleugh that George Clerk Maxwell and his mining partners took out a tack with the Earl of Hopetoun in 1758 (Chapter 8). In addition, since George Irving's son Alexander is known to have been involved in mining in that area, it would come as no surprise at all to find that George Irving himself was interested in the mines either on his own property or on neighbouring holdings where he may have been involved in a partnership. We know so little about the second George Irving that we simply cannot say either way, but if he did have mining interests he would have spent rather more time at Newton than otherwise.

As to an Edinburgh residence, we find him in the available directories for only one year, 1775, when 'Irvine, George Esq.' was at Milne's Court, in the Old Town, at the top of the Lawnmarket. In the previous year, it was his brother Thomas, the surgeon, who was listed at that same address. After George's death, his widow was in Bristo Street in 1786, listed as 'Irvine Mrs. of Newtown', while her sister Jean, listed as 'Irvine Mrs. Dr.' was elsewhere, at Sommervile's Close in the Canongate. Bristo Street, as it was then, lay just beyond the city wall, close to George Square. The houses there had appeared piecemeal even before the New Town expansion. In 1775, therefore, the Irvings were still pretty much in the Old Town, but by 1786 they were in the vicinity of George Square, the alternative New Town development touched on in Chapter 9. After the second George Irving died in the winter of 1782, Alexander, his heir, would have come into Newton in his own right by 1787, the year before he qualified as an advocate. Listed under his mother's name and in due course his own, for nine years the Irvings are to be found at Bristo Street, after which they moved 'round the corner' to 5 Buccleuch Place, shown as location 7 in Figure 1.1, a mix of recently built town houses and tenements adjacent to George Square. In this tenement, comprising four storeys entered by a common stair, they remained until 1805 with Alexander as the head of the family. In 1806, however, Alexander Irving of Newton, advocate, now joint Professor of Civil Law at Edinburgh University, was to be found living at Heriot Row, listed as number '8 on the west side'. In 1811, when the numbering changed to the present-day system, this became 27 Heriot Row, the same address at which James Clerk Maxwell called for his schoolfriend, Lewis Campbell (Chapter 2).

As an advocate and as a judge in the Court of Session (Plate 16), the character of Alexander Irving, later Lord Newton, seems to have been the antithesis of his contemporary, Lord Eldin, to whom he was related by his half-sister Janet's marriage to James Clerk HEICS. Although he completely lacked Eldin's force of character, he was particularly successful as a judge in his own particular sphere of activity.[10]

In 1814 he married, at the age of forty-seven, his first cousin Bethinia Irving, who was the daughter of his father's brother, Dr Thomas Irving (see Family Tree 6). In 1815 they had their only child, a son they named George Vere Irving[11] (the co-author of Irving & Murray, 1864). Alexander used his wealth to build a large mansion on his estate at Newton (Plate 17), the site of which can be seen in Figure 11.1. It was here that the Clerk Maxwell family would stop over on their journeys between Glenlair and Edinburgh.

Alexander died on 4 March 1832 after bravely enduring what must have been an excruciating operation to remove stones from his urinary tract, a trial he bore with great fortitude. His great-nephew James Clerk Maxwell was then only nine months old. Alexander's widow lived on at 27 Heriot Row until 1842, the year her son married, and thereafter she decamped to 2 George Square.

It appears that James Clerk Maxwell would have known his great aunt Bethinia when he first came to live at Heriot Row. She died about 1854, and is buried alongside her husband Alexander and his father and mother, George Irving and Mary Chancellor, in St John's and St Cuthbert's churchyard.

• • •

Alexander and Janet Irving had two younger brothers: John, born in November 1770, and Thomas, as shown in Family Tree 6. John Irving, along with William Clerk, second son of John Clerk of Eldin, was one of Sir Walter

Scott's closest friends. They were of much the same age and were neighbours, Scott in George Square and Irving in nearby Bristo Street;[12] both would have gone to the High School. They inevitably spent much of their youth together, from time to time going on long wanders, despite Scott's lameness, as far as Penicuik House; and we know that John Irving was there with Scott and William Clerk on an impromptu visit paid to Sir John and Lady Mary Clerk (Chapter 9). At the time of that visit, Janet Irving had already married Sir John's younger brother, James Clerk HEICS.

Unlike Walter Scott and William Clerk, however, John Irving's legal career was not as an advocate. His older brother Alexander having already taken that route, it was his lot to become a Writer to the Signet, which he accomplished in July 1794. Ten years to the month later, he became the second Irving to become connected with the Clerks of Penicuik, albeit somewhat distantly, through his marriage to Agnes Clerk Hay, a great grand-daughter of George Clerk Maxwell, 4th Baronet; and they had a large family.

Their fourth son, John, was born in February 1815. We mention him first because he was to achieve fame something akin to that of the heroes that accompanied Captain Scott on his Antarctic expedition, but alas it did not come in his lifetime. After attending Edinburgh Academy as one of the first intake of pupils at the newly opened school (Henderson & Grierson, 1914) he went to Portsmouth Naval College where he excelled and took the second prize silver medal in mathematics before passing out as a lieutenant in the Royal Navy (Bell, 1881). After several years' service during which he failed to get promotion, in 1837 he left for Australia in fulfilment of a plan to become a sheep farmer there. Nevertheless, while on a trip to Sydney in 1841 to sell wool, he happened to spy HMS *Favorite* lying in the bay and, having encountered some of its officers whom he knew well, he was persuaded to return to the Navy. Having done so, he joined the ill-fated Franklin expedition that set out to navigate the North-West Passage in 1845.

Despite the two ships *Erebus* and *Terror* having been strongly built, strengthened with iron plates and equipped with steam power[13] (Bourne, 1852, Appendix: p. i), they became lodged in the ice in Baffin Bay, and so it was that he perished, along with the rest of the crew, in the vicinity of King William Island sometime between 1848 and 1849. There his remains lay frozen and undisturbed until they were discovered in 1880 by a search party led by Lt Frederick Schwatka of the US Army (*Sydney Morning Herald*, 1881). His remains were identified by his silver medal, which his comrades had placed beside him in his shallow grave before they too met their fate. After their discovery, his remains were returned to Edinburgh and buried with full honours in the Dean Cemetery to the west of the New Town on 7 January

1881. The funeral procession, led by his brother Alexander and sister Mary, who was by then Mrs Scott Moncrieff (see note 29 to Chapter 5), departed from Mary's home in Great King Street, another fine Second New Town address just to the north of Heriot Row. His monument in the Dean includes a bass relief plaque depicting the expedition (Plate 18).

John and Agnes had several other children, the eldest of whom was George (1805–1841), named after his paternal grandfather. He followed in his father's footsteps, first by becoming his apprentice and then, in November 1828, a Writer to the Signet. He died at the age of thirty-five in February 1841 without either having married or having had the opportunity to fulfil his potential to the sort of extent that some of his siblings managed to achieve.

The second son was Lewis Hay (1806–1877), named after his maternal grandfather who had been a colonel in the Royal Engineers. He trained for the ministry of the Church of Scotland and by 1830 had become the minister of Abercorn church on the Hopetoun estate near Queensferry (Scott, 1915, vol. 1, p. 191). During the 'Disruption of 1843' (Scotland.org, 2014; Prebble, 1971, pp. 324–325) he chose to come out with the seceders and joined the Free Church, and by the end of that year he was inducted to a charge in Falkirk. There he founded the Garrison Church in 1844 (Falkirk Archives, 2012) and in the course of time two others followed. Having encountered many examples of serious poverty and terrible social deprivation amongst his flock, he was soon labouring to provide schools, of which two were built, and to bring about lasting social change (*The Falkirk Wheel*, 2011). Having first married Isabella Carruthers who died in 1836, in 1840 he married Catherine Caddell (1817–1890) who was from a family with whom his mother had some connection. Lewis Hay Irving is buried in Camelon cemetery near Falkirk and the present Camelon Irving Memorial Church is named after him.[14]

The next son, Alexander Irving (1813–1882), was a soldier who rose to the rank of major-general in the Royal Artillery. He served in the Crimean War and was appointed a Companion of the Bath (Bell, 1881, p. 5; Henderson & Grierson, 1914).

Archibald Stirling Irving (1816–1852) was born on 18 December 1817, the namesake of a mining manager at Leadhills who is mentioned in the final section of this chapter. Although like his brother John he attended Edinburgh Academy, he had a delicate constitution that left him unsuited to any formal career. Supported by a pension from his father, he lived in various parts of Scotland, drawing on them as inspiration. An avid scholar of the classics and literature, he also wrote many songs and poems which were published anonymously as *Original Songs*, in two volumes, by the Rev. William Murray (Irving, 1841). In the last year of his life he married Helen Laing who cared for him until his death on 29 October 1851 (Rogers, 1856, pp. 235–236).

David Williamson Irving (1819–1892) may have been named after David Williamson, the judge Lord Balgray. Having been much influenced early in life by his elder brother John, David went with him to Australia where, unlike his brother, he decided to stay. He settled, married, had a successful career, and eventually became a police magistrate in Tamworth, New South Wales (Bell, 1881; Henderson & Grierson, 1914).

John Irving, the father of these illustrious children, died at about the age of eighty on 26 May 1850. His wife Agnes died long before him in 1823 (*Gentleman's Magazine*, 1823, p. 647) and he was buried by her side in the Canongate churchyard. His sons George and Archibald Stirling, who both died in his lifetime, are buried with him. Buried with Agnes are their daughters Barbara and Agnes who both died in infancy. It is curious to note, however, that on Agnes' headstone, which is clearly the older of the two, the first lines of the inscription are to someone apparently unconnected with the family:

> To the Memory of
> CHARLES ALSTON M.D.
> Professor of Botany
> in the University of Edinburgh
> Died 22nd November 1760

We shall soon, however, find the connection between the Irvings and the man whose name was given to the genus of trees Alstonia by the botanist Robert Brown (mentioned in Appendix 5 as the discoverer of random 'Brownian Motion' in tiny particles).

A Case of Multiple Connections

James Clerk Maxwell, James Stirling, John Napier and Sir Isaac Newton: these four great names are connected one to the other by the subject of mathematics, and while two of them were natural philosophers, a different pairing was very interested in metals, specifically lead and gold. On top of that, three of the four were related to each other! However, let us straightaway eliminate the obvious guess that the Irvings of Newton might have some connection with Sir Isaac Newton (1642–1727). While it has been mooted that Alexander Irving named his house near Elvanfoot in honour of the great Sir Isaac, any such claim is specious because, as explained in Chapter 11, Irving's estate had been known as Newton long before Sir Isaac came on the scene.

Let us, however, examine other possible connections from Newton himself to the rest of our famous quartet of mathematically orientated geniuses. Apart from being the premier physicist and mathematician of his era, Newton for a time was deeply involved in alchemy (O'Connor & Robertson, 2000). Although the subject is now thoroughly discredited, such ideas were no further from the truth than many other serious scientific ideas of the day, for example, those regarding the basic elements and the principles of medicine. His interests were therefore scientific rather than mystical and his experiments in this area should be seen as early probings in the subject of metallurgy. Given that he was Master of the Royal Mint from 1699 to 1727, any rumours concerning the possibility of creating gold from base metals would have prompted thorough investigation. Metals and mining are, of course, intimately related.

In the early eighteenth century, Newton encountered the much younger James Stirling (1692–1770) (Fraser, 1858, pp. 91–102), the Scottish mathematician who gave us Stirling's asymptotic approximation for N factorial, the Stirling numbers, and central difference formulae (O'Connor & Robertson, 1998). As a young man, Stirling had travelled to Venice where he discovered the secrets of making Murano glass, and so he was not simply a mathematician, he was also a 'hands on' enquirer with a technological bent. Newton would have recognised a common bond with Stirling not only in mathematics but in this sort of early scientific interest. But above all, the young Stirling must have deeply impressed Newton with his work on cubic curves, for in a paper 'Linae Tertii Ordinis Newtonianae' published in 1717, he essentially completed Newton's own efforts on the subject. Despite the age gap, a bond seems to have formed between the men and they became firm friends (Carlyle, 1898).

But where does this get us? In 1734–35, when he was back in his native Scotland, James Stirling undertook some studies on the lead mines around Leadhills where, in 1736, he subsequently took up residence, presumably as some sort of scientific and management consultant to the mine manager. He was then appointed as the manager in 1739, a role in which he proved surprisingly adept (Harvey, 2000). His house at Leadhills, built for him under the architect William Adam by the Scots Mining Company, still stands. Being of a similarly wide-ranging capability as Newton himself, Stirling took to the task in hand with great accomplishment. On his death in 1770, he was succeeded in the post by his nephew Archibald Stirling (1738–1824).[15] Archibald may not have been at the same level of competence as his uncle, for in 1800 Alexander Irving of Newton was appointed to assist him

(Waterston & Shearer, 2006). But why did Alexander Irving, then an advocate and professor of law, get involved? He had lands in the area of Leadhills, specifically at Shortcleugh, which had mining potential (Harvey, 2000); more than likely, he had invested in the mines and made money out of them when the going was good, and it may even be the case that his father, George Irving, also had some involvement in his time.

Alexander Irving was sole manager of the mines until 1820. Since he was a joint professor of civil law with John Wilde, it could be thought that this latter arrangement would have given him more time to devote to running the mine, but the reason for it was that John Wilde had become unfit for the job, and Alexander had been appointed to the post on the understanding that he was to stand in for Wilde and carry the entire workload of two courses, one on the Institutions and the other on the Pandects.[16] As we have already seen, Scottish professors and judges had virtually six months of the year to attend to their own business, but during the academic term Alexander would have had to rely on correspondence with someone on site upon whom he could rely. When some years later the owners became dissatisfied with their returns, they appointed an assistant for him, John Borron, whose role was ostensibly to support him, but in fact he took over the role completely in 1824. No doubt there was too much on Alexander's plate to allow him to focus on the job in the single-minded manner that James Stirling had been able to demonstrate.

Alexander Irving, as James Clerk Maxwell's paternal grandmother's half-brother, is a connecting link between Maxwell and James Stirling, and as we have already seen, John Irving named one of his sons Archibald Stirling Irving. Nor should we overlook the fact that Stirling was at Leadhills during the time that George Clerk Maxwell and his father the Baron were frequently in the area and in particular, as discussed in Chapter 8, George and his partners had a tack on land nearby, in fact adjacent to George Irving of Newton's property of Shortcleugh. In addition, William Adam, the Baron's architect for Mavisbank, had built James Stirling's house at Leadhills. In such a desolate area where mining was the greatest ongoing concern, George Clerk Maxwell, James Stirling and the second George Irving of Newton, must have been at least acquainted with one another. But what is there to suggest that the connection went any further than that?

A study of Family Trees 4(a) and 4(b) shows that James Stirling and the Baron's second wife, Janet Inglis, had a common ancestor, James Stirling of Keir (d. 1588). Through the line of Stirling of Garden, James Stirling the mathematician was the fifth generation after James Stirling of Keir by direct descent (Fraser, 1858, pp. 35–44, 91–102). Through the marriage of Margaret

Stirling, the latter's daughter, to Sir John Houston of Houston (d. 1609), Anne Houston was also his fifth generation descendant, and she was the mother of Janet Inglis, and consequently the grandmother of George Clerk Maxwell. Families such as these knew their lineage intimately, and the Stirlings and the Houstons would have been well acquainted with the details of their interrelationship. Perhaps the surprising thing that arises out of this family tree is that James Stirling of Keir was also John Napier of Merchiston's father-in-law! Curiously, at the time of their marriage, the Napiers owned Wright's Houses (Grant, *c.* 1887, vol. 5, pp. 32–34), the property just outside Edinburgh purchased by the first John Clerk of Penicuik about 1650.

There is therefore no real mystery concerning the name of Newton, and any connection between James Clerk Maxwell and Sir Isaac Newton amounts only to the coincidental facts that they both made outstanding advances in fundamental physics; both were fellows of Trinity College, Cambridge; Newton knew James Stirling and, as seems very likely, so did Maxwell's forebears, George Clerk Maxwell, the Baron and George Irving of Newton.

While it is not impossible that the Baron met Sir Isaac Newton on one of his many visits to London, there is no mention of it in his memoirs. When the Baron was eventually elected a fellow of the Royal Society of London in 1727, Newton had been dead for over six months. In fact, when the Baron first visited Sir Hans Sloane earlier in that year, Sir Hans had only just taken over the presidency from Newton. The Baron had arrived in London 'about the begining of Aprile' 1727, coincidentally just a few days after Newton's death on 27 March. Given that fact, if the Baron had known Newton at all, he could not have failed to mention it.

• • •

We now return to the curious presence of the name of Dr Charles Alston (1683–1750) on the gravestone of Agnes Clerk Hay, the wife of John Irving. A professor of botany at the University, Alston also held the associated post of curator of the Edinburgh Physick Garden.[17] The reason his memorial is above that of Agnes Clerk Hay lies in the fact that John Irving's mother, Mary Chancellor, was related on her own mother's side to the family of Birnie of Broomhill (Family Tree 6). Two generations earlier, Elizabeth Birnie had married Thomas Kirkaldie, and Charles Alston was their grandson, making him John Irving's second cousin twice removed. An additional link was formed in 1741 when Dr Alston married his second wife, Bethia Birnie, who was John Irving's great aunt. Although John's and Agnes' gravestones are of matching design, Agnes' stone is clearly much older, and so it is possible that it stood there for fifty years with only Charles Alston's name on it before the first of the Irvings' children was interred. Perhaps John Irving had inherited it, or by some other means managed to acquire the rights to the lair for his own family's use. Death was a frequent visitor, and burial space was in short supply.

If there is any disappointment at there being no familial connections with Isaac Newton, at least we have serendipitously found that James Clerk Maxwell and Professor Charles Alston MD share a common ancestor, the Reverend Robert Birnie (1608–1690). That Maxwell was also related to the mathematicians John Napier of Merchiston and James Stirling through the Baron's mother-in-law seems quite incredible!

CHAPTER 13

Dr Thomas Weir and Anderson's Patent Pills

In this chapter we explore in more detail the family connections and dealings of Dr Thomas Weir, father-in-law of the first George Irving of Newton, for they reveal a surprising and rather interesting story. Dr Thomas Weir, great-grandfather of Janet Irving, and thereby three-times-great grandfather of James Clerk Maxwell, was from 1686 the possessor of the secret recipe for Anderson's Pills, a favourite patent remedy and 'cure-all' that endured both in the United Kingdom and overseas for over 200 years. Family Tree 6 will be helpful in following the connections involved in this story.

It has been mentioned before that Dr Thomas Weir married Bethia Blackwood, a daughter of Robert Blackwood of Pitreavie. They married sometime before 1692, for they had a child who died in that year (Paton, 1902, p. 681); sadly, two more died in 1697 followed in just a few weeks by Bethia herself. However, they did have at least one surviving child, a daughter by the name of Sarah, and it was she who in 1711 married George Irving of Newton. Although a connection has not yet been established, the author suspects that George's father-in-law, Dr Thomas Weir, was probably related in some way to George's uncle, John Weir of Newton, but whether or not that was the case is not really germane. Now, on Bethia's side, the Blackwoods of Pitreavie are not directly related to the Weirs of Blackwood, rather they are connected with the Blackwoods of Ayr (Paterson, 1847). There is a possible connection, however, between all three, for William Weir of Blackwood, who was a co-defendant of John Weir, was also known as Robert Blackwood when it suited him (see Chapter 11 and note 11 thereto).

Patrick Anderson's Legacy

We now take a step back in time; Corley (2004) informs us that Patrick Anderson (*c.* 1580–1660) was a Scottish doctor, author and manufacturer of patent medicine. By 1618 he was well-known in Edinburgh, and in 1625 he was appointed as a physician to Charles I, which may mean no more than he had the right to use the royal warrant in association with his profession. In 1635 he published a treatise *Grana Angelica* as a means of promoting his own patent digestive pills (Wootton, 1910), which were reputedly an excellent remedy for hangovers. Of 'walnut' size and made with aloes, colocynth and gamboge, he claimed to have registered the formula in the Rolls House at Edinburgh after bringing it back from Venice around 1630, but the record seems to be no longer extant.[1] Eventually known as 'the Scots Pills' or 'Anderson's pills', they were also apparently favoured by a subsequent monarch, Charles II.

There are several versions of what happened to the pills after Patrick Anderson, but the most detailed account is given in *Chronicles of Pharmacy* (Wootton, 1910, pp. 168–170). After Anderson died in about 1660, the formula and rights went to his daughter Katharine who continued to sell the pills in Edinburgh, while about the same time another Edinburgh doctor, Thomas Weir MD, took over Anderson's medical practice and in 1686 also bought from Katherine the formula and the wherewithal to manufacture and distribute the pills.

Chambers (1980, p. 27), however, states that Thomas Weir was probably the son of a second daughter of Anderson's by the name of Lilias and so, Katherine being his aunt, he came by the rights through that relationship. This would also explain why Weir had become involved in Anderson's practice in the first place, for Patrick Anderson would then have been his grandfather. Whatever the circumstances, Dr Weir obtained royal letters patent from King James VII for Anderson's Scotch pills in 1687 (Scotland, Privy Council, 1687) but, as a matter of asserting its authority over the monarchy, the English parliament soon thereafter revoked all royal warrants that had been granted without its specific approval. Dr Weir was therefore forced to seek such protection as he could for his monopoly from the new monarchs, William and Mary (Scotland, Privy Council, 1694) and from Edinburgh Town Council, which he obtained by 1694.

Unfortunately, the protection that Dr Thomas Weir managed to obtain was not wholly watertight against others who also claimed to have the rights to the recipe. One, Mrs Isabella Inglish, was said to have stolen the recipe when she had been a servant in Weir's household; not only was she selling the pills in London with apparent impunity, she was even advertising the fact in the *London Gazette* (Inglish, 1689):

DR THOMAS WEIR

Advertifements,

These are to give Notice, That Dr. Anderfon's or the Scotch Pills, (which is a convenient and effectual Medicine both by Sea and Land, have been much abufed fince the Death of Mrs. Catharine Anderfon) and are now faithfully prepared and fold only by Mrs. Ifabel Inglis, living at the Hand and Pen near the Royal Bagnio in Long-Acre, London.

Even though she was libelled (alleged) as a counterfeiter in Edinburgh in 1690, Mrs Inglish continued to trade the pills in London and was selling them at the Unicorn in the Strand, *c.* 1707–09. For long thereafter, her son James, and followed in turn by his son David, were continuing the trade from more or less the same location and still regularly advertising them in the *London Gazette* as being the 'true pills' right up to the start of the nineteenth century (Inglish, 1800).

Another 'pretender' to the pills was someone by the name of Thomas Steill, who claimed to have got the recipe from Anderson himself through a Mrs Hastie:

> In the action anent Anderson's Pills, betwixt Weir and Steill, Mrs Hastie, who gave the secrets thereof, and gave it to Steill, depones, she had it from Mr Anderson; and so both were allowed to sell the same. (Fountainhall & Scott, 1822, p. 237)

These pills, selling at the princely sum of one shilling per box in 1748, were therefore highly in demand; at about £6 in today's terms, they were the basis of a lucrative business.

Milne's Court

Thomas Weir conducted his Anderson's pills business from a house in Miln's (now spelled as Milne's) Court (Chambers, 1825, pp. 255–257; 1980, pp. 27–28), a tall tenement on the north side of Edinburgh's Lawnmarket facing the head of the Upper Bow (Plate 19a and b). The building still stands and dates from 1690, just about the time when Thomas got protection for his remedy from the Town Council. The entryway is just two doors down from the long-established Ensign Ewart pub, where Thomas' customers may well have desired to wash down the bitter pills they had just bought with a glass of ale.

From Thomas Weir, the recipe and rights went in 1711 to his widow (implying that Dr Weir had married a second time) and from her they went in 1715 to his son Alexander, followed in 1726 by his sister Lilias (namesake of Dr Anderson's daughter Lilias, who was probably her grandmother). But, of course, Dr Weir had another daughter, Sarah, who it was that had married the first George Irving of Newton, and so it turned out to be George and Sarah's youngest son, Dr Thomas Irving (c. 1725–c. 1804) who, in 1770, eventually inherited the recipe and rights from his aunt Lilias. It was Thomas' elder brother who became the second George Irving of Newton, and great-grandfather of James Clerk Maxwell.

In August 1747, Dr Thomas Irving was appointed as surgeon to the 14th Regiment Hamilton's Dragoons serving in Ireland. Until his return to Edinburgh in October 1774, he was based mainly at Lisburn, in the north (Hamilton, 1901, p. 18; Grant, 1944, p. 110 [see under Irving, Alexander]).

By 1774, not only were Dr Thomas Irving and his wife Jean Chancellor[2] doing business at the same place in Milne's Court, they lived at that very same address alongside Thomas' elder brothers George and his second wife, Mary Chancellor, and their family. At least they seemed to have had reasonable accommodation, for when the house was put up for sale nearly a century later it was described as comprising seven rooms plus a garret and a cellar.

The pills must have been very successful, for not only was there strong competition, the pills were also being sold much further afield. The *Cumberland Chronicle* (1777) advertised: 'Doctor Anderson's Pills (made by Thomas Irving, surgeon of Edinburgh) are sold in Whitehaven by J. Dunn, the printer of this paper.' If that is at all surprising, by the middle of the eighteenth century the pills were even considered to be one of the eight essential medicines for use within the American colonies (Griffenhagen & Young, 1959, pp. 155, 162).

Wooton (1910) has it that after Thomas Irving died, the secret recipe and rights passed 'to his widow, Mrs. Irving, 1797', who would continue to make the 'only genuine pills'. The Anderson's Pills advertisement in the *Edinburgh Advertiser* of 20 July 1798, however, informs us that he died some months earlier in that year. Until about 1805, moreover, the same advertisement was being placed to reaffirm Dr Irving's very recent death![3] There was some form of subterfuge; either he was alive and lying very low, e.g. incapacitated, or his widow was trying to keep his close association with the pills going in spite of his death.

However, it seems that she temporarily gave up the house cum shop at Milne's Court in 1804, towards the end of her husband's rather extended cycle of 'deaths'. Nevertheless, she remained in the pill business through numerous agents who sold the pills both in Edinburgh and beyond. During these years, she probably lived at her son James' house in Chessel's Court in the Canongate (Storer & Storer, 1820), which is towards the other end of the Royal Mile (Plate 20).

James Irving was then a colonel in the Royal Irish Guards and was mostly away from home until 1822, although he may have been back in Edinburgh often enough to take some part in the pill making business, for Wootton maintains that his mother assigned him the rights in 1814. She was then seventy-six and may have been concerned about the possibility of dying when her son was away from home and the secret then being lost or stolen.

Decline and Ruin

After her son returned to Chessel's Court, Mrs Dr Irving went back to Milne's Court and the following classified advertisement was placed in a Glasgow newspaper:

> GENUINE ANDERSON'S PILLS, CONTINUE to be prepared as formerly by Mrs. IRVING, Widow of the late Dr. IRVING, the sole proprietor of the Genuine Receipt, and may be had, Wholesale and Retail, at the original house, Milne's Court, head of the West Bow, Edinburgh, where they have been constantly sold for nearly 150 years [...] The public may also be supplied with this Medicine by [...] most of the respectable Druggists and Medicine Venders throughout Scotland. (*The Glasgow Herald*, 1832)

According to this, she was back at Milne's Court and still very much in business and continuing to supply the pills once more at the age of about ninety, which must have been exploited as a testament to their efficacy. Very similar advertisements had also been running in *The Scotsman* (1823, 1831, 1834) with an even longer list of druggists and agents selling the pills for her in Edinburgh. Judging by the scale of her advertising and network of agents, it would be quite unlikely that she was doing it all on her own. Nevertheless, bearing in mind what had happened during her husband Dr Irving's latter years, she probably wanted to make it look as though she was still the one in charge of it, all for the sake of the brand continuity and the 'secret recipe'.

What also points to this is that Mrs Irving and her son were not the ones actually selling the pills at Milne's Court. Chambers (1825, p. 256) describes the situation as it was about 1825:

> The Pills continue to be sold here [...] by Mr James Main, Bookseller who is agent for Mrs Irving [...] Portraits of Anderson and his daughter are preserved in this house: the physician in a Vandyke dress, with a book in his hand; the lady, a precise-looking dame, with a pill in her hand about the size of a walnut, saying a good deal for the stomachs of our ancestors.

Mr Main's Post Office directory entries for 1827 and 1828 concur with this. A possible explanation for his involvement is that by then Mrs Dr Irving's involvement in the pills may have been on a parallel with Colonel Sanders' current involvement with fried chicken. By 1829, however, Mr Main had moved to a shop at 52 George Street in the New Town and was no longer mentioned in connection with the pills. Curiously, the advertisements in *The Scotsman* reveal that a druggist by the name of Gardner had already been dealing with the pills from that same address, and had been doing so since 1823.

Mrs Irving also had the daughter named Bethinia who married her cousin, Alexander Irving of Newton, in 1814 (Chapter 12). It seems that the intention had been to call her Bethia, which was not an uncommon name at the time and had been the name of her great-grandmother, Bethia Blackwood the wife of Thomas Weir, but there was a slip of the tongue on the part of the minister who christened her, a thing apparently irrevocable (Irving, 1907, p. 7). Lord Newton's entry in *The Faculty of Advocates in Scotland* (Grant, 1944) indicates that Bethinia was born in Lisburn, just outside Belfast. We can be sure that Bethinia was indeed born in Ireland because that is also recorded against her entry in the 1841 census. From that same source we have her year of birth as being about 1780.[4] As previously mentioned, in 1747 her father Dr Thomas Irving was appointed surgeon to the 14th Dragoons stationed in the vicinity of Belfast, but returned to Scotland in 1774 and was back at Milne's Court by 1775. About ten years later there was a period of absence, for he was not listed there between 1784 (the next available directory year) and 1789. This implies Dr Irving was back in Ireland with his wife around 1780. Indeed, we find later that he claimed to have been physician to the County of Antrim Hospital (*Edinburgh Advertiser*, 1798), a civilian appointment that he may have taken up in the hope of settling down there. However, we find that from 1786 'Mrs Dr Irving' at least was back again in the Canongate, not far from Chessel's Court.

Mrs Irving seems to have lived out the rest of her years with her son Colonel James Irving and his family, first at Chessel's Court in the Canongate and then in the New Town at 8 West Maitland Street where she died in 1837. Our story, however, is not yet complete because an advertisement in *The Scotsman* of 22 September 1838 informed the public that her son, named only as 'Mr Irving', was now making the pills from the same, original, secret recipe, notwithstanding the fact that the basic recipe of all the competing brands had by then been openly published in a current pharmacopoeia (Rennie, 1837, p. 27):

> ANDERSON'S PILLS. Five grains each, made with ʒiv of aloes, ʒss of jalap, ʒij of scammony, xxx drops of oil of anise. Several purgative pills are called Anderson's, but are all similar to these.

One might have thought that as a senior military man James would not deign to dabble in the manufacturing and selling of the pills himself, rather he would have employed someone else to do it for him. Instead, he carried on the family tradition of making and selling them just as before, keeping Milne's Court as the focus of the business and remembering to advertise that they were still 'just as his mother used to make them', and at the same time thinking it better to refer to himself as plain Mr Irving rather than by his military title of Colonel.[5] Now considering Mrs Irving's great age, one would not expect her son to outlive her by long, and given that his final entry in the Edinburgh directory was 1841, we could guess at this being around the year of his death, but this was not so.

The story as revealed almost twenty years later is entirely different (*The Scotsman*, 1860). 'Mr Irving' was making £400 a year from the pills when he took over making them, a substantial income at the time and more than three times his full-pay army pension, but he soon got into debt. The business was by then in decline; there were many competitors and his only remaining advantage was the historic family brand. Furthermore, he had let things slip; the way of selling had not changed. To add to his woes, there had been family troubles even before his mother had died. He complained that his cousin cum brother-in-law Alexander Irving, now the judge Lord Newton, had practically removed him from his house at Chessel's Court, probably in the latter half of 1831, and put him and his elderly mother into a rented flat at West Maitland Street in the New Town. It may well have been a problem concerning a debt, or a disputed family inheritance, which Lord Newton wanted to sort out while he was still able to do so.

Debt indeed seems to have been the likely problem, for about the time of his mother's death, and certainly before 1840, James Irving had to set up a trust administered for the benefit of his creditors by an accountant. He blamed Lord Newton once again for his woes by saying that they all started when he had to leave Chessel's Court, and in particular Lord Newton had promised to pay his rent in return for getting him out, but then had gone and died. His new rent was £130 a year all told, quite a considerable sum given his net army pension was £120, and so he was dependent on the pill business to cover the difference plus all his other expenses. By the year his mother died, he had found it expedient to quit West Maitland Street and by 1839 he had moved far to the opposite side of the town to Wheatfield, a farm house just beyond the cavalry barracks at Piershill (Wallace, 1987, p. 111). It would not have been very convenient for the pill business at Milne's Court but rather more convenient for the Abbey debtors' sanctuary (Figure 13.1), where it would have been be possible to take cover from his creditors when the need arose. As a retired senior officer fallen on hard times, perhaps he was able to find some sort of employment at the barracks. In the end, however, he remained listed at Wheatfield for only three years, after which any further trace of his whereabouts evaporates.

From then on, things only worsened; the pill business was still going downhill and so, ironically, the now General Irving felt the need to change his agent, which he did about 1850. He appointed a Mr Cotton (possibly George) who had a shop at number 231 on the High Street, which had previously been owned by the Edinburgh tobacco magnates James and John Gillespie of Spylaw.[6] It is now marked by a commemorative plaque bearing James' image in bas-relief (Edinburgh Museums, 2013).

Mr Cotton does not seem to have taken the pill business as seriously as General Irving would have liked, for the General claimed to have got little money back from him over the next ten years. In fact, he suggested that Mr Cotton was cheating him by selling instead his own counterfeit version of the medicine. Whatever the truth of the matter, the lack of any substantial income eventually left him with debts amounting to nearly £3,300, an enormous sum that could have purchased a very substantial house. His creditors therefore took action, leaving Irving in very straitened circumstances.

The secret recipe, however, was still being held under lock and key by Mr Gibson, accountant to his creditors' trust. Strangely, Gibson did not offer it up to the creditors, perhaps because he knew it to be worthless, being no different from what had already been published in the pharmacopoeia and therefore better off being kept 'secret'. Despite Irving's difficulties, there is no evidence to suggest that the pills themselves had waned in popularity, the real problem was that the number of sources of them was increasing, all claiming

Figure 13.1 (opposite) Holyrood Abbey Sanctuary, Piershill Barracks, and St Margaret's Church, Restalrig
Wheatfield, not shown, is a little to the east of Piershill Barracks.

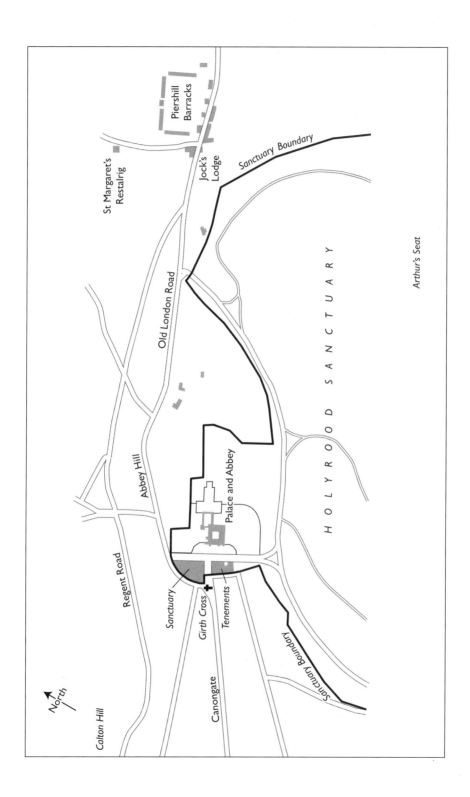

to be the genuine article. In the end, realising that Irving would never clear his debts, his creditors sued for bankruptcy.

During the bankruptcy hearing, General Irving again maintained that he could set things right by getting back the wherewithal to merchandise the pills and starting anew, but by now he was severely deaf and nearly blind, and so the court did not press him further. In 1863, however, things came to a resolution with both the rights to manufacture the pills and the house at Milne's Court being put up for auction. Meanwhile, a few days before the auction, someone placed a notice in *The Scotsman* claiming that the recipe for the true Anderson's pills was 'worth so much waste paper' (*The Scotsman*, 1863).

Whatever was the outcome of the auction, General Irving survived only to the following February, dying 'from old age and reduced circumstances'. His death was reported by his son George, who lived in Portobello. Clearly he had not done much to help his father in his difficulties, nor had his sister Bethinia, Lord Newton's widow, who had lived until 1855.

Survival

Returning to the question of others having got into the Anderson's pills market for themselves, Raimes, Blanchards & Co., predecessor of the Edinburgh pharmaceutical company Raimes and Clark, were by 1842 manufacturing and selling copies of the pills, supposedly under licence (but from whom?). However, they eventually bought out the rights, such as they were, in 1876 in order to reduce costs in the face of the competition (Cumming & Vereker, c. 2008). Wootton (1910) confirms this date, and informs us that it was then a Mr J. Rodger who sold them, and so perhaps it was he who had picked them up about twelve years previously, when they were auctioned.

Raimes and Clark were still making the pills in small quantities well into the First World War period, and one source mentions that they could be had in some places as late as 1956 (Ellis, 1956). Contrary to Chambers' interpretation of Anderson's portrait, however, the information held by Lindsay & Gilmour, the present owners of the archived materials, shows that the pills are black and more the size of a pea than a walnut, and so in his description Chambers appears to have confused the pill-box for the pill itself! Also in the archive are some of the pills, their authentic packaging, trademarks and early sales literature, examples of which are shown in Plate 21 (Cumming & Vereker, c. 2008).

CHAPTER 14

The Cays and Hodshons

We now come to James Clerk Maxwell's mother's side of the family, the Cays. Frances Cay's parents were Robert Hodshon Cay and Elizabeth Liddell. Robert took his first name from his grandfather, as was the custom, but his middle name came from his mother's family, the Hodshons of Lintz. The origins of the Cays and the Hodshons were not Scottish, for Robert's parents were from the north-east of England. We will begin with their rather interesting background and how they came to settle north of the border.

The Cays of Charlton Hall

Although this family were known as the Cays of Charlton Hall (Burke, 1834–38, vol. 1, pp. 384–385), the association with North Charlton in Northumberland came about only around the end of the seventeenth century. In their earlier days, when the spelling of the name was Kay or Key, they nevertheless seem to have been living in the same general area as North Charlton, in fact near Alnwick, an ancient market town lying just to its south and within thirty miles of the Scottish border at Berwick-upon-Tweed (Figure 14.1). They therefore lived on the English side of the Eastern March of the turbulent border region, which until the seventeenth century was effectively under the separate rule of the March Wardens and beyond the direct rule of the separate monarchs of Scotland and England, as discussed in Chapter 6. As with many of the families that lived in these parts, the Keys were drawn into the frequent ructions that took place. Indeed, they settled their scores, as others did, in the time-honoured manner, and so at some early date a duel took place at Alnwick between one of the heads of the local Key family and some adversary now unknown. Key slew his opponent and made off south to Newcastle-upon-Tyne to escape the reach of the March law, according to which he had committed an act of treason. His estate and possessions were subsequently forfeit and so he had to remain in straitened circumstances, sequestered in

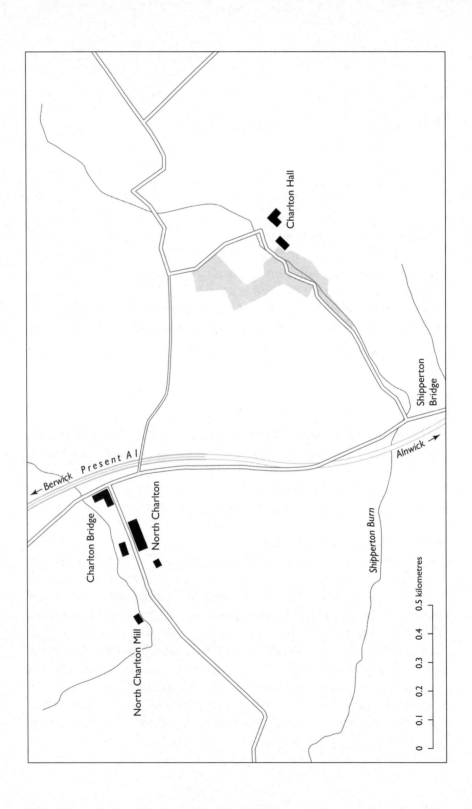

Newcastle for the rest of his days (Burke, 1834–38, vol. 1, pp. 384–385).

Key's family stayed on in Newcastle and in subsequent generations they managed to build up a brewing business of significance, which was then in the hands of Thomas Cay, who was a baker and brewer in Newcastle from 1594, the year he was apprenticed, until January 1623, when he died (Bateson, 1895, pp. 297–300; Cay et al., c. 1947). Following Family Tree 7, it may be seen that Thomas' business passed to his oldest surviving son, the first recorded John Cay, who was born in 1605. When John died, the business was then carried on by his widow, Isabel Wilkinson, until it eventually passed down to their second son, Robert Cay (1634–1682). When this took place is not known, but by the time Robert Cay died, he owned lands in Tynemouth, was a free baker and brewer of Newcastle, and was a prominent non-conformist, most likely a Puritan. His widow, Barbara Carr, whom he had married in 1665, expanded the business and was able greatly to improve the family fortunes. Thereby this lady was able to acquire much property in and around Newcastle and, significantly, to educate her four sons, Jabez, John, Robert and Jonathan, and to give them houses. The upward trend was still continuing by June 1682 when her eldest son, Jabez, was at Edinburgh University. Again according to Burke, it was her third son, Robert, who eventually brought the business to ruin and in the aftermath took off to Londonderry to escape the consequences. All of this appears to be consistent with the Cay Family Tree (Cay et al., c. 1947), which notes that it was this Robert who was in partnership with his mother.

This catastrophe with the family business seems to have occurred after 1704, for Barbara's son Robert was admitted as a free hoastman[1] of Newcastle in that year and therefore had not yet taken flight. Reduced in circumstances[2] but undeterred, the brave widow Barbara started all over again, and once again met with success. She carried on until 1723, in all probability with the help of her grandson Robert Cay, John's son, who went into the brewing and baking business and was the one who eventually carried it on. But before moving on to his part in the Cay family history, we must go back to the stories of his father John and his three uncles, Jabez, Robert and Jonathan.

Figure 14.1 (opposite) Charlton Hall and Shipperton
North Charlton lies just off the present A1 road about twenty-five miles south of Berwick-upon-Tweed, while Charlton Hall is just over a half mile to its south-east, on the other side of the A1. Note the local placename Shipperton, corresponding to Shepperton, after which Robert Dundas Cay named his house on the outskirts of Edinburgh.

Jabez Cay was sent to continue his education at the university in Edinburgh where, in November 1682 (by which time his father was already dead) he was unlucky enough to be expelled for taking part in a student romp that involved burning an effigy of the Pope. While it was the sort of traditional rag that the majority of Edinburgh citizens would have wholeheartedly approved of, this took place under the chancellorship of the Roman Catholic Duke of York, later James VII, who had been appointed as Lord Chancellor in Scotland (Chapter 11) by his brother the king, Charles II.[3] Jabez and his fellow rompers were duly brought before the Privy Council and, with all due severity, banished from Britain. Faced with banishment, Jabez was at least lucky enough to be able to count on his mother's financial support, and so it was that he took himself to Europe not in poverty, but able to go off to Padua and continue his studies there and eventually qualify as a physician.

Some years later, after James VII's abandonment of the throne, Jabez was able to return to Newcastle where he married Dorothy Gilpin (1668–1708). She was the sister of William Gilpin of Scaleby Castle (1657–1724), a cleric and antiquarian, who in 1718 became the recorder of Carlisle.[4] Now, here we have a connection to the Clerk family, for William Gilpin was a fellow antiquarian of the Baron, Sir John Clerk 2nd Baronet of Penicuik, who corresponded with him and visited him shortly before he died (Chapter 4; BJC, pp. 116–119). Jabez Cay was likewise an antiquarian, and at one time possessed the ancient Roman stone from Benwell to which a curious story is attached.[5] William and Dorothy's younger brother, John Gilpin (1670–1732), returned the compliment by marrying Jabez' sister Hannah Cay (b. 1675) in 1699; he later became a wealthy and respected merchant in the Virginia tobacco trade at Whitehaven in Cumberland (Symson, 2002). Through the Gilpin family, Mary Dacre, wife of Sir John Clerk 5th Baronet of Penicuik (Chapter 9) was also related to the Cays, for her grandmother was Susannah Gilpin, Dorothy Gilpin's niece (Family Tree 5).

Jabez Cay, like his mother, prospered at Newcastle and he was able to purchase a moiety (half) of the estate of North Charlton in August 1695, which he managed to accomplish with the assistance of Jonathan Hutchinson MP (*c*. 1662–1711), each simultaneously buying one half of the property (Hayton et al., 2002b; Burke, 1834–38, vol. 1, p. 384). It was in this way that North Charlton came to be long associated with our Cay family.

As well as being a keen antiquarian, Jabez was also an avid mineralogist, as revealed by his letters over the period 1696–1702 in which he wrote to the London physician and naturalist Martin Lister (1639–1712).[6] However, it was through his interest in antiquities that he became the correspondent and friend of Ralph Thoresby (1658–1725), a merchant in Leeds who later in life spent most of his time on antiquarian studies and kept up a prodigious correspondence with many of the learned men of his time. He also kept a voluminous diary, which, along with his correspondence, has been published (Thoresby, 2009; Colburn & Bentley, 1832). Amongst the diary pages we find:

> [15 April 1695] [. . .] had Dr. [Jabez] Cay's, of Newcastle, company, viewing collections, &c. with several other friends at dinner; with whom spent most of the afternoon amongst the coins.
> [6 Jan. 1702] [. . .] I heard also that my kind friend, Dr. Cay, of Newcastle, is very weak, if alive.
> [8 February 1703] Visited cousin Whitaker, who told me of the death of my kind friend and benefactor to my collection of natural curiosities, Dr. Cay, of Newcastle; sense and seriousness filled his last hours, as Mr. Bradbury's expression was. He died 22nd January, Lord sanctify all mementos of mortality!
> [21 May 1703] [. . .] we had the convenience of seeing the Hell-kettles,[7] the best account of which, is in my late kind friend Dr. Jabez Cay's letter, inserted by Dr. Gibson in the new edition of the Britannia, p. 782.

On Jabez Cay's death in January 1703, his brother and successor John Cay (1668–c. 1730) continued to correspond with Thoresby (as mentioned in the latter's diary). However, not only did John succeed to Jabez' moiety of North Charlton, he was subsequently able to take over the whole estate by buying out Hutchinson, which certainly shows that he had significant wealth. In fact, he was able to make a match to a wealthy heiress, Grace Woolfe of Bridlington, Yorkshire. He was perhaps able to live as a gentleman, for there is no note of him actually being involved in the family brewing business, which part seems to have been taken over by his brother Jonathan.

When Grace Woolfe's father died *c.* 1709, he remembered John, his daughter and their family in his will:

> I, Henry Woolfe of Lay Yett, near South Shields [. . .]
> To my Son in Law John Cay & Grace his wife my daughter [. . .]
> To my Grandson Robert Cay [. . .] Robert Cay, Messuage [i.e., a dwelling house] & five salt pans held from Dean and Chapter [. . .] twentieth part of Elswick Colliery [. . .] Farm in Harton lately bought of Thomas Watson.
> Dated April 25th, 1709 [. . .] Executors, John Cay & Grace his wife.
> (Phillips, 1894, p. 210)

His grandfather's bequest made good provision for Robert (1694–1754), who was then but a boy, for he duly became a merchant and salt manufacturer at Laygate (given as Lay Yett in the will) in South Shields, which is on the south bank of the River Tyne and closer to the sea than Newcastle itself. The will also informs us that he was left a share in a coal mine at Elswick, which neighbours Benwell, the origin of the Benwell stone given to his uncle Jabez by Mrs Shafto.

According to family tradition, Robert philanthropically published a letter that appeared in the *Newcastle Courant* of 5 January 1751 (Enever, 2001; Thomson, 2005) suggesting the institution of an infirmary at Newcastle. However, the only direct evidence for him having been the writer is that it is signed with the initials B. K., and there is an opposing story that it was in fact written by a Newcastle surgeon, Richard Lambert, Two other sources, however, say the writer's initials were 'K. B.' (Mackenzie, 1827; Baillie, 1801, p. 322). Now, the true identity of the philanthropist who suggested the idea was probably unknown even then, for the ward named to commemorate the instigation of the hospital was simply designated as the 'B. K. Ward'. Robert Cay did have a plausible claim to the initials B. K. inasmuch as he may have been known as Bob Key.

Like his uncle Jabez, Robert was a considerable antiquarian. He corresponded with John Horsley, and not only contributed to his magnum opus *Britannia Romana*, which was a work well known to the Baron (BJC, p. 258), but at Horsley's behest he filled in many of the gaps in the text and relieved him of the burden of revising his manuscript and preparing it for printing. According to Chambers (1840, vol. 3), Cay was 'an eminent printer and publisher at Newcastle'.[8] Certainly, Horsley wrote to Cay requesting advertisements he wanted placed in the *Newcastle Courant* in terms that suggest that Cay may have been in some way involved in the publication of that very newspaper (Hodgson, 1832, p. 444n). It also appears that in 1740 there was a William Cay who described himself as a bookseller in Newcastle and published an edition of the works of Allan Ramsay and some other 'bards' under the title *The Caledonian Miscellany* (Ramsay et al., 1740).

If Robert Cay was indeed in the publishing business then he may well have been related to this William Cay. Burke, however, does not mention a brother William, nor does William feature on the Cay Family Tree anywhere even near this time. While neither of these sources can be considered as fully complete and accurate, the notion of Robert Cay actually being a printer and publisher, as well as antiquarian, brewer and salt manufacturer, could simply have been an error on Chambers' part. The fact that Chambers and his brother were publishers themselves, and would have known something about who was who in the business of publishing and bookselling, makes the question all the more intriguing.

Even if Robert Cay was not involved in actually publishing *Britannia Romana*, he was still deeply involved in its detailed preparation, effectively as both its general editor and its copy editor. He likewise contributed to Horsley's great map of Northumberland, and even gets a mention in the original citation:

> A Map of Northumberland, begun by the late Mr John Horsley, F. R. S., continued by the Surveyor he employed [George Mark], and dedicated to the Right Honourable Hugh, Earl of Northumberland. By R. Cay, A. Bell, Sculpt. Edinburgh, 1753. (Hodgson, 1832, pp. 445–457)

In 1726, Robert Cay had married Elizabeth Hall, daughter of Reynold Hall of Catcleugh, which lies just four miles south-east of the Scottish border at Carter Bar. In 1754, he was succeeded by his eldest son, John, who was born on 16 April 1727. Part of John's education may have taken place at Edinburgh University, where his uncle, John, and great-uncle, Jabez, had studied law and medicine respectively. An Edinburgh connection would certainly explain a sudden move to Edinburgh later on. Having taken up the law, by 1756 he was a barrister at Middle Temple, London (Hodgson, 1832, p. 220), which is about the time of his marriage to Frances Hodshon, daughter of Ralph Hodshon of Lintz in county Durham (Cay et al., *c.* 1947, pp. 10–11), of whose family we will find out more in the following section.

John and Frances' first son, Robert Hodshon Cay, was born in 1758, as far as we know, at North Charlton (Brown, 1910, p. 298). While John had been successful to the extent that his position in society was recognised by his appointment as a justice of the peace and then a deputy lieutenant[9] for the County of Northumberland, he entered into a lawsuit with the Duke of Northumberland that was his undoing. The lawsuit concerned the mining rights under his property and there would have been a reasonable chance of the rights being worth something since there were already many lucrative coal mines in the Newcastle area.[10] The Duke, at least, thought they might be worth taking for himself, for it was often the case that mining rights were reserved by the feudal lord. The upshot was that John Cay had to go to court to protect what was rightly his, and by the judgment given in August 1761 he did in the end succeed. The cost of the litigation, however, had put him in debt, for he had subsequently to flee north to Scotland and take refuge at the Abbey sanctuary at Holyrood (Figure 13.1; Brown, 1910, p. 241). As we shall later discuss more fully, his father's brother and own namesake[11] was a steward of the Marshalsea, the infamous debtors' prison in London, from 1750 to 1757. John Cay would have known only too well what his fate would have been had he ended up at a place like that, for a large proportion of its inmates simply did not survive the appalling conditions.[12] In addition, he would have been in the jurisdiction of the English courts, and would no doubt have been compelled to sell off North Charlton to pay his creditors. We may guess he had some prior knowledge of the Edinburgh sanctuary, and the possibility that it afforded him both to stay out of prison and the clutches of his creditors. The rules of the sanctuary also offered him the chance to earn a living, and so that is where he decided to go.

At the time, the sanctuary at Holyrood Abbey was an area that extended some way around the edge of the royal park in Edinburgh that is now called the Queen's Park. Some of the old sanctuary buildings to the north-west of the Palace still exist, and the sanctuary boundary is now marked out by brass letters 'S' embedded in the causey stone setts of the Abbey Strand. In ancient times it was a holy sanctuary available to all sorts of offenders but, by the latter half of the eighteenth century, it applied only to debtors. One of the more noteworthy recipients of its protection was James Tytler, editor and writer-in-chief of the second and third editions of *Encyclopaedia Britannica* (Henderson, 1899), whose stay there may well have been contemporaneous with John Cay's.

Through his self-enforced exile in Edinburgh, Cay managed to hold on to his lands at North Charlton, but they were certainly mortgaged. His financial problems are referred to in his correspondence with William Ord (*c.* 1715–1768), the High Sheriff of Northumberland and wealthy landowner; to whom he was in debt.[13]

John Cay died in 1782. There is no record of any Cay in the Edinburgh directories during his lifetime and so either John Cay remained in the sanctuary for the rest of his life or he did not seek to publish an address that would give his creditors the opportunity of catching up with him. Frances survived her husband John by some twenty-two years. By the time of her death in July 1804 at Fisherrow, a coastal village just to the east of Edinburgh, she had many grandchildren. Being a Roman Catholic, she could not be buried at any of Edinburgh's parish churches and so she was interred

Figure 14.2 (opposite) Lintz and Lintz Hall
The River Derwent flows north-east from Lintzford to reach the River Tyne some six miles downstream, and from there the Tyne Bridge in Newcastle is only a further four miles downstream.

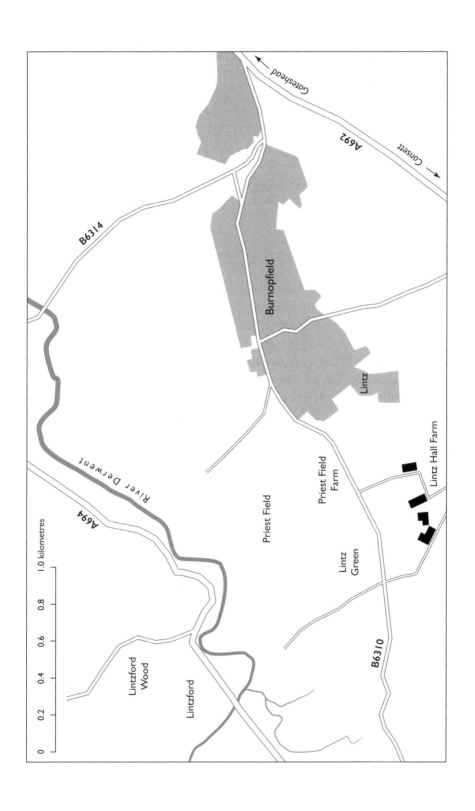

alongside her husband at St Margaret's Church in Restalrig, shown in Figure 13.1 just to the north-east of the Abbey sanctuary. The gravestones in the family plot there are still clearly legible today.

Given the pressures involved in having to live in a debtors' sanctuary and the problems that a Protestant–Catholic marriage would bring on them, John and Frances Cay would have had no easy time of it. We must therefore look into the origins of the Hodshons of Lintz to investigate how a Cay, being from a puritanical Protestant background, came to marry a Hodshon, from a line of highly devout Roman Catholics.

The Hodshons of Lintz

At first sight of the name, one could be forgiven for thinking that Lintz was a European place, and indeed Linz, pronounced in the same way, is the principal city of Upper Austria. But although Hodshon also seems a bit unusual, it does not look a quintessentially Germanic name; it is in fact frequently given as the more familiar Hodgson.

But think no more of Europe, because Lintz is a small village in county Durham that is only ten miles south and west of central Newcastle. The name Lintz was previously written variously as Linz, Lynths and Lynce, after the name of the early landowners there, of whom some snippets survive: 'given by William de Linz to Adam, fil. William de Linz [...] In 1350, Richard de Lynce held the vill of Lynce by homage [...]' (Surtees, 1820).

By the second half of the sixteenth century, Lintz was the property of Nicholas Hedley, who married Elizabeth Hodshon. If there is anything at all European about the Hodshons, it is that they staunchly held to their Roman Catholic faith throughout the Reformation, the Civil War, and the reigns of the houses of Orange and Hanover. They were a large family (Family Tree 8), with many branches that had long been settled south of the River Tyne, generally around Gateshead. In 1549, we find that a Richard Hodgson was Sheriff of Newcastle, and between 1555 and 1580 he had been the Lord Mayor on no fewer than three occasions. He died in 1585 leaving four sons and a daughter Elizabeth, who it was that married Nicholas Hedley of Lintz, which came thereby to be associated with eight generations of Hodshons.

As a result of Nicholas Hedley and Elizabeth Hodshon having no children, Nicholas took the step of entailing his property, with the first line of succession going to Elizabeth's brother William, followed by the third brother Richard (Surtees, 1820). In the event, it was Richard who succeeded, and from this gentleman, Richard Hodshon of Lintz, it passed down through three generations to Richard Hodshon of Lintz Hall in 1669. From there, Lintz passed in turn to his son, grandson and then great-grandson, each one named Ralph Hodshon of Lintz.

The first Ralph married Mary Killingbeck and died before 1715 leaving his widow Mary, his son and heir Ralph, and three daughters, Mary (d. 1760), Elizabeth (1704–1783) and Catherine (c. 1706–1771) who all became nuns.[14] After Ralph's death, Mary Killingbeck remarried to a man called Hay who was a benefactor of the Bar Convent in York. While there may have been an element of wishing to provide for the young women by placing them in convents, it is clear that the family were not only devout Roman Catholics, they were overtly supporting their Church in times of religious persecution.

The second Ralph Hodshon of Lintz married and had at least three children: his successor, the third Ralph Hodshon of Lintz (c. 1730–1773), had two daughters, Mary Agnes (d. 1782), who joined her aunt Catherine as a nun at Gravelines, and Frances (1730–1804), who took an entirely different course and married John Cay of North Charlton.[15]

Having placed three of Frances' aunts and one of her sisters in holy orders, a family as devout as the Hodshons of Lintz would naturally have wanted Frances to marry a Roman Catholic. How she came to know and marry John Cay can only be guessed at. The Cays and the Hodshons both had interests in coal mining, and Benwell, the origin of Jabez Cay's Roman monument, happens to be the one place in the vicinity where it is possible that they both may have had business.[16] Unfortunately, we have no more 'evidence' than that to offer as to why John Cay, Protestant, married Frances Hodshon, Roman Catholic; nevertheless, it would be reasonable to speculate that the marriage was the result of an elopement. Rather than become either a nun or a housekeeper to her parents, Frances may have preferred to run off with John Cay despite the long shadow it might cast over their future.[17]

John's parents may well have wanted him to make a marriage that would have helped to secure the family future, for example, by marring his cousin Grace (1722–?), eldest daughter of the aforementioned John Cay, Judge of the Marshalsea. But by his marrying Frances Hodshon, they would have been disappointed in their wish. Though scant, there is however some evidence to suggest that the true story is not as we would have expected, and that by marrying a Roman Catholic, John may have actually stood a chance of *improving* his circumstances. We shall look into this possibility at a later time, but on the face of it John and Frances had placed themselves in a difficult situation.

The Edinburgh Cays

By 1784, two years after the death of his father, Robert Hodshon Cay (1758–1810) was living in Buccleuch Street, just to the east of Edinburgh's George Square. Despite his financial embarrassment, John Cay had been able to send his son to Glasgow University in 1778 to study law. By 1780 Robert had qualified as an advocate (Grant, 1944) and in 1788 he was appointed one of the four commissaries in Edinburgh. His parents had wished him to marry his cousin Catherine Hodshon who, like his own mother, was a Roman Catholic. This fact appears to compound the mystery surrounding why John and Frances would have chosen to marry across the religious divide, but in fact it explains it. In his family book, his descendant Albert Cay made a marginal note:

> R. H. Cay above mentioned was heir presumptive of the estate of Lintz until Miss Catherine Hodshon was born. The family arrangement was that he was to marry her (his first cousin). She was a papist. (Cay et al., c. 1947, p. 15)

Catherine Hodshon (1761–1826) was the daughter of Ralph Hodshon of Lintz, Mrs Frances Cay's brother. In this we have a clue to the likely answer of our conundrum: in this, the next generation, the parents *wanted* their Roman Catholic daughter to marry a Protestant. It was clearly a scheme to keep Lintz in the family at all costs by getting round the draconian anti-Catholic laws then in force.[18] We may guess that it had been the same sort of motivation that led to John Cay marrying Frances Hodshon, for it seems that Frances must have been the presumptive heir to Lintz at the time of their marriage. In turn, their son Robert Hodshon Cay, born in 1758, was next heir presumptive until 1761, when Catherine eventually entered on the scene. The family then seems to have decided that the best remedy for this diversion of their tactics was to marry off Catherine to her cousin Robert, whereupon the original plan would be back on track. But it was not to be, for in the end the match was unmade and they went their separate ways.

As asserted in note 19, Catherine Hodshon must have married about 1780, while Robert Hodshon Cay married much later, on 26 October 1789, Elizabeth Liddell, daughter of John Liddell of Dockwray Square in North Shields. Elizabeth was a talented amateur pastellist (Jeffares, 2014), and it was she who did the portraits of herself (Colour Plate 12), her sons John and Robert Dundas, and her daughters Frances and Jane, that are now on display at 14 India Street. On their marriage, Robert and Elizabeth lived at 1 South George

St (renumbered as 2 George Street in 1811). Their house is described as being:

> the fourth and fifth storeys of the tenement [...] consisting of a dining room, drawing room, kitchen, two bed chambers, and a bed closet on the first floor, five bed rooms and three light closets [...] two ceiled garrets, and one unceiled above, with four cellars and other conveniences. (Harris, 2011)

It was a north-facing gable-end building where St Andrew Square meets George Street, the entrance to which faced towards where the statue of James Clerk Maxwell now stands (Youngson, 1966, p. 208 [plan 28]). Unfortunately, like many of the First New Town's original buildings, it was torn down to make way for a twentieth-century redevelopment, in this case a new building for the Caledonian Insurance Co., later known as the Guardian Exchange. A very fine building the new one may be, but any idea of what the Cays' house was like can only be gauged from the one or two remaining old gable-end buildings in the area.

It would have been here that all but the youngest of Robert and Elizabeth's children would have been born. On the last day of 1800, Robert was appointed Judge Admiral of the High Court of the Admiralty in Scotland (Colour Plate 13), a significant promotion with an annual salary of £400, four times his salary as a commissary (Cay et al., *c.* 1947, p. 15). The Cays, although not gentility like the Clerks, were apparently a well-to-do family that could now afford to move to a grander abode in the Second New Town, the development of which had more or less just begun. And so it was, for in the Post Office Directory of 1804 we find them at 'Heriot's Row', thereafter mentioned as 11 East Heriot Row. After renumbering in 1811, 'East' was dropped but the house number stayed the same. The house is very similar to 31 Heriot Row (shown in Colour Plate 4), but lacks the top floor, which was added in later times. Instead, it has a dormer window, which is how Jemima Blackburn recalled number 31 being when she and her brother Andrew Wedderburn used to smoke 'out on the leads' there of a summer's evening (Fairley, 1988)! The location of the house can be found in Figure 1.2.

Robert and Frances' children were:

John (1790–1865), friend of John Clerk Maxwell since their schooldays. He was Sheriff of Linlithgow and district and, like his friend John, a fellow of the RSE;
Frances (1792–1839), who married John Clerk Maxwell and was, of course, the mother of James Clerk Maxwell;

Robert (1/12/1793–1/9/1804), a midshipman on HMS *Atlas*,[19] who died three months short of his eleventh birthday when his ship was off Demerara (Guyana);

Albert (1795–1869), originally a merchant who dealt in wine, tea and stocks and shares, who later became a news and advertising agent;

Jane (1797–1876), James Clerk Maxwell's maiden aunt, who became a mother figure to him after her sister Frances' death;

Robert Dundas (1807–1888), a Writer to the Signet and Registrar to the Supreme Court of Hong Kong; he was named in memory of the brother who died in 1804.

Two others, George and Elizabeth, died in infancy.

Elizabeth Liddell and Robert Hodshon Cay were therefore James Clerk Maxwell's maternal grandparents. From 1809 onwards, John Clerk Maxwell and John Cay both lived in Heriot Row and, being of a similar age and ideas, became close friends. It is not hard to see how this would lead to John Clerk Maxwell and Frances Cay becoming acquainted. The real question is, what took a relationship between them so long to develop and blossom? Jane Cay never married; instead she took a particular interest in her nephew James, especially after her sister's death, and they remained close throughout his lifetime. According to C&G (p. 14), Jane inherited her mother's artistic talent, to which her two watercolours on display at 14 India Street amply testify. One is of her and the other of young James, each taking their tea in her sitting room at 6 Great Stuart Street.

While the Cays lived in St Andrew's Kirk Parish, they did not worship there, for unlike the Clerks, including John Clerk Maxwell, they were Episcopalians. They were associated with the Charlotte Chapel in Rose Street and thereafter its successor, St John's Church in Lothian Road, seen in Plate 22 (Historic Scotland, 1970b; Harris, 2011).

After Robert Hodshon Cay's death in 1810, his widow Elizabeth stayed on in Heriot Row until her death in 1831, when her grandson James Clerk Maxwell was just a month old. Although the family were ostensibly comfortably-off, on his death at the age of fifty-two, Robert's estate was valued at only £900 (Harris, 2011). His father had never managed to clear the mortgage on North Charlton, and any hopes that Robert would be able to do so also went unrealised. The problem passed on to his eldest son to wrestle with, and it may have been a factor in why, of the two daughters, Francis married late and Jane never married.

John Cay, Sheriff of Linlithgow

The eldest son, John Cay FRSE (1790–1865), followed his father into the legal profession. He qualified as an advocate in 1812 (Grant, 1944), and from 1822 he was the Sheriff for Linlithgow and the surrounding area, some distance to the west of Edinburgh. John married Emily Bullock in May 1819 (see Chapter 9), and from 1822 to 1831, they lived at 5 SE Circus Place, a very short distance from Heriot Row.

Following the death of his widowed mother in 1831, John Cay took possession of the family home at 11 Heriot Row and, presumably, rented out 5 SE Circus Place, for he returned to that address in 1847 and a variety of names appeared there over the intervening years. John was one of the first members of the vestry of St John's Episcopalian Church in Lothian Road, which is very likely where he and Emily wed, for Bishop Sandford conducted the wedding and it was his charge.[20] Although Emily died at the young age of thirty-six, she and John had several children (Cay et al., *c.* 1947; Harris, 2011; Henderson & Grierson, 1914):

> John (1820–1892), eldest son and heir, became solicitor to the General Post Office in Edinburgh;
> Robert (1822–1888) went to Australia to become a sheep farmer, and died at Brisbane;
> William (1823–1840) died of influenza;
> Edward (1825–1870) went to Australia with Robert to become a sheep farmer. Died at Melbourne;
> Emily (1826–1904) married R. Robertson of Auchleeks, Perthshire;
> Elizabeth (1828–?) married George Alexander Mackenzie of the Applecross family;
> Lucy (1829–1883?) married the Hon. Sir Montagu Stopford;
> Frances (1831–1832) was born just two months after James Clerk Maxwell and died aged one;
> Thomas (1834–1868) died of cholera at Rosario de Santa Fe, in central Argentina;
> Francis Albert (1836–?) was the last.

Lewis Campbell described John Cay thus:

> though not specially educated in mathematics, was extremely skilful in arithmetic and fond of calculation as a voluntary pursuit. He was a great favourite in society, and full of general information. (C&G, p. 14)

John Cay was just two months older than John Clerk Maxwell, and they very probably attended the Royal High School together. They trained as advocates together, and from 1809 to 1818 they lived just a few doors away from each other in Heriot Row; their seemingly lockstep progress continued when in 1821 they were both elected fellows of the RSE within a month of each other. They were great friends and had many common interests, particularly in the progress of science and technology. John's brother, Robert Dundas Cay, recalled that in 1821 or 1822 they were both: 'engaged [. . .] in a series of attempts to make a bellows that should have a continuous even blast' (C&G, pp. 7–8). Campbell goes on to say: '[John Clerk Maxwell's] acme of festivity was to go with his friend John Cay (the "partner in his revels"), to a meeting of the Edinburgh Royal Society.' He later quotes from John Clerk Maxwell's diary entry for Monday 2 March 1846: 'Return [from the law courts] with John Cay, called at Bryson's and suggested to Alexander Bryson my plan for pure iron by electro-precipitation from sulphate or other salt' (C&G, p. 74n1); while, in a letter to his son in February 1854, John Clerk Maxwell mentioned: 'I am going to dine with John Cay, and with him proceed to the Royal Society. I may perhaps catch Prof. Gregory about the microscopist' (C&G, p. 207).

Campbell also gives several illustrations of John Cay's interest in science, and of helping to encourage his nephew James Clerk Maxwell in these interests. For example, in October of 1844 James wrote home to his father: 'I was at Uncle John [Cay]'s, and he showed me his new electrotype, with which he made a copper impression of the beetle. He can plate silver with it as well as copper' (C&G, p. 67).

James mentions him again in a letter to Lewis Campbell of July 1849 that further illustrates his scientific interest:

> Perhaps you remember going with my Uncle John Cay ([when we were in the] 7th Class), to visit Mr. Nicol at [4] Inverleith Terrace. There we saw polarised light in abundance [. . .] Well, sir, I received from the aforesaid Mr. Cay a 'Nicol's prism', which Nicol had made and sent him. (C&G, p. 123)

Modern times were emerging; railways, electric machines, electroplating, and photography. John would obviously have been interested in this last new wonder and so he joined, possibly even helped to found, the Edinburgh Calotype Club. The benefit to us is that for the first time we have a photographic record of our subject. He is seen seated, flanked by his two sons Robert and Edward, in a photograph taken by members of the Club (Plate 23). The picture was probably taken between 1843 and 1844, for the Club started in 1843, and there is also a photograph of his brother Robert Dundas Cay (q.v.) who left for Hong-Kong in 1844.

For some reason or other, in 1848 John and family departed from 11 Heriot Row and returned to their original domicile at 5 SE Circus Place. By this time, however, John's youngest child had reached twenty years of age and his daughter Emily had married in that same year, and so one of the factors may simply have been a case of 'downsizing' as a result of the family branching out. But it was also about this time that his brother Albert's business foundered (see next section), and John may well have invested either in the business itself or in the stocks and shares he was dealing in, hoping thereby to help pay down the debt on North Charlton. The general economic situation had been poor, and it may simply have been the case that the demise of Albert's business made it impossible to repay the debt. Back in 1832, John Cay had said of North Charlton:

> [it] has been a struggle for my grandfather and father to retain, chiefly in consequence of a grievous law plea with the Earl of Northumberland [. . .] I fear [it] will one day quit the family, for it is now heavily burdened. (Bateson, 1895, p. 297)

Rather than using Charlton Hall as his own summer residence, he had already been renting it out to help service the debt (Parson & White, 1828, p. 389). And so it was that, late in 1847 or early 1848, 11 Heriot Row was given up and not long thereafter, in September of 1849, the bulk of the North Charlton estate was put up for sale. It was advertised as being over 2,000 acres in area, bringing in an annual rental of some £1,280, and possessing a bed of coal of excellent quality (which clearly the Cays did not have the means to exploit to sufficient advantage). Charlton Hall and Shepperton were purchased by a Mr William Spours of Alnwick, while John held on to those parts that remained (Bateson, 1895, vol. 2, p. 298).

After the Cays moved back to 5 SE Circus Place, 11 Heriot Row was occupied by a firm of solicitors. John Cay died in 1865 leaving an estate worth £10,000 (Harris, 2011), and so although his sacrifice of the family estate and the house in Heriot Row must have been bitter pills to swallow, he had not only stabilised the family's financial situation but eventually left them fairly well-off (his father's estate amounted to only £900). Of his children, John jnr qualified as a Writer to the Signet in 1851 and was Solicitor for the Post Office in Edinburgh. He married his sister-in-law, Geddes Elizabeth Mackenzie of the Applecross family in 1857; she was the sister of a Liverpool merchant George Alexander Mackenzie 12th of Applecross, who had just the year before married John's youngest surviving sister, Elizabeth. John jnr and Geddes lived at first with his father at 5 SE Circus Place. After his father's death, however, John and Geddes moved to their lodge-house, called 'Charlton' in memory of the lost Cay family estate, which was at the east end of Strathearn Road, not at all far from where John's uncle Robert and aunt Jane lived from about the same time. After a year there the couple moved back into town, staying from 1867 in Alva Street, just off the west side of the Queensferry Road, and this later Charlton was either sold or rented out. There they stayed there for the rest of their days, with John dying in 1892 and Geddes following him in 1909. The remainder of the North Charlton estate was sold during John's lifetime.

John Cay snr's next three sons Robert, William and Edward, spent only a few years at the Edinburgh Academy, leaving at about the age of twelve. Robert was in the same class as Hugh Blackburn, who later married Jemima Wedderburn (Chapter 9). William went for only one year and Edward attended for two years in the same class as Charles Mackenzie (see note 32 of Chapter 2). William died in 1840, and a few years later Robert and Edward, seen in Plate 23 with their father, left for Australia to seek their fortunes as sheep farmers, for by then it was clear to John Cay that he could hardly provide a decent inheritance for even his eldest son. Some of the background to this comes from John Cay's comments he made on North Charlton in Bateson's *History of Northumberland* (1895, p. 297): 'its owner [John Cay senior] has too numerous a progeny to admit of his making a wealthy squire of the eldest'. In addition, in a letter he wrote in the autumn of 1839 to his dying sister Frances,[21] perhaps looking for small talk to avoid the painful subject of her dire situation, he said,

> Bob [...] is working at English wool which he says is mostly dirty stuff & is constantly jagging his fingers with the thorns in it. He says he is now a very fair sorter of Botany Wool & has a pretty good knowledge of it. His mind is plainly fully occupied with his businefs & his prospects & I have every hope he will do well, poor fellow.

Robert, at seventeen, was not heading for any career that called for intellect, rather something more practical, wool trading. He was learning the basics of his trade somewhere in England, and perhaps emigration to Australia was already in mind. Why 'poor fellow'? Was it that John Cay felt sorry for his son having to learn a dirty trade? Or was it that he could not bestow on him the sort of future he would have hoped for?

Robert and Edward were certainly both about the age of twenty[22] when they set out to be sheep farmers in Australia, and they seem to have been well established there by about 1850. Robert married Anne Montgomery from Melbourne in 1851 at Mount Fyans, while in 1855 Edward married Anne Burdock, also in the state of Victoria. Furthermore, both men were appointed territorial magistrates for Avoca and the Loddon respectively, in 1852 (*The Argus*, 1852).

The youngest son, Francis Cay completed only one year at the Academy and later followed his elder brothers to Australia (Henderson & Grierson, 1914, p. 141), but he may not have survived as there is no other mention of him. Edward died at Melbourne in 1870 at the age of only forty-five, while Robert died at Brisbane, having recently moved there from Victoria. Robert and Edward between them had twelve offspring to carry on a new generation of Cays that were firmly settled in Australia, and their descendants were still living in Queensland at the start of the twenty-first century.

All Sheriff John Cay's sons went to Edinburgh Academy save for Thomas, who ended up in Argentina, as narrated on his memorial at Restalrig. The younger sons of John Cay may not have had the opportunity or the inclination for professional careers, rather they were pioneers and adventurers. If Robert and Edward had been intrepid in setting out for the Antipodes, at least it was populated in the main by their own countrymen. South America was an altogether different proposition. Prospects were there for certain in timber, cattle and many other commodities, but Thomas would have found it quite a culture shock. Rosario de Santa Fe lies on the navigable lower reaches of the Parana River about 200 miles inland of Buenos Aires, and is now one of Argentina's largest cities. He was one of a dozen to die of cholera there in the height of the summer of 1868 (January) and who were buried at the Methodist Episcopal Church (Howat, 2013). Thomas must have made some success of his brief life in Argentina, for although he was just thirty-three years old at the time of his death, the sale of his property fetched £2,000 net, paid to his brother John.[23]

Albert Cay, Entrepreneur

Sherriff John Cay's younger brother, Albert (1795–1869), is not much mentioned in the usual biographies that touch on the Cay family, perhaps because, rather than belonging to one of the elite professional institutions, he was a businessman. Worse than that, he had the misfortune to have been made insolvent. By late 1847, his wine, tea and stockbroking businesses were foundering and as a consequence his assets were sequestered by the courts on 1 February 1848 (*Edinburgh Gazette*, 1848). Until then he had had salerooms at 99 George Street (Figure 1.2), and, at one time, a counting house in the Port of Leith.

Several factors had caused a significant economic crisis at this time. Firstly, the potato famine of 1845 ran on, affecting not only Ireland but much of Great Britain. This, compounded by a failure of the corn harvest, caused the price of basic foodstuffs to soar. Lastly, the over-inflated prices of railway stocks collapsed. By 1847, an economic crisis was in full swing and record numbers of businesses failed, Albert's amongst them (Figure 14.3). In spite of this, he had managed to set up and hold onto a partnership, Cay and Black, that traded as advertising and newspaper agents from another address in George Street. It continued thereafter under his business partner until it was eventually taken over by Keith's advertising and newspaper agency, which lasted into the early 1900s (EPOD, 1887–1911).

Albert must have been a larger than life fellow and had his fingers in many pies, earning himself the soubriquet *Albertus Magnus* (Frey, 1888). In addition to his businesses, he was a member of the Royal Company of Archers (like his brother John he won many of their prize competitions) and was director of the Highland Club of Scotland (*c.* 1825) and of the Church of England Life and Fire Institution (*c.* 1844). As to properties, from 1846 he had Fixby Cottage in Corstorphine;[24] from 1849 Slateford House; and from around 1853 Ratho Villa. This does not mean to say he owned the properties.[25]

When John took over 11 Heriot Row as his family home, Albert moved with his younger brother Robert and his sister Jane to 6 Great Stuart Street, a property comprising some flats entered from a common stair (Colour Plate 3; Figure 1.2), and lived there till 1835, the year of Robert's marriage. It is understandable that this event necessitated further residential changes. In the same year, Albert had moved his wine business from 37 to 99 George St, where it is possible that he also had a flat, for he is listed only at that address until the demise of his business. Thereafter he listed himself either at his out of town address or at Cay and Black. In his later years his eyesight began to fail; he moved in with Robert and Jane and eventually became blind.

Dyce and Cay

Robert Dundas Cay WS (1807–1888), the youngest child of Robert Hodshon Cay and Elizabeth Liddell, was born at the family seat of Charlton Hall (Cay et al., *c.* 1947, p. 26) and brought up at 11 Heriot Row where the Cay family had lived since 1804. As a boy he must have been familiar with the Clerk family

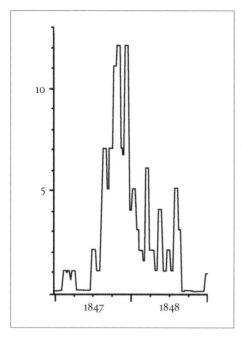

Figure 14.3 Business failures between 1845 and 1850
Some the economic problems of 1844–47 may have played a part in reducing the circumstances of several prominent members of the Cay family, thereby making the eventual loss of the already burdened North Charlton estate inevitable. The graph depicts the weekly number of business failures that followed in the wake of the failure of the corn harvest of 1847. (From Campbell, 2011)

at 31 Heriot Row, and later with the Wedderburns who moved there in 1821, but if not, they would soon have come to his attention when John Clerk Maxwell started courting his older sister Frances, whom he married in 1826. In 1831, the year James Clerk Maxwell was born at 14 India Street, Robert would have been in his early twenties.

William Dyce RA (Barringer, 2004; Ferguson, 2006) was born in 1806 and therefore of a similar age to Robert Dundas Cay. He enters our story possibly because Robert had some gift for drawing and painting, as did his mother and youngest sister (C&G, p. 14; Cust, 1888).[26] For whatever reason, he and William became firm friends. William was born into a well-off Aberdeen family. His parents were the distinguished physician Dr William Dyce FRSE of Fonthill and Cuttlehill and his wife Margaret Chalmers, of the Westburn family.[27] Educated at Marischal College in Aberdeen,[28] William graduated with an MA at the age of only sixteen. Though talented in drawing and painting, he at first followed his father's wishes by studying medicine, but tiring of that he turned to theology with the aim of becoming a priest, with much the same result.

In the meantime, however, William not only continued to study art but sold enough of his paintings to afford his passage to London in 1825. He contrived to meet the president of the Royal Academy, who was impressed enough with him to intercede with Dr Dyce to allow him to study at the RA as a probationer. For William, however, even such good fortune as this was not enough, and soon thereafter he accepted the offer of Alexander Day and William Holwell Carr, both painters and art collectors over thirty years his senior, to visit Rome, where he spent nine months during 1825 to 1826 studying, in particular, Titian and Poussin before returning to Aberdeen. The following year, at just twenty-one, he exhibited at the RA[29] after which he returned, as a nascent pre-Raphaelite, to Rome.

While in Rome, William's 1828 painting of *Madonna and Child* made such an impression on the members of the German art community there that they offered to finance him to stay on by buying it at a handsome price. Considering that he had taken an MA at sixteen and had then flung himself in at the deep end as an artist, within half a dozen years he had reaped the critical acclaim of both the Royal Academy and the European cognoscenti. Nevertheless, the painting may have been unfinished later that year when William was back home in Aberdeen. All too predictably, the straight-laced inhabitants of the Granite City did not find much enthusiasm for his chosen style, and even less for his subject matter, Madonnas. Amazingly, however, he spun once again on his heel and headed back to Marischal College to study the emergent subject of electromagnetic theory! This was no immature flight of fancy, for he came away with the Blackwell Prize for his essay entitled 'The Relations between the Phenomena of Electricity and Magnetism', a concept that was only just then beginning to be appreciated. But the change of direction was short-lived and he turned again to art, this time as a portrait painter, which in contrast to his attempts with Madonnas, was showing some early signs of success.

In order to pursue this as a career, in 1830 he moved to Edinburgh, perhaps staying at first with his sister Margaret, who had married an Edinburgh solicitor by the name of James Ross some years before.[30] He then bought 128 George Street[31] where he lived with his younger sister Isabella, who was no doubt sent to look after him. It is said that Dyce's portraits of ladies and children were much admired and, by his own account, he had no shortage of commissions. Dyce was elected to the RSE on 2 April 1832, at not yet twenty-five years of age, a considerable accolade for one so young.[32] This could hardly have escaped the attention of John Clerk and John Cay, who were already fellows of some years standing and regular attenders at its meetings.

In the summer of 1832, the Clerk Maxwells took young James with them for a vacation at Nether Corsock, which must have been his first visit there. The couple were clearly already on friendly terms with Dyce, for they invited him to come along. He did some of his earliest recorded watercolours there, none of which survives (Pointon, 1979, p. 21).

Dyce evidently enjoyed the experience, roaming the hills and moors and letting his artistic eye take in the wild scenery, visiting Threave and, by walking a total distance of some forty miles, the abbey at Dundrennan, which he sketched and later painted. It is believed that Dyce hung onto these early works for sentimental reasons, but they were all lost after being sold off after his death (Pointon, 1979, p. 22).

Sometime between 1832 and 1834, Dyce drew a charcoal sketch of James Clerk Maxwell as Puck in the same style as his original version from 1825 (Plates 25 and 26). A copy is on display at 14 India Street. In the will of Mrs Barbara Wallis,[33] James is given as being three in the sketch, but he looks much younger. While Dyce was at Nether Corsock in 1832, he would surely have favoured his hosts with some sort of memento? The age of the child in the sketch could be fifteen months, consistent with a date of 1832.

It was by 1834 at the latest that William Dyce had got to know Frances' brother Robert Dundas Cay. They became close enough for him to be able to ask Robert for help in opening a bank account, for which he needed, even at the age of twenty-eight, two guarantors. His problem as a fledgling portrait painter was that he was perpetually short of money and had to resort to 'dunning' his clients too frequently for his own good standing (Dyce & Dyce, aft.1864, pp. 116–119). One presumes that Robert did vouch for Dyce, for he had taken a fancy to Isabella, the artist's sister, and married her almost exactly one year later.

While Dyce probably became acquainted with Robert and the rest of the Cay family as a result of having already met John Cay and John Clerk Maxwell, and then Frances Cay, it has been suggested that they may have come to know each other through the Edinburgh Calotype Club (Melville, 2006, p. 92), of which Robert's elder brother John was a member. This, however, is erroneous, for the two men were corresponding by the autumn of 1834 and the club was not founded until, at the very earliest, late 1842. On 14 October Dyce wrote to Robert: 'I daresay Isabella has told you of my having been in this town for three or four weeks enjoying a little repose after my London campaign' (Dyce & Dyce, aft.1864, folio 113).

The 'London campaign' refers to his exhibition of three paintings at the Royal Academy, among them his portrait of Sir George Clerk of Penicuik. But it is the reference to Isabella that is significant, for we learn that Robert is now corresponding with William's sister. She also appears to have been back in Aberdeen at this time, for William mentions that he is going to do her picture in pastels before he returns to Edinburgh: 'She means to give you an agreeable surprise [. . .] My mother and father and all here join in kindest remembrances and *one individual in something more warm*' (Dyce & Dyce, aft.1864, folio 114, my emphasis).

There was a similar cryptic message for Robert in his letter of 22 October, which certainly seems to imply that there was more than mere letter writing going on between Robert and Isabella!

That winter found Dyce commissioned by Robert and Jane to paint Frances sitting with her young son James (Colour Plate 2), for which he was paid £31 10s in March 1835 (Dyce & Cay, 1888). Dyce had begun the painting probably when Frances and John were on their winter visit to Edinburgh, but he took ill before it was finished and in the interim the Clerk Maxwells had to return to Nether Corsock.

Dyce had not exhibited at the Royal Academy that year. Dyce's son James suggests it was to do with Isabella's wedding,[34] but Isabella had no wedding plans just then. At that time a serious illness could take quite some time to recover from. On top of this, his father had died just a few months earlier, in March. It is much more likely that, due to these untoward circumstances, he got behind in his commissions, for he was not able to get to Nether Corsock to finish the portrait till the late summer.

It was on this visit that he did the watercolour which did survive, *Glenlaer* (Melville, 2006, pp. 90–91), shown in Colour Plate 14. He was still there on 18 September when a letter arrived from Robert announcing his engagement to Isabella. William recounts this in his reply to Robert in jubilant tones, telling him that the laird, John Clerk Maxwell, had left his mantle of earnestness aside for long enough to send up a rousing three cheers, leading in the hip-hip-hoorays himself. Jane Cay must have been there also, for it was a 'Miss Cay, rosy as the morn' who brought him the letter.[35] He ends by saying the portrait had gone remarkably well, and Robert's sister (Jane or Frances?) had hopes of William himself beginning to do better in 'matters pertaining to the fair sex'.

During his stay in Edinburgh, William Dyce got to know some of the trustees of the Board for Manufactures, whose secretary was then James Skene of Rubislaw,[36] a fellow Aberdonian and devotee of art and design, and so began to get involved in the design activities of its drawing school (Chapter 8). If this seems an odd thing for a portrait painter to be involved in, Dyce was in many respects a polymath, hints of which have already come out in his volte face from art to electromagnetic theory and back again. He also wrote essays on an eclectic range of subjects from the garments of Jewish priests, to the Jesuits and church music. While it may be recalled that Baron Sir John Clerk and his son George Clerk Maxwell had both been trustees of this Board for Manufactures, by now its aims had had been redirected from stimulating the growth of ordinary manufacturing and fisheries in Scotland, to promoting a drawing school with the emphasis on turning out skilled artists, draftsmen and craftsmen who would go on to conceive and design goods and works of the highest quality. Dyce became in effect a consultant to the Board, in which capacity he eventually became very active. During his last year in Edinburgh, he submitted his ideas to Lord Meadowbank (Dyce, 1837), which then led to him being asked to take up a similar but more wide-ranging role in England.

Dyce moved back to London in 1838 on his appointment as director of the School of Design at Somerset House, but he did not marry until 1850. His wife, Jane Brand (1830–1885), nearly twenty-five years his junior, was from a Kinross-shire family that had settled south of the Thames at Balham, Surrey. They had two daughters and two sons. Dyce was now setting up design schools for the Board of Trade, whose vice-president in 1841 was William Ewart Gladstone;[37] the two men became close friends.

Not only did he continue to paint, he designed coins and medals and made frescoes for galleries and, in particular, for the Queen's Robing Room; he wrote on theological matters and was at the fore of the Anglican High Church movement; he was a talented organist and composer, and founded the Motet Society; and he published *The Book of Common Prayer with the Ancient Canto Fermo set to it at the Reformation*. In all of these areas, and others, he wrote numerous essays and pamphlets, much of which was transcribed and kept by his son James Stirling Dyce (Dyce & Dyce, aft.1864) whose name suggests that Dyce also had maintained an interest in and admiration for mathematics.

After his death at Streatham on 14 February 1864, Dyce's friend Gladstone said of him that he had exhibited the 'very ideal of the profession of an artist'. Such praise does not go far enough: wonderful and enduring artist though he was, he was nevertheless a man of considerable talent who could have achieved great heights in many fields. Compared with his art, his pioneering efforts in the setting up of the design schools in Edinburgh, Somerset House, Spitalfields and Manchester over the years 1837 to 1848 are now little remembered, but they are equally important in terms of their long-term benefits to the nation. Laying down the key principles of applying design to technology and the furnishing of schools that could turn out skilled designers to embody these principles provided the foundation for a multitude of inspired Victorian creations that are still marvelled at today. While the Forth Bridge (1893) is a prime example that reaches the very apex of these ideals, it is only one of countless bridges, buildings, steamships, steam-engines and so on, all the way down to cups, saucers and coinage, where the benefits of these principles were demonstrated. Technology has moved on, but the principles remain.

Robert and Isabella

Robert Dundas Cay and Isabella Dyce, seen in Plates 26 and 27, were married on 29 October 1835 in Aberdeen by the Reverend John Murray, 'minister of the North Parish, Aberdeen'.[38] The witnesses were the bride's uncles on her mother's side: Alexander Brown, bookseller and sometime Provost of Aberdeen, and David Chalmers, a printer. The Dyce and Cay families were Episcopalian, yet they were married by a Presbyterian minister. Dr John Murray of the North Parish was not only a Presbyterian, he was one of those principled ministers who came out for the Free Church in 1843 and took his congregation with him. But he officiated at the marriage because he was one of the family: he was married to Isabella's cousin on her mother's side, Margaret Brown, of the same family as the Lord Provost. We may adduce that the Dyce side of the family were Episcopalian because a number of William's siblings were baptised in that faith.[39] On the other hand, it seems likely that the Chalmers side was Presbyterian, for it would have been unlikely for a Presbyterian minister to have married someone not of his own denomination. A separate indication that the marriage of William Dyce MD and Margaret Chalmers was a unison of the Episcopal and Presbyterian faiths is that William himself was baptised by a Presbyterian minister.[40]

What this clearly demonstrates, however, is that even though matters of religion were still taken very seriously, and the principle 'each unto his own' was still very much the order of the day, by the early nineteenth century Episcopalians and Presbyterians were not only getting on well with each other socially, they intermarried, as did, for example, John Clerk Maxwell and Frances Cay, and Isabella Clerk and James Wedderburn (Fairley, 1988, p. 98), and here we have two Episcopalians, Robert and Isabella, being married by a Presbyterian minister. Although discrimination against Roman Catholics was still endemic, the days when the nation was close to tearing itself apart over the issue of episcopacy were largely over.

Having qualified as a Writer to the Signet in 1833, Robert was a clerk to the Court of Session before being promoted to Keeper of the Rolls in the year of his marriage; he would therefore have had a comfortable income. He left 6 Great Stuart Street to live with Isabella at 18 Rutland Street,[41] which turns off Lothian Road opposite the entrance of St John's Church, and here they started to raise their family of eleven children:

Robert Hodshon, baptised 27 October 1836;
William Dyce, born 28 March 1838;
Elizabeth, born 14 April 1840;

Charles Hope, born 26 April 1841;
John Frederick, born 2 November 1842;
Alexander Dyce, born 7 May 1844, died 22 February 1853;
Albert, born 1846 in Hong Kong (as were the children that followed);
George d'Aguilar, born 1848, died 1881 (he was named after
    the retiring Lieutenant Governor of Hong Hong);
Margaret, born 1849, died 1921;
Isabella, born 1850, died 1934;
Dundas, born 1851 and died in the same year.

The year 1844 brought a momentous change, for Robert secured a position as a civil servant in Hong Kong. It was an adventurous step to take, particularly since Hong Kong had become a British colony only a few years before. It was halfway round the world and foreign in all senses of the word; language, culture, religion, climate and food being just the obvious ones. His motivation for going may have been to improve his prospects, perhaps to help restore the family fortunes, but the loss of the office of Keeper of the Rolls, which had lapsed in 1841, may also have been a setback he needed to recover from. But he had a wife and five young children to consider. Unlike the situation that prevailed in India, they would not be able to rely on the benefits of a well-established colonial settlement. Nevertheless, it showed enterprise on his part, for in gaining a toehold in the legal system there, he would be playing a key role, and perhaps in the long-run he would have ample opportunity for advancement.

Hong Kong and its dependencies were declared a separate colony by a royal charter of 1843 that established full legislative powers, and the law courts to enforce them, under a plenipotentiary governor, Sir Henry Pottinger HEICS, who had already been Superintendent of Trade and Her Majesty's Plenipotentiary since the formal cession of Hong Kong to Britain by China at the end of the Opium War in 1841. He was formally appointed as the first governor on 5 April 1843. While Robert had started preparations to move a whole world away, the organisation of a proper colonial administration in Hong Kong had only just begun to get under way. In the first couple of years there was much that needed to be done and too few adequately qualified people to do it all. Mistakes and misunderstandings contributed to the stresses and strains of this crucial period, and so when Sir Henry put forward his resignation it was accepted, no doubt as much to his own relief as that of his officials and a grateful British government. More manpower was needed to fill the gaps in the administration, and so it was that Robert sailed for Hong Kong in early 1844 as one of the fresh appointees sent out to build up the manpower of the administration and to get things going.

Given the low expectation of the situation they were likely to find on arrival, Robert went on his own,[42] for we find that some months after he set off, his son Alexander Dyce was born at Rutland Street and was baptised in St John's at the end of June.[43] This child was born on the very day that Robert sailed into Hong Kong harbour on HMS *Spiteful*. Isabella and the children would come out later.

Robert had set off early in 1844, sailing to Bombay in stages via the Mediterranean and Red Sea, and crossing Suez over land.[44] He then departed from Bombay on HMS *Spiteful* in the company of the newly appointed governor, His Excellency John Francis Davis, who was to relieve the incumbent Sir Henry Pottinger, and Frederick Bruce, the new Honorary Foreign Colonial Secretary (Norton-Kyshe, 1898, p. 47). They arrived on 7 May and were sworn in by the Legislative Council of Hong Kong the following day. By October things must have been going well for Robert, for he was appointed as a Commissioner for Taking Affidavits and was also provided with a deputy, Mr Smith.

There were many teething troubles in the new administration, and things were exacerbated by a rift between the governor and his Chief Justice, John Hulme. There was also a serious problem with the mail, for the local Postmaster General proved to be not only incompetent but highly irrational. After being suspended from duty in July 1845 he ended up committing suicide (Lim, 2011, pp. 114–115). Problems with the administration were one thing, but everything depended on the mail. Everyone would have been frustrated by the mail debacle for not only did it hamper their work, communications with home were difficult enough at the best of times. Robert must have longed to know whether Isabella and his new child were alive and well, but five months after the child's birth his name was left blank on the baptismal certificate. A letter would no doubt have been sent to Robert in all haste following the child's birth, but even so Isabella would have been lucky to hear by Christmas that Alexander Dyce was agreed as the name.

When it was decided that it would be safe for Isabella and the children to come out, preparations were made for a departure in the late winter or early spring of 1845, for she arrived in Hong Kong in the following July having brought with her two female servants, three sons and a daughter (Lim, 2011, p. 270). Now, while the daughter could only have been Elizabeth, at the time of Isabella's departure she had *five* sons: Robert, William, Charles, John and finally Alexander, who was only about nine months old. Nevertheless, Lim is correct for, as we shall soon discover, the two eldest sons were left at home.

Safely reunited, they all moved into a house called Rosehill within the security of the administrative quarter. By and by the establishment of the administration was progressing and when, in January 1847, the formation of the Admiralty Court was announced, Robert was appointed as its registrar, following which, in 1848, he was appointed Master in Equity, that is to say, he could handle cases relating to the valuation of land and property in his own right. In the meantime, more children came along: Albert, George, Margaret, Isabella and finally Dundas, who died in infancy. Unfortunately, Alexander, who journeyed out to Hong Kong as an infant, developed spinal weakness and was sent home in 1848, where he died in 1853.

Otherwise, things seemed to be moving along on a level keel until 1852, when Robert's world was suddenly shattered by Isabella's death in a carriage accident at Victoria: 'The Registrar of the Supreme Court, Mr. Robert Dundas Cay, on the 21st June, had to deplore the loss of his wife whose death took place at Victoria, on that date' (Norton-Kyshe, 1898, p. 325).

She was buried at Wong-We-Chung, Happy Valley Cemetery, Victoria, alongside her infant son Dundas, who had died the year before, 'in the centre of the largest plot for an individual grave in the entire cemetery' (Lim, 2011, p. 270). The inscription on the memorial is simply:

SACRED
TO THE MEMORY
ISABELLA DYCE
1811 – 1852
BELOVED WIFE OF
ROBERT DUNDAS CAY
ROSEHILL, HONG KONG
AND THEIR INFANT SON
DUNDAS 1851   (Lim & Atkins, 2011)

After burying Isabella, Robert must have wondered, in his grief, what he must do. He no doubt had servants and maids who could look after his children in the short term, but there was no mother figure, and no help from family and close friends, as there would have been at home in Edinburgh. In the face of this woeful situation, Robert wrote home. On the assumption that he would get a reply back from Edinburgh by about January the following year, 1853, he had a few weeks to set things straight in Hong Kong, get permission for a substantial leave of absence, and get a passage home: 'Mr. Cay, the Registrar of the Supreme Court, proceeded on leave of absence on the 24th March [1853]' (Norton-Kyshe, 1898, p. 332).

With nine children and some maids to look after them, it might have been difficult to book a passage for all of them at one time, both for financial reasons and availability of berths; we do not know how he managed, save to say they all got safely home. About a year after he set foot on home soil sometime in the summer of 1853, it was recorded that: 'on the 29th of [...] [August 1854] it was announced that Mr. Cay, the Registrar of the Supreme Court, had obtained an extension of twelve months leave of absence' (Norton-Kyshe, 1898, p. 353), while nearly two years later, the report was:

> In consequence of the resignation by Mr. Robert Dundas Cay who had now been on leave for some considerable time, of his office of Registrar of the Supreme Court, on the 29th April 1856 [...] The exact reason for Mr. Cay's resignation of his appointment was never known.[§] (Norton-Kyshe, 1898, p. 383)

The footnote [§] to the above passage includes a comment quoted from the journalist W. Tarrant (1862, p. 87): 'Mr. R. D. Cay, the first Registrar of the Court, was a Scotchman with a large family, which is about all we know of him, nor can we tell why he threw up such an excellent appointment.'

Robert and Jane

By 1855, that is to say within two years of his return to Edinburgh, Robert had moved into 42 Northumberland Street, whereupon his sister Jane left 6 Great Stuart Street, where her nephew James Clerk Maxwell had spent many a happy day, to be with him and look after his children.

Just as had been the case with James, she doubtless became more than just an aunt to Robert's younger children. But, before moving on to this part of the story, we still have the mystery of what happened to the two boys Isabella left behind when she went out to Hong Kong to be with Robert. Charles, John and the baby Alexander were the three boys that Isabella took with her. From Cay, et al. (c. 1947) and elsewhere, we find that William had been placed at an Episcopalian boarding school in Peebles run by the Rev. William Bliss (Williamson, 1895), where he was enrolled in 1844, the year before his mother left. The accommodation for this long established burgh school was in the Duke of Queensberry's old residence, or Dean's House, on the High Street of Peebles, a county town some twenty miles to the south of Edinburgh. Charles was also placed at the school but much later (Venn, 1898, p. 349) when it was under the charge of Rev. Thomas Dwyer, who became the new incumbent at Peebles on Bliss' death in 1850 (Bertie, 2000, pp. 180, 499).

Robert was the other son who stayed at home, and we must therefore presume that he too was at the Bliss School with William. It was not uncommon for officers in the army, HEICS or the emerging colonial services, to take their wives abroad with them and leave their children of school age in a boarding school at home; even when travel became much easier, the practice persisted. Of Robert we know very little other than that he was named after his grandfather Robert Hodshon Cay, baptised on 27 October 1836, read arts at Glasgow University, and died on 13 May 1860 at his father's home in Lauriston Lodge of an infection following a surgical operation.[45] He had been a humble merchant's clerk.

Returning now to Alexander Dyce, as previously mentioned, he had indeed travelled with Isabella to Hong Kong but was sent back in 1848 due to a severe illness, a 'spinal weakness', which implies a paralysis. The burial records for Restalrig (Grant, 1908) reveal that he died of dropsy at 20 Albany Street on the first day of March 1853, a month before Robert set off on his return voyage to Edinburgh.[46] However, the place he died touches on another little mystery that we have not yet resolved: who took care of Alexander when he was sent back to Scotland in 1848? The occupant listed at that address in that year is James Ross SSC, husband of Margaret Dyce, Robert's sister in-law (EPOD, 1852–53). When the decision was made that Isabella would not take Robert and William with her to Hong Kong, it seems that Margaret and James Ross were left *in loco parentis* of the two boys, and on his return in 1848 Alexander also came under their care.

Three months after Alexander's death, however, the Rosses moved to Mauricewood House at Penicuik, which was where they would have been living when Robert arrived home. Incidentally, this house was later owned by Joseph Bell, the esteemed Edinburgh surgeon who, with his power of insightful deductive reasoning, was Sir Arthur Conan Doyle's study for Sherlock Holmes.

From 42 Northumberland Street, Robert, Jane and the children moved in 1856 to Lauriston Lodge.[47] From his return to the time of this move, however, Robert had had no regular employment; he said that illness was one of the factors in this (Cay, 1857), and it may have also been the case that they could live more cheaply there than in the town proper. Luckily, as a result of his past experience he was able to become an agent for the Oriental Bank of Hong Kong in St Andrew Square at least by the following year (EPOD, 1858–59).

By Whitsun 1860, Robert and Jane had moved again, to Greenhill Gardens where houses had been built on the fringes of the old burgh moor, or Boroughmuir as it is now called. Finally, they moved to a smaller house on Blackford Road; it was named Shepperton Cottage after part of the much missed North Charlton estate, and it was here that they spent the rest of their days. Their nephew John Cay made a similar move to Charlton Lodge in the same area of town, and so it is possible that after the death of his sheriff brother the Cay family fortunes were in much better shape. All the houses that Robert and Jane lived in are still in existence.

Robert's eldest surviving son William Dyce Cay went to Edinburgh University from 1853 to 1856 but, like his cousin James Clerk Maxwell, he did not stay for his final year; he had a brilliant mathematical mind and became a civil engineer of some standing. Perhaps through a similarity of technical minds, he was quite close to his cousin and on leaving university in 1856 it was James who persuaded him to go to Belfast to be apprenticed to James Thomson (Hamilton, 1896), an engineer who was the older brother of Maxwell's good friend and correspondent, William Thomson, later Lord Kelvin.[48] Not only did James recommend William, he accompanied him to Belfast for the interview (Harman, 1990). Furthermore, in 1858, it was William who was best man at James' wedding to Katherine Dewar. A few years later, in 1861, he put his training for his future career into practice, designing for James the stone access bridge over the Urr, as part of the improvements he was making at Glenlair (see Chapter 2). One of William's major works during his subsequent career, however, was the construction of the new harbour breakwater at Aberdeen, where he became the resident engineer.[49]

The second son, Charles Hope Cay, was also a brilliant mathematician who took the Mathematical Medal in his final year at Edinburgh Academy and afterwards graduated as sixth wrangler in the Mathematical Tripos at Cambridge, where he then became a fellow of Caius College. He later married and became a mathematics master at Clifton College. Also on close terms with his cousin James, he was to die before he reached his thirtieth year (Cay et al., c. 1947; Forfar, 1992). The two men corresponded regularly (C&G, pp. 324–343), and it is clear, from James' occasional advice to his younger cousin, that he was under some sort of strain, perhaps depression, at Clifton, which their mutual friend Lewis Campbell hints may have been a factor in his untimely death:

> He died in 1869, at the early age of twenty-eight, the most devoted of teachers, one of the purest-hearted and most amiable of men. If he could have listened to his cousin's gentle warnings against excessive zeal, perhaps his services to Clifton College [. . .] might have been continued longer. But who knows? They whom the gods love die young. (C&G, p. 324)

John Frederick Cay was another of the family with a good head for mathematics, for he also took the Mathematical Medal in his final year at Edinburgh

Academy. He firstly went into banking, originally in India, before becoming private secretary to the renowned international banker, Baron Rothschild. He then married and became a director of a company in the booming business of making axles and wagons for the railways.[50]

Albert, named after his uncle, was one of the children born in Hong Kong. After attending Edinburgh University, he went into business as a glass manufacturer in Birmingham, later becoming a Justice of the Peace and the High Sheriff for Warwickshire; he was also a lay canon in Coventry (Henderson & Grierson, 1914). He married Annie Jaffray, daughter of John Jaffrey, 1st Baronet of Skilts, a Scottish journalist from Stirling who settled in Birmingham and became the founder and owner of two newspapers (Cay et al., c. 1947, p. 43; Jaffray Family, 1836–1926).

Elizabeth Cay married Thomas Dunn in 1870, a house-master at Clifton College where her brother Charles had been mathematics master since 1864 until his early death. Through the Dunns, the Cays have many descendants, including the Giles family (Cay et al., c. 1947, pp. 37–41).

After her father's death, Isabella Cay went to Paris where she studied drawing and painting. She then lived in Venice until her failing health brought her home. It was she who painted the fine copy of Raeburn's portrait of her grandfather, Robert Hodshon Cay in his judge's robes. While this copy now hangs in Maxwell's birthplace, 14 India Street, the original Raeburn is in the Houston Museum of Fine Arts (Raeburn, n.d.). She never married and died at Bournemouth (Cay, et al., c. 1947, pp. 44–45).

Full Circle

Jane Cay had stayed at 6 Great Stuart Street, first with her brothers Robert and Albert, and then, for a spell, on her own. When Robert returned from Hong Kong with his children, she no doubt felt it was her place to be by his side, helping recreate a family unit. One by one Robert's children got married and went their separate ways, but Jane stayed with him the rest of her days, which came to an end on New Year's morning of 1876, at Shepperton, with her brother by her side. She had been suffering for some years from a tumour in the nasal sinuses that was accompanied by severe facial neuralgia.[51]

Lewis Campbell described her as being:

> one of the warmest hearted creatures in the world; somewhat wayward in her likes and dislikes, perhaps somewhat warm tempered also, but boundless in affectionate kindness to those whom she loved. (C&G, p. 14)

She had appointed Robert as her executor, and so in the days after her funeral he accordingly set about taking care of things. On 10 January, one of the letters that he wrote in that capacity was to his nephew James Clerk Maxwell, who was then at Cambridge. He tells James that the will has been read and his aunt Jane has left him her sketch of a picture of some cows from the original, at Penicuik House, by Nicolas Berghem (1624–1683). We may guess, first, that it had some special significance for James and, secondly, that the original had probably been brought back from Holland by either the Baron or one of his sons on their return from Leiden. Jane also left him six of her sketches, which he could choose for himself; and lastly, but far from least,

> the Picture by Dyce of your mother and yourself (this last was a mistake, *as the picture was painted for me*, and then when I married I gave her [Jane] the liferent of it; but after all she has done for me and mine, I will have great pleasure in fulfilling her wishes and sending it to you; which indeed will only be letting you have it a very short time sooner, as I always intended to leave it to you). (Dyce & Cay, 1888)

But just why, back in 1835, had Robert wanted that portrait done? After all, it was an expensive thing to commission, and for people in his position, that is to say, a young unmarried man not yet wealthy, quite unusual. In fact, it seems that he even had to get Jane to share in the expense. We find out a little more in a later letter about the picture; on this occasion the letter was to Robert from Andrew Wedderburn, now Wedderburn Maxwell, offering him the return of the portrait. Having come to Wedderburn Maxwell on James' death, the picture was then hanging in Glenlair House. When, at the age of thirty-five, James thought it about time to draw up his will, he had been advised by his cousin Colin Mackenzie WS, who acted as his lawyer, to try to keep the house and its most significant contents together for the benefit of his successors.[52] We must infer that James' widow Katherine had the liferent of Glenlair House, for Wedderburn Maxwell did not write to Robert Dundas Cay until shortly after her death in 1887, which left no-one with a closer connection to the picture than James' uncle Robert who had had the picture painted over fifty years earlier. Having had no sentimental connection with the picture himself, Andrew Wedderburn Maxwell now generously offered it back. Robert's reply, dated 29 March 1887, is as revealing as it is heartfelt:

> My Dear Wedderburn Maxwell,
> It is difficult for me to explain how very grateful your kind offer of the picture of my Sister and James has made me. It so happens that it has

been almost woven in with my life. There was first the extreme difficulty I had in inducing Fanny [Frances] to sit – then Dyce took ill before it was finished – the patients [sitters] could not remain in town – the Artist had to go to Glenlair to finish it there – then all this time I was pushing my fortune with his sister my dear wife. When I had with great difficulty got it home the next thing was I had to go to China – as I did not wish to take it with me to be knocked about in foreign parts I left it with my sister Jane to take care of. When I returned home I lived with her and so it remained so long on her walls that she seems to have forgotten it was not her own and left it in her will to James.

After her death I did not wish to assert my claim to it especially as I was her executor and had benefitted so much from her care of my children and in other ways so I sent it to him stating the facts. And now by your kindness it is to return to me again as your present for which accept my most hearty thanks.

– I am so nearly blind now that I hope you will excuse any mistakes I have made in this [. . .]. (Dyce & Cay, 1888)

However, we have to take this letter in context with the original receipt for the painting:

> Edin.^r March 1st 1835
>
> £31.10
> — Received from *Robert Dundas Cay Esq & Mifs Cay* the sum of Thirty one Pounds Ten Shillings, Sterling, being the payment of the *Portraits* of Mrs Clerk Maxwell & child.
>
> William Dyce
> (Dyce & Cay, 1888, my emphasis)

This reveals, firstly, that Robert seemed to forget that Jane also paid towards the painting (there was no other Miss Cay at the time, and he refers to Frances separately as Mrs Clerk Maxwell). Secondly, this was not just for one portrait; perhaps there was also a smaller one or a preparatory sketch. Taken together with the preceding letter, it reveals a possible motive for the painting of the picture. Robert Dundas Cay had befriended William Dyce shortly after Dyce's move to Edinburgh. We know that Dyce had encountered the Clerk Maxwells by 1832, the year Dyce was invited to join the Clerk Maxwells at Nether Corsock (Pointon, 1979, p. 21). In due course, Isabella and Robert met and, by October 1834, they were becoming close. And here we have it from Robert himself, in his letter of thanks to Andrew Wedderburn Maxwell: 'There was first the extreme difficulty I had in inducing Fanny to sit [for the picture] [. . .] then all this time I was pushing my fortune with his [Dyce's] sister.'

We may infer that sitting for the portrait began in the early months of 1835, when the Clerk Maxwells were back in Edinburgh and most likely staying, as became customary, with Isabella Wedderburn at 31 Heriot Row. The receipt was given in March 1835,[53] before the painting was completed, but Dyce mentioned to Robert that he frequently had to ask his clients for early payment for reasons of cash flow, and no doubt Robert was more than willing to pay up. Work on the painting had to stop when William took ill, and when he recovered, he decided to show his work for that year at the Royal Scottish Academy in Edinburgh rather than travel to London. By the time the exhibition was over, the Clerk Maxwells would have been at Glenlair, and he had to make the journey there to finish the long awaited picture.

In the end, William was very happy with the portrait, despite the fact that James, aged four, was bored with sitting and would just not keep still. The reason for the engaging pose is mainly due to Frances having to clasp her hands round the boy in order to keep him in place, and he seems only to be looking in the painter's direction for an instant, for his body is facing in a different direction, the one that he would run off in, could he only wriggle free.

Within days of the portrait being finished, the engagement of Robert and Isabella was announced. The painting had therefore fulfilled Robert's true purpose, and now, many years after Frances, Isabella, William, Jane, James and Katherine had all gone, it had come back to him to comfort him in the last year of his life. He died on 19 March 1888 at 'Shepperton', his home on Blackford Road.

CHAPTER 15

James Clerk Maxwell's Scottish Homes: 14 India Street and Glenlair

It has already been revealed that James Clerk Maxwell was born at 14 India Street in Edinburgh's Second New Town on 13 June 1831. It had been the house his father and mother had lived in after they married, and although they had started living at Nether Corsock on John's estate some two days' journey away, they returned to their town house in Edinburgh each winter until James was in his early infancy. In Chapter 9, we discovered that James' grandmother had come to live at the newly built house in India Street after having lived for many years just round the corner at 31 Heriot Row. We now complete the story of how and why 14 India Street came into being, and what happened to it after the Clerk Maxwells moved permanently to Nether Corsock and the house there that eventually came to be known as Glenlair.

The Origins of 14 India Street

Building in India Street commenced when Heriot Row was more or less finished. In fact, William Wallace,[1] who was responsible for building much of India Street, had started work long before in Heriot Row where he had a building or yard listed at the 'Back of Heriot Row' (nowadays called Jamaica Street South Lane), a mews lane from which the stables and other outbuildings belonging to the houses in Heriot Row were accessed. The Kirkwood map of 1817 shows only the outline of the street and the corner house adjoining to Heriot Row, whereas a later map of 1819 shows just two houses, built on the west side (Kirkwood, 1819). Building work had slowed due to the financial constraints occasioned by the recent Napoleonic wars, but in 1820 James Clerk Maxwell's father and grandmother were looking for a new house.

They were then living at 31 Heriot Row, but it was a large, indeed grand, house and, apart from possibly Mrs Clerk's stepmother Mrs Irving and some of her stepsisters, there was no other family living there on a permanent basis. By May 1820 a decision had been made, and on the seventeenth of that month

John Clerk Maxwell got a charter (James Clerk Maxwell Foundation, 1820) from the trustees of Heriot's Hospital for a house just then being built close to the top of India Street, on the plot allocated for the house that is now number 14 (Figure 1.2). Since the houses were being built from the top of the street downwards (away from the town), it would have been the closest then available to Heriot Row, in fact just six doors up to the top of the street and then just twelve eastwards to 'Old 31', as it was called by the family. A sasine of 1 June 1820 records the fact that John Clerk Maxwell had duly obtained his charter and contracted for the house with 'William Wallace, Architect', who had acquired the rights to build on the land in question but was in fact more of a builder than architect as such:

> (23,859)   Jun. 1. 1820
> JOHN CLERK MAXWELL, Advocate, Seizes, May 17. 1820, – in an Area of ground with the Buildings thereon, on the west side of INDIA STREET, Edinburgh:– on Feu Ch. by the Governors of George Heriot's Hospital, with consent of William Wallace, Architect, Edinburgh, May 16. 17. 1820.   P. R. 881. 222.
>
> By Courtesy of Registers of Scotland

While Kirkwood's plan of 1819 shows that building on India Street proper was not then started, Brown's plan, around 1823, shows that it was by then substantially complete. It also reveals that the original intention to overbuild the old road running north-west from Edinburgh to Stockbridge was abandoned, leaving the northern terminus of India Street not only a cul-de-sac, but railed off at a height of some 5m above the street below. This means that the north-facing ground floor windows of the corner house at the north end of India Street look out onto North West Circus Place from two floors up!

An entry for 14 India Street in the Edinburgh Post Office directory first appears in the year 1821, under 'Clerk, Mrs.', while at this time her son is still listed under 'Maxwell, Clerk esq. advocate, 31. Heriot row'. Until 1826, John Clerk Maxwell continued to give his address as Heriot Row, a clear indication that this was his residence rather than a business address, for Heriot Row is what he gave as his address at the time of his marriage.[2] Not until 1827, the year after his marriage, was John Clerk Maxwell's name listed at 14 India Street, by which time his mother had been dead for five years.

John never actually moved with his mother to 14 India Street; the notion that John must have lived there because he purchased the house is mere surmise. Certainly, his mother had not purchased 31 Heriot Row, it was actually John's brother George who had done so; living there only occasionally, he had never bothered to list himself at that address. John had been in full possession of his Middlebie estate since reaching his majority in 1811, and from this he drew a reasonable income in rent. The building of the house at India Street may have been his part of the deal to facilitate the sale of 31 Heriot Row, and for him it could have represented both an addition to his portfolio of properties and an amenable place for his mother to live nearby, in all likelihood with her step-mother and some of her half-sisters.

That is how it seems to have been. In 1804 John Irving married and left the Irving family home at 5 Buccleuch Place to set up house at 73 Princes Street. In the following year, his brother Alexander Irving also decamped and moved to 27 Heriot Row, and so it is fairly certain that their mother, Mrs Molly Irving, would have accompanied him to keep house, and some of their sisters may have lived with him and some with John. After Isabella's marriage in 1813, however, it would appear that only Mrs Clerk and John Clerk Maxwell remained at number 31. It seems only natural that some the Irving ladies would then have been lodged with Mrs Clerk, who was just doors away. But when Alexander married in the following year and by 1815 had started a family of his own, it is likely that the majority of the Irving ladies would have been accommodated at number 31. The probable situation, therefore, is that from 1821 the same arrangement would have been continued at 14 India Street.

Even when Mrs Janet Clerk died in March 1822, John Clerk Maxwell did not list himself at 14 India Street, which surely he would have done had he actually been living there. In fact, after James Wedderburn's untimely death later in the same year, John Clerk Maxwell would have felt inclined to remain close to his sister and his nieces and nephews. Indeed, no-one else was listed at 14 India Street until 1827. The most likely possibility is that Mrs Irving and some of her daughters continued to live there until about 1825, when she re-emerged just a very short distance away at number 38 in the very same street. It cannot be ruled out that Mrs Irving had lived there for some time before choosing to list herself in the directory; alternatively she may have chosen to retire to the family seat at Newton, where in fact some of her daughters were living as late as 1841.[3] Only when we reach the year 1827 can we be certain about the residents of 14 India Street, for it was then occupied by the recently married John Clerk Maxwell and his wife, Francis Cay.

Before we leave this intriguing question, however, let us recall Lewis Campbell's apparent mistake in saying that 'Mrs Clerk died there in the spring of 1824', when in fact we have found definitive evidence that she actually died on 25 March 1822. If the elderly person still living at 14 India Street was Mrs Irving she may well have been mistaken for Mrs Clerk. Had Mrs Irving left there in the spring of 1824, this would clearly tie up with her listing at the number 38 having appeared at the following Whitsun, 1825. It may simply be a case of mistaken identity some sixty years after the fact on the part of Campbell's informant.

The House

We now turn our attention to the house itself. Like all the even-numbered houses, number 14 lies on the west side of the street (Colour Plate 1). Built in 1820 and occupied by 1821, it is of the same Georgian neo-classical style as all the other houses in the upper part of the street, which slopes fairly steeply to the north, intersecting firstly with Circus Gardens on the right and then Gloucester Place on the left, before coming to an abrupt end at North West Circus Place below, the only direct access to which is via a steep flight of stairs.[4]

The street itself, like many others in the New Town, is cobbled with hard-wearing causey stones, with the pavements about two feet higher, so that beyond the kerb there is a sloping skirt of cobbles that leads down to the gutter channel, hewn out of the same hard stone as the causeys. The New Town pavements were originally made of Caithness stone, dressed flat and square, and while some of these pavements remain, in the vicinity of 14 India Street they have been replaced by utility concrete paving slabs that are hardly so attractive. Near each house there is a lighting step that projects beyond the kerb; though redundant now, they allowed people to get in or out of their carriages easily, and without stepping into the probably very dirty street. Given the ladies' fashions of the time, and the torrents that can flow down the gutters after heavy rain, it was just as well.

By 1820, gas lighting had come to Edinburgh and by 1826 most of the New Town was equipped with it (Shakhmatova, 2012). The original gas lamp, though now electrified, still stands in front of number 14. The house itself is on four levels. Built on the original ground level, there is a basement, with doors to both the back and front, which housed the kitchen and domestic staff. The rear door gives access to the garden, which would have been used as a drying green. The garden, also at more or less the original ground level, stretches westward to a small building that was once the coach house and stable. It now comprises a mews flat above, and a garage below, both of which face onto Gloucester Lane. The front door of the basement opens onto a paved 'area' with several storage cellars built under the pavement level. A staircase on the north side of this area allowed servants and tradesmen access to the street outside, while its south side shelters under a sort of bridge formed with massive stone slabs that, after a few steps up, leads from the pavement to the main front door of the house. Here an ancient and faded brass plate advises callers not to ring the pull-bell 'unless an answer is expected'; but why else would one ring?

Despite its degree of elevation above the level of the street in front and the garden to the rear, we will refer to the floor entered from the main door as the ground floor, and in keeping with British usage the floor above that is the first floor and so on. This is mere nomenclature, for the first floor, with the ground floor and basement beneath, is the third level of the house! After a spacious entrance hall with its high ceiling decorated with a large stucco oval moulding, two Ionic columns form an entrance to a lofty stairwell, off which are the doors to the rooms ahead and to the right; while to the far left a wide, carpeted open stairway hugs the walls on its way up to the first and second floors. This stairway is illuminated by an oval glazed cupola in the ceiling that admits a good deal of daylight, but at a price; the grandness and airiness of the stairway requires it to be both high and wide, so that it takes up a space that could have been devoted to three additional rooms.

The high-ceilinged ground floor rooms comprised the dining room (now the exhibition room) to the front and a small office to the rear, the former looking out on the street and the latter over the garden. While the dining room is visible from the street, its elevation above street level, and its distance back from the pavement, gave the occupants relative privacy, and at night time, when the rooms were lit, panelled wooden shutters, which are in use to this day, could be folded out to keep warmth in and prying eyes out. An unusual feature of rooms to this design is that the wall facing the window is cylindrically curved so that the entrance door and the matching 'press' door on the opposite side of the room are likewise curved, which looks handsome but must have been costly to make. On the north wall of both rooms are marble fireplaces, with the one in the dining room being large and, although handsome, not ostentatious.

The two main rooms on the first floor are the drawing rooms. They are as large and as high-ceilinged as the rooms below; they also have grand white marble fireplaces, in particular the front room. The front drawing room is the entire width of the house and may be entered either from the landing on the central stairway or from the adjoining rear drawing room, to which it is connected by a wide double door which may be swung wide open so that both rooms can be used as a single, large, L-shaped area. One can imagine a party taking place in which the carpet would be rolled up for dancing in the large drawing room, with space for the musicians and for people to sit out in the other. The rear room has a window that reaches from floor to ceiling, where one can step out onto a cast iron balcony, such as those that can also be seen on the front of some of the houses in the street.

The rooms on the upper floor are somewhat smaller and would have provided, say, three bedrooms, a dressing room, a nursery and perhaps a water closet, but the attic space above was never converted for additional accommodation. One quirky feature of the upper floor is that of the three window spaces to the front, only those on the right and left are glazed, and the central one is merely a dummy. Formed by recessing the ashlars to look like a window space, it allowed two similar sized rooms to the front of the house without unduly affecting the fairly regular frontage of the street. Not only that, it saved on window tax!

For internal lighting, many such houses would have used oil-lamps supplemented by candles, but it was not long before gaslight, already used for street lighting, came inside. According to MacIvor (1978), however, no gas at all was used at 14 India Street until the twentieth century. He attributes the fairly unspoiled original nature of the house to the fact that, as it was rented out until 1891, there was no resident owner hankering to make significant alterations, and little incentive for the landlord to do so either. The occupants were mostly elderly, and so keeping the status quo probably suited them.

The houses were not always provided with some of today's essential requisites when they were first built and as already mentioned some of the main access roads, such as Howe Street, were not properly built up. Such problems also caused delays in getting a connection to the common sewers and town water supply. Even some of the earlier houses in the Second New Town did not have provision for an inside water pipe or water closets; water was originally brought in to the First New Town houses by a daily army of water caddies (Youngson, 1966, p. 241). Perhaps, after centuries of being accustomed to the circumstances of the Old Town where water was fetched from communal wells and waste was despatched from any convenient window directly into the street, the early designers and inhabitants of the houses in the New Town were slow to catch on, but the truth is that the houses simply advanced much faster than did innovations in plumbing. But according to Rodger (2004, p. 65) even if there were cases of delay in getting connected up, by 1809 most New Town houses were being built with at least basic sanitation and a water supply. We may assume, therefore, that as far as 14 India Street is concerned, they were there from the start.

As to the water supply, it seems that a spur taken from the main supply pipe in India Street fed a cistern in one of the outside cellars off the front basement area. From there, the water would generally be fetched and carried to wherever it was needed within the house itself. Nevertheless, there had also been a pump, the purpose of which is not clear, but perhaps it was used later on to pump water into a cistern within the house itself. While there were likely to have been water closets, for example, located where the present toilets are on the ground and first floors, they could not originally have been flushed by a self-filling cistern for there was no internal water supply; they would generally have been flushed manually from a jug of water kept at hand and replenished by the servants. Some houses collected rainwater from the roof for the purpose, but whether this made any difference to the method of flushing is not known; at least it would have provided the possibility of fitting one of the early forms of flush toilet that was then becoming available.

A fairly minor feature of these houses, but which was nevertheless of great interest to James Clerk Maxwell, was the system of call bells (still in the basement) used for summoning the servants. The brass bells themselves hung where servants could both hear and see them, and after a bell was rung they would have had to watch for which one was still bouncing on its spring in order to be able to tell from which room it was rung. The mechanism was a system of wires, cranks and pulleys that ran from each bell-pull down to the basement, usually hidden within channels in the walls. Lewis Campbell refers to the story of the very young James being fascinated by the bells at Glenlair (C&G, p. 27) and if, as we suspect, he last lived at 14 India Street when he was just about two years old, it is highly likely that, as an inquisitive and active toddler, his early fascination with them first came to light there.

The Clerk Maxwells and Glenlair

When in 1809 John Clerk Maxwell came to 31 Heriot Row he was a young man of eighteen or nineteen. Frances Cay was then about seventeen and lived with her family just a block away at 11 Heriot Row. Nevertheless, it was Frances' elder brother John who first knew John Clerk Maxwell (Chapter 14) and their friendship continued for many years before any prospects of marriage arose between John and Frances; perhaps John had some other prospect in mind before he set his heart on Frances, but at any rate they would have known each other for some considerable time before they were married on 4 October 1826 in Frances' own church, St John's Chapel, Edinburgh.[5]

It seems that it did not take long before Frances suggested that John should try to do something with part of the Middlebie estate that he had inherited from his grandmother, Dorothea Clerk Maxwell. While it did bring him an income, he had as yet no house on it that he could call a country seat; the project would be a considerable challenge, but they were game for it:

> The pair soon conceived a wish to reside upon their estate, and began to form plans for doing so; and they may be said to have lived thenceforth as if it and they were made for one another. (C&G, pp. 4–5)

And so it was decided that they might live at Nether Corsock,[6] which was then a piece of farmland in need of improvement and with nothing more than a few buildings on it. What it was that turned their attention to this particular portion of land can only be guessed at. It may have been that Nether Corsock had already become their place for the customary summer vacation, or perhaps the lease was due for renewal, or it offered some potential for development; or its situation was particularly appealing; or, even more likely, any combination of such things. Their idea of moving there to embrace an agricultural life tends to corroborate the idea that John was not going to be a great advocate, he was merely: 'doing such moderate business as fell in his way, and dabbling between-whiles in scientific experiment' (C&G, p. 4).

When the couple were staying at 14 India Street during the winter of 1830, John received a telling letter from his uncle, John Irving WS, informing him:[7]

> There is a case of Mr Stirlings in $2^{\underline{d}}$ Dec$^{\underline{r}}$ tomorrow. I have feed by my brother's advice Mr Arch$^{\underline{d}}$ Bell to attend to it, as you and [he] may be concerned in some other cases of Mr Stirlings you might attend also although *I am sorry I cannot fee two counsels.* [My emphasis]

John Clerk Maxwell was clearly still taking the opportunity to practise at the bar when he and Frances were overwintering in Edinburgh, but the wording of this passage seems to have been carefully chosen to avoid embarrassment by letting him down gently, for he was not the first choice of counsel, nor was it worth feeing him as a second. John Irving tactfully insinuates that the recommendation of Mr Bell as counsel was not his, but his brother's. Since Alexander Irving was by then the judge Lord Newton, his recommendation could not be disregarded. By way of consolation, however, he reminds John that he and Frances have not yet let him know when they can come and dine with him.

Like his grandfather George, it turns out John Clerk Maxwell was a practical man, but in place of his grandfather's impetuousness he came to decisions slowly and carefully, and had a talent for managing things well, and in much detail:

> In matters however seemingly trivial – nothing that had to be done was trivial to him – he considered not what was usual, but what was

best for his purpose. In the humorous language which he loved to use, he declared in favour of doing things with judiciosity. One who knew him well describes him as always balancing one thing with another exercising his reason about every matter, great or small. (C&G, p. 8)

He therefore must have been far from being the kind of dyed-in-the wool, cut and thrust advocate that his cousin once removed, Lord Eldin, had been. Not only did John Clerk Maxwell have a tendency to reason slowly and deliberately, he inclined in an altogether different direction from the minutiae of the law; he was more interested in the very latest things in science, manufacturing and machinery (C&G, pp. 7–8).

While most of the landed gentry had both a town house and a country estate, Nether Corsock was inconveniently far away from Edinburgh, more or less two days distant, and as we know, it did not then boast a dwelling house befitting a country laird and his family. Thus, a project was formed that John could throw his heart into; in 1830 he started building a decent, practical house. Albeit a far cry from the refinement of Heriot Row and India Street, it was *to his own design*. Once they had settled into the house that for the time being they called Nether Corsock, the journey back to Edinburgh would be contemplated only when it was strictly necessary, for example, the winter vacation in Edinburgh, and matters of business or social necessity. The railways were yet to come to this area and travel by horse and carriage was a great inconvenience. At least there was a new road, built by Telford in 1824, from Dumfries to Glasgow, which would get them as far as present-day Abington, but according to Lewis Campbell the journey was still grim indeed.

During 1828 and 1829, John Clerk Maxwell did not list himself at 14 India Street in Edinburgh. All the same, the house remained his, and while no-one else listed themselves there, it is not to say that it remained empty; that we simply cannot tell. In the December of 1829, however, the Clerk Maxwells lost their first child, a girl called Elizabeth who died at just one day old. It may be inferred that the child was born in Edinburgh for she is buried in Restalrig Cemetery, just to the east of the city, in the tomb of John Cay (Grant, 1908). At any rate, when they realised that Frances was once again pregnant, they were very probably already back in Edinburgh, for John had relisted his name at 14 India Street for the year 1830, and continued to do so until 1833.

In view of the unhappy outcome of Frances' first pregnancy, the couple would have been keen to minimise the risks. Frances therefore remained in Edinburgh while John made the journey to Nether Corsock as and when he needed to collect rents or progress the building work. In fact, he placed a time-capsule[8] containing plans of the house and other mementoes, under the floor on 25 March 1831, celebrating the laying of the foundations just eleven weeks before James was born.

In between times, when they were in Edinburgh, the community of tenant farmers, servants and other workers on the estate would carry on just as their laird would have expected them to, irrespective of his absence. By the same token, while they were both at Nether Corsock, there would have been someone keeping an eye on 14 India Street. Exactly how the couple apportioned their time between their country and town residences is uncertain, but from 1834 onwards 14 India Street was rented out and 'Old 31' became the stopping-off point in Edinburgh on any future visit or vacation.

Rented Out

After James was safely brought into the world, the family used 14 India Street as their Edinburgh winter residence until the spring of 1834. Lewis Campbell does not say exactly when they gave up living there, and John Clerk Maxwell's diaries, to which he often refers, seem to have been lost after Campbell's biography was written.[9] The first mention of them being settled back at Nether Corsock comes in April 1834 when James had reached the age of two years and ten months (C&G, p. 26). Also, in the following month, we find 14 India Street listed in the Edinburgh directory under the name of John McFarlan, Surgeon, and thenceforward there is no further reference to John Clerk Maxwell or to his son James ever again living there.

The Register of Sasines for Edinburgh shows that John Clerk Maxwell did not sell the property, and the Edinburgh directories for the ensuing years show that he must have rented it to several different tenants. Of course, they would all have had to be of the requisite social standing, and gauged able to pay the rent on time and to keep up the strictest moral standards (or at least a good appearance of it). They included such people as John McFarlan, Colonel and Mrs Thomas Wardlaw, and a Mrs Simson and her daughters.[10]

The ownership of 14 India Street eventually passed out of the hands of James Clerk Maxwell and his heirs in 1891. We will now address the question of who actually owned it up until that time.

Ownership after John Clerk Maxwell

The Register of Sasines reveals what became of the house after John Clerk Maxwell's death. In June 1856 we find the entry to be just as we would have expected:[11]

> JAMES CLERK MAXWELL of Middlebie, as heir to John Clerk Maxwell, Advocate, his father, Seised,– in an Area or Piece of ground with the Buildings thereon, on the west side of INDIA STREET, Edinburgh.

When James Clerk Maxwell was in his forty-third year and presumably thinking about securing the financial future of his wife in the event of his predeceasing her, we find this disposition of 14 India Street in her favour:[12]

> Disp. by JAMES CLERK MAXWELL of Middlebie, to Katherine Mary Dewar or Clerk Maxwell, his wife, exclusive of his rights [...] with the Buildings thereon, on the west side of INDIA STREET, EDINBURGH. Dated Dec. 26, 1873.

Following James' death in November 1879, his executors filed a notary instrument concerning a trust disposition on behalf of his widow appears and, again, the subject is 14 India Street.[13] Finally in 1887, following Katherine's death on 12 December 1886, we have her own disposition to the trustees followed by their ensuing disposition to the beneficiaries:[14] 'Louisa Helen Mackenzie and Jean Charlotte Mackenzie, both residing at No. 25 Ainslie Place, Edinburgh'.

The Mackenzies

But who were the Misses Louisa and Jean Charlotte Mackenzie? To answer that question we must briefly investigate yet another branch of James Clerk Maxwell's family. We already know that James Clerk Maxwell's first cousin, Janet Isabella Wedderburn, had married James Hay Mackenzie; and Colin Mackenzie, whose name has appeared frequently in our story, was their son (Chapter 9). Colin may have been the one in line to inherit 14 India Street, but he had died just a few years after James.

The main clue given in the record as to the origins of Jean Charlotte and Louisa Helen Mackenzie is their address, 25 Ainslie Place, which cropped up in Chapter 9 as the residence, from 1861 until their deaths, of James' aunt, Isabella Wedderburn, his cousin George Wedderburn, and his cousin once removed, Colin Mackenzie. After Colin's death in 1882, his two sisters Louisa Helen and Jean Charlotte were indeed listed there from 1884 until the year 1897.

Lewis Campbell (C&G, p. 60) brings Colin into the story through one of James' letters written home in 1844 from 'Old 31'. James recounted to his father a visit to his cousin Janet Isabella at her new house, called 'Marine Villa' or 'Silverknowe', near the coast just a few miles to the north-west of the New Town. James played on the shore with some other boys, and Colin, then just three years old, was there with them. James informed his father that Colin was known by the baby name 'Coonie', which James wrote in Greek letters as 'κυνη'.[15]

During James' final days, 'his cousin, Mr. Colin Mackenzie, who acted the part of a brother at the last, as he had done many a time before', came to Cambridge to be by his side (C&G, p. 411). Just before the end, James called on him for assistance: 'Colin, you are strong, lift me up' (C&G, p. 416). Since James and Katherine had no children of their own, James may have considered leaving some of his property to Colin. The estate and the house at Glenlair, being tied in with the Middlebie entail, were ultimately destined for another cousin, Colin's uncle, Major Andrew Wedderburn, later Wedderburn Maxwell (1821–1896), who had made his career in the Madras Civil Service. His older brothers having predeceased him, Andrew Wedderburn was next in line as heir of entail to Glenlair and Middlebie. Colin would be fully aware of the situation, for he was a Writer to the Signet and he not only acted as James' solicitor, James had appointed him as one of his executors. When Colin advised James about those issues he needed to be aware of in drawing up his first will in 1866, he raised the question: 'What of the house in India Street? Is Mrs Maxwell or your heir to get that?'[16] James' answer was, surprisingly, that the house should not go to his wife; rather, it should go to his heir.[17] But his rationale for that decision may have been that Katherine had never lived in Edinburgh and would have been lonely there. His strategy was to will her his moveable estate, i.e. household goods, money and so on, and the lease of their current house in London. For whatever reason, he had had a change of mind when he put 14 India St into Katherine's hands in December 1873.

Notwithstanding that the house was now Katherine's, any chance of Colin Mackenzie eventually succeeding to 14 India Street perished when he died less than three years after James, at which time Katherine was still alive. His demise is mentioned by Lewis Campbell when finishing the foreword of his book in August 1882:

While the last sheets were being revised for the press the sad news arrived that Maxwell's first cousin, Mr. Colin Mackenzie, had died on board the Bo[th]nia, on his way home from America. There was no one whose kind encouragement had more stimulated the preparation of this volume, or whose pleasure in it would have been a more welcome reward. (C&G, p. x)[18]

At his own request, Colin was buried beside his uncle, George Wedderburn, in Edinburgh's Dean Cemetery. George had died of consumption in 1865 (Chapter 9), and we may surmise that being unmarried and childless that he would have left his house, 25 Ainslie Place, to one of his nephews. In the event it was Colin who appears to have either inherited it, or had the liferent of it, for in the year 1865 his name was the only one listed at that address. It is exceptional that this was so, for George Wedderburn died on 1 May of that year, whereas Colin would have had to submit the revised directory entry some months before that date. Colin, who had just turned twenty-four at the time of George's death and had only recently qualified as a Writer to the Signet, may well have been living at 25 Ainslie Place since early on in that year, for his father, James Hay Mackenzie, had died in February, just months before. George must have known his own fate was looming by then, and perhaps he had the generosity to make available to Colin what was soon going to be his in any case. This would help to explain how George and Isabella Wedderburn's listings were so speedily replaced by Colin's.

Interestingly, the Blackburns (that is to say, the in-laws of Jemima and Jean Wedderburn), who were also related to the Wedderburns from two generations previous, chose not to attend Colin's funeral. According to Fairley (1988, p. 77) there had been 'some whiff of scandal' about it. Neither Colin nor George had ever married, and they had been frequent companions. When George was diagnosed with consumption in 1860, he was told by his doctors to take a long holiday in a dry climate. He therefore went on a visit to Egypt taking not only his mother with him, but also Colin Mackenzie, then about the age of twenty (Fairley, 1988, p. 162). It is not possible to apply today's attitudes to the situation, because platonic relationships between men, and between women, were then seen as being quite regular. Nevertheless, in the Victorian era it would only have taken even the vaguest hint, true or untrue, that things were not quite as they should be to bring a swift condemnation. Fairley seems to be suggesting that the Blackburns had indeed caught such a whiff.

Returning now to the inheritors of 14 India Street, Katherine's trustees did, as we have seen, alight upon Colin's two surviving sisters, Jean Charlotte and Louisa Helen of 25 Ainslie Place, who therefore came to inherit 14 India Street, an address they never cared to live at. Instead, they stayed on where they were and simply collected the rent from Misses Anne and Mary Simson until 1891, when they sold it to them. Mrs Simson appears to have been Mary Shepherd (1794–1871), widow of Henry Simson, minister of the Chapel of Garioch, Aberdeen (Cowper & McIver, 1993). They had two daughters, Anne (1829–1915) and Mary (1839–1914).[19] It could be that Louisa Helen died in 1897 or 1898, by which time there were no Misses Mackenzie at 25 Ainslie Place. As we can see from the gravestone she shared with Colin, Jean Charlotte Mackenzie lived until February 1926. Curiously, there is no mention of Louisa. Their parents, James Hay Mackenzie and Janet Isabella are buried at St Cuthbert's and St John's cemetery, little more than a mile away, but Louisa is not mentioned there either.

Had Colin Mackenzie lived, he would have had precedence over his sisters and would probably have inherited the house his cousin was born in. Curiously, 25 Ainslie Place is on the corner where it abuts with 6 Great Stuart Street, where Isabella's sister-in-law, Jane Cay, had once lived.

14 India Street after the Misses Mackenzie

Louisa Helen and Jean Charlotte Mackenzie were the last of James Clerk Maxwell's family to own 14 India Street. In 1891 they sold the house to Mrs Simson's daughters, Mary and Anne. Mary moved out on Anne's death in 1914, after which the house went unlisted until 1919, possibly having been used in support of the national war effort. According to the Register of Sasines,[20] from then on the house had four different owners and sadly ended up being altered for renting out as separate flats. In 1978, however, it was rescued from further depredation by Iain and Marion Macivor, who carefully restored the house to very near its former state (MacIvor, 1978). They also repurchased the garage and workshop to the rear of the house, but the flat above is still in separate hands.

Finally, in commemoration of James Clerk Maxwell, 14 India Street was acquired for preservation by the James Clerk Maxwell Foundation,[21] who turned the ground floor dining room into an exhibition room open to visitors; the study to the rear was turned into a library with books concerning James Clerk Maxwell and his works; the first floor drawing rooms were made into a presentation suite accommodating forty people; and finally, the unused rooms on the upper floor, the basement flat and the garage, are all presently rented out to secure an income to cover the upkeep of the house, with any surplus going to charitable causes associated with education in science. From 1994 to 2010, the rented out accommodation was home to the International Centre for Mathematical Sciences (ICMS).

Epilogue

What happened to 14 India Street after James Clerk Maxwell's death is now clear, but of his wife Katherine we hear only that until her death on 12 December 1886 she continued to live in Cambridge at 11 Scroope Terrace, the lease of which James had left to her, and that she was buried beside him within the walls of Old Parton Kirk.

After Andrew Wedderburn succeeded to Glenlair and Middlebie in 1879, he undertook several alterations at Glenlair. He enlarged the rear section of James Clerk Maxwell's new wing to Glenlair House in order to provide a billiard room and extra bedrooms. As he needed more servants' quarters, he also added to the external buildings adjoining the house and converted the attic space in the original part of the house built by John Clerk Maxwell. The change in its external appearance of the front elevation was subtle, betrayed by some new dormer windows.

From Andrew Wedderburn Maxwell, Glenlair passed to his son Major James Andrew Colvile Wedderburn-Maxwell in 1896, and from him in 1917 to his son Brigadier John Wedderburn-Maxwell, a soldier who saw service in both World Wars. After the First World War, the estate would not pay its way, but one means of making some money towards its upkeep was to let it out to wealthy businessmen for shooting parties that were in demand for two to three months in the late summer and autumn of each year (Wedderburn-Maxwell, 1985). And it was during such a let in 1929 that the heat from a fierce kitchen fire caused a beam in the converted attic above to catch alight. In such isolated houses, there was no easy way to tackle the blaze and, when a fire appliance did eventually arrive on the scene, they could not get water.[1] For whatever reason, they were not able to prevent the fire spreading along the roof and into the newer wing. Had there been the equivalent of a fire door in the connecting passageway, the newer wing might have been saved. As it was, it probably fared worse.

After the fire, the partially gutted 1860–90 wing was left to the elements, while Brigadier Wedderburn-Maxwell was obliged to reclad the roof of the old house in order to provide what he may at first have thought of as temporary living quarters. This suggests that the ground level floor and the internal walls may have been left sufficiently undamaged to leave a basis upon which to build the structure seen in later photographs.[4] Executed in makeshift fashion with corrugated iron and wooden boards, a loftier roof space was formed under a single lengthways ridge.[2] This roof outline was certainly not there in 1928,[3] the year before the fire, as a photograph taken in that year clearly shows. Interestingly, at the time of writing, and for reasons unknown, several US corrugated iron suppliers refer to the roof of the resurrected Glenlair House in testimony to their product almost as though it had been one of the original design features![5]

That the owners of Glenlair would live under such conditions seems surprising, but although they had the land, they had no great family fortune. Following the Great War there had been years of economic difficulty in the UK, and in 1929 the world was hit by the Great Depression, so that any roof over one's head was better than none.

Given that he had disentailed the estate in 1922, it may be that Brigadier Wedderburn-Maxwell sold Glenlair even before the fire, but there were two main factors against this. The first was the legacy of the Great War, which not only badly depleted the male workforce, especially in the countryside, but also changed social attitudes towards the landed classes to an extent that made running a country estate a burden rather than a benefit. Secondly, as already mentioned, there was a dismal economic climate in which there were far many more prospective sellers than buyers of such properties. He did, however, manage to sell Glenlair Mill (just over six acres) in 1930, but it was not until after the Second World War that the remainder of the estate was sold, to John Liston Dalrymple, on 5 September 1946. In 1949, it was broken up and sold in lots, the last one of which, Glenlair Home Farm (127 acres), was purchased by the father of the current owner, Captain Duncan Ferguson, in February 1950.

Captain Ferguson, who was brought up at Glenlair, established the Glenlair Trust, which has been responsible for restoring the old house and the vestibule of James Clerk Maxwell's wing, and it is a work in progress (Glenlair Trust, 2012). He lives with his wife Henrietta in the same gardener's cottage that John Clerk Maxwell and his wife Frances Cay lived in nearly two centuries ago before John had built their new house that he was to call Glenlair. The Clerk Maxwells made it their home, the home in which the young James Clerk Maxwell was raised between the years of two and ten, and where he came back every summer in the years thereafter. After his father's death, the lower right front room on the ground floor[6] became the study in which he worked on many of his papers. From 1866 to 1871, Glenlair was his main residence, during which period he fulfilled his father's original concept of an additional wing to the house. It was unfortunate that he had no son of his own to benefit from the adventures and freedoms that he himself had enjoyed there as a boy, or from the sort of caring and devoted relationship he had with his own father, the creator of Glenlair.

APPENDIX 1
Maxwell's Innovative Method of Drawing True Ovals

The well-established method for drawing an ellipse is shown in Figure A1.1.

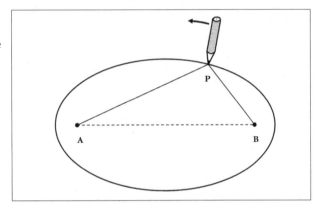

Figure A1.1 The conventional way of constructing an ellipse

Maxwell's ingenious idea involved wrapping the thread back on itself, as the rope does in a block and tackle, so that when the pencil moves, more thread will now pass round it on one side than on the other. The simplest form, shown in Figure A1.2, has the thread tied to the pencil point P then looping round pin B and sliding round P again before being finally tied at A.

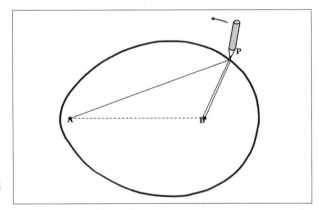

Figure A1.2 Maxwell's innovative method of constructing a general oval

APPENDIX 2

Awards and Commemorations for James Clerk Maxwell

Prizes and Honours (C&G, pp. 431–432)

1846 Successful Competition in Geometry, 5th Class Prize Medal, Edinburgh Academy
1847 Mathematical Medal, 6th Class, Edinburgh Academy
1851 Smith's Prize, University of Cambridge Mathematical Tripos (shared)
1856 Fellow, Royal Society of Edinburgh
1857 Adams Prize, University of Cambridge
1860 Rumford Medal, Royal Society of London
1861 Fellow, Royal Society of London
1866 Bakerian Prize Lecture, Royal Society of London
1869 Keith Medal, Royal Society of Edinburgh
1870 Honorary LLD, University of Edinburgh
1874 Foreign Honorary Member, American Academy of Arts and Sciences of Boston
1875 Member, American Philosophical Society of Philadelphia
1875 Corresponding Member of the Royal Society of Sciences of Göttingen
1876 Honorary DCL, Oxford University
1876 Honorary Member, the New York Academy of Science
1877 Member of the Royal Academy of Science of Amsterdam
1877 Foreign Corresponding Member, Mathematico-Natural-Science Class of the Imperial Academy of Sciences of Vienna
1878 Volta Medal and Honorary Doctor of Physical Science, the University of Pavia

Commemorations (various online sources)

Commemorative plaques at his residences in Edinburgh, Aberdeen, London and Cambridge, and at Old College, Edinburgh University, and the Cavendish Laboratory
Commemorative postage stamps by San Marino and Mexico, and a souvenir sheet by the Republic of Mali
Glenlair Preservation Trust, Glenlair (restoration of Glenlair House)
James Clerk Maxwell Foundation, 14 India Street, Edinburgh
James Clerk Maxwell Medal, IEEE, RSE
James Clerk Maxwell Memorial Window, Corsock Church
James Clerk Maxwell Prize for Plasma Physics, American Physical Society
James Clerk Maxwell Statue, George Street, Edinburgh
James Clerk Maxwell Telescope, Hawaii
James Clerk Maxwell: The man who changed the world. BBC TV documentary (2015)
Maxwell: an asteroid
Maxwell: a crater on the Moon
Maxwell (abbreviation Mx): the CGS unit of magnetic flux
Maxwell's displacement current: a concept which completed electromagnetic theory
Maxwell's electromagnetic stress tensor
Maxwell's equations: for the electromagnetic field
Maxwell–Boltzmann distribution: of the velocities of molecules in a gas
Maxwell's Demon: a well-known thought experiment originated by Maxwell but named as such by Lord Kelvin
Maxwell Montes: a mountain range on Venus
The James Clerk Maxwell Building, King's Buildings, University of Edinburgh
The James Clerk Maxwell Building, Waterloo Campus, King's College, London
The James Clerk Maxwell Centre, Edinburgh Academy
The Maxwell Building, Science Park I, Singapore
The Maxwell Centre for the Physical Sciences, New Cavendish Labs, Cambridge
The Maxwell Gap: in the rings of Saturn
The Maxwell Institute for Mathematical Sciences, Edinburgh and Heriot Watt Universities (jointly), Edinburgh

APPENDIX 3

James Clerk Maxwell's Legacy to Science and Mankind

As we have seen, Maxwell made many great contributions to science, but if we had to choose the one that has had the greatest impact on our lives today, we must single out his work on the foundations of electromagnetic theory. It eventually resulted in the development of a technology that almost every person today finds indispensable. In addition, it enabled Albert Einstein to make his great leap forward with the theory of relativity. The history of the world since about 1900 would have been completely different were it not for these two great pillars of modern physics, electromagnetic theory and the theory of relativity.

By the eighteenth century the subject of mechanics was well developed, and while there had been some study of electricity and magnetism, but little that could be put to practical use. By the early nineteenth century, work on electromagnetic theory, that is to say, on the magnetism that is produced by electric currents rather than by magnetic materials, was just beginning, and by the middle of the century it was still incomplete. On the other hand, thermodynamics, the theoretical study of how heat can be put to work, was making great progress and being applied to the further development of the key invention of the industrial revolution, the steam engine. Maxwell's friend William Thomson was at the forefront of this research, from which came some fundamentally important physical revelations, for example: work can be readily turned into heat whereas the same amount of work cannot be recovered from that heat; perpetual motion is impossible; heat only flows from a hot body to a colder one, and never vice versa without work being done (see 'Maxwell's Demon' in Chapter 2). By the end of the 1850s electromagnetic theory had some catching up to do, but James Clerk Maxwell had already been at work on the ideas of Michael Faraday.

Electricity and magnetism had long been known as separate subjects until the year 1820, when Hans Christian Oersted discovered that an electric current creates a magnetic field. So began the study of electromagnetic phenomena. In 1832, Michael Faraday had also discovered that a changing magnetic field gives rise to an electric one. By then, the working knowledge of electromagnetism was such that William Sturgeon had invented the first practical electrical motor (Day & McNeil, 1996, pp. 1178–1179) and Hippolyte Pixii the first dynamo, a type of electrical generator (Day & McNeil, 1996, p. 970). James Clerk Maxwell would have been inspired by seeing an exhibition of such things in 1842 when he was a boy at school in Edinburgh. A theoretical understanding of the subject was only then nudging along in piecemeal fashion with major contributions from Andre-Marie Ampére, Jean-Baptiste Biot, Félix Savart and again Michael Faraday, all of whom have given their names to various aspects of the subject.

Although it did not appear in print until 1864, James Clerk Maxwell's first essay on the subject, 'On Faraday's Lines of Force', was given in December 1855 and February 1856 to the Cambridge Philosophical Society, when he was just twenty-four years old (Maxwell, 1864a). The paper opens with the observation, 'The present state of electrical science seems peculiarly unfavourable to speculation', which of course indicated that the situation was particularly favourable for further enquiry by Maxwell himself! For him it was just the beginning, and by 1861 he had published the first part of a major four-part paper, 'On Physical Lines of Force' (Maxwell, 1861–62), in which he drew together the known laws of electricity and magnetism as they stood at the time and, by way of a mechanical analogy, attempted to explain how a material (generally called a medium) could carry and transmit electrical charges, currents and forces. Although this formulation of the theory was able to support the existence of electromagnetic waves and, indeed, to identify light as being a form of electromagnetic wave, his mechanical analogies were taken too literally, and he therefore decided to abandon them and republish his findings based on the concept of electromagnetic fields which, being abstractions, could not be denounced as being improbable or unrealistic. In 1864, almost nine years to the day after presenting his first paper on the subject, he read 'A Dynamical Theory of the Electromagnetic Field' to the Royal Society of London (Maxwell, 1865). The major equations and conclusions were the same as before, but this time the paper was accepted and in due course it became the cornerstone of modern electromagnetic theory.

But if Maxwell had merely drawn together the laws of electromagnetic theory as they stood at the time, he would not have achieved this at all. In picking out the distinct laws that are essential to the theory, he chose in particular to consider what constitutes an electrical current. It was then held that an electrical current was the result of a flow of electrical charges around a circuit, something which must be familiar to almost everyone today, when we can flick a light on with a switch that completes a circuit to let the current flow; we can also see the effect of the current on the light-bulb filament, which glows bright from the heat created by the frictional effect of the filament wire on the current flowing through it. But now Maxwell was thinking more generally, and took the view that any motion of electric charge would constitute a current, irrespective of whether or not it was able to move around a complete circuit.

His rationale was that even insulating materials like glass or amber contained electrical charges imprisoned, as it were, within them, for charge could be taken away from their surface simply by friction, say by rubbing them with a cat's fur. If such an insulator were placed in an electric field, for example by placing it near some charge that had been collected in this way, the charges trapped within the insulator would be attracted or repelled and, despite not being able to flow freely to the surface, they would nevertheless be displaced to some small degree. But any motion involved in that displacement is only a momentary thing, so that the displacement itself reaches a limit and cannot therefore constitute an ongoing current. However, what would happen if the applied electric field were then to be reversed in direction? By that time it was well known, from Faraday's law of induction, that a rotating magnet would not only produce an alternating magnetic field along a given direction, that is to say one that continually reverses in direction in time with the rotation of the magnet, but also an alternating *electric* field. Maxwell reasoned that his displaced electrical charges within the insulator would follow the alternating electric field and would be displaced first in one direction and then the other, so forming an alternating current: 'This displacement [itself] does not amount to a current, because when it has attained a certain value it remains constant, but it is the commencement of a current, and its variations constitute currents' (WDN1, p. 491).

In a normal steady current, or direct current, charges flow continuously, for example, round a loop of conducting wire. In an alternating current, however, they need only vibrate back and forward without giving rise to a continuous flow. Maxwell called his new sort of current a 'displacement current', and he must have had great satisfaction in noting that, like any other current, it would give rise to a magnetic field. This was a new phenomenon that no one before Maxwell had ever considered, and so it necessitated a change to the electromagnetic equations as they were then known. The entire concept came to him as a result of thinking deeply about what must happen when molecules in an insulating body are subjected to an electric field, and even although his resulting insight was simple, it was a turning point; so much so that it was in effect the key to the mystery: what is light?

While it was known that light was likely to be some sort of wave, until then there had been no clear idea as to what sort of wave it might be. In 1846 Michael Faraday had speculated that it might be some sort of 'vibration in the lines of force which [...] connect particles' (Faraday, 1846), which was at least suggestive of an electromagnetic wave and may have given Maxwell some clue as to what he was looking for. Having found the key, it did not take him long to unlock the answer: electromagnetic waves.

The discovery of the displacement current allowed Maxwell to manipulate his electromagnetic equations into what is called a 'wave equation'. Wave equations were already well known, and whether they describe sound waves, the vibrations of a string, or the waves on a pond, they all have the same simple, characteristic form. To this extent, at least, electromagnetic waves are no different. Once the wave equation has been found, the velocity of the wave in question is readily determined. In the case of sound vibrations, the velocity of the wave, that is to say the speed of sound, is found from the ratio of the stiffness of the material κ to its density ρ, and in fact it is equal to the square root of that ratio,

$$\sqrt{\frac{\kappa}{\rho}}$$

The velocity of electromagnetic wave turns out very similarly, except that the velocity is equal to

$$\frac{1}{\sqrt{\varepsilon\mu}}$$

in which the constant ε indicates how much of Maxwell's displacement is induced in the medium by a given electric field, whereas μ relates to the degree of magnetisation induced in it by a magnetic field. Seeking the data for air as the medium, Maxwell was profoundly struck by the fact that Kohlrausch & Weber (1857) had not only already measured both ε and μ, but had gone so far as to note that

$$\frac{1}{\sqrt{\varepsilon\mu}}$$

was equal to 311 million metres/second, which more or less matched Fizeau's most recent result for the speed of light in air, 315 million metres/second. Now, nothing in the ordinary world travels at such a speed. The speed of sound in air is only some 300 metres/second, and it may have been known that in metals and fused quartz it could reach 6,000 metres/second, but even so the speed of light is some 50,000 times greater, making it unique. Surely it could not be the case that the speed of electromagnetic waves was so close to the speed of light and yet there be no connection between the two? Whereas Faraday had speculated on the possibility, Maxwell was now convinced that light must be an electromagnetic wave, a thrilling conjecture that in the course of time has become a matter of fact.

By bringing all of the then known electromagnetic equations together

APPENDIX 3: LEGACY TO SCIENCE AND MANKIND

with his particular new idea of electrical displacement, James Clerk Maxwell laid the foundations of a set of four equations now known after him as 'Maxwell's equations'. It fell to four significant others to complete what he had started: Oliver Heaviside, Heinrich Hertz, Hendrik Lorentz and Albert Einstein. Heaviside reduced Maxwell's somewhat ungainly original equations to pretty much the same form that is in use today; in 1887 Hertz provided the experimental verification that electromagnetic waves did indeed exist; and by the beginning of twentieth century Lorentz had given Maxwell's theory a proper microscopic basis, the very thing that Maxwell had attempted to explore with his mechanical analogies of 1861. Finally, Einstein showed that Maxwell's equations were in complete agreement with special relativity.

As to the practical benefits to follow from Maxwell's work, although Sir Oliver Lodge was granted a US patent on signalling with radio waves in 1894, it was Guglielmo Marconi who made wireless telegraphy a practical reality in the final years of the nineteenth century, nearly forty years after Maxwell had first mooted the electromagnetic wave.[1] For Marconi, it was just the beginning, for it had occurred to the British admiralty that if wireless telegraphy could be made to work over large enough distances then it would revolutionise their ability to command and control their huge fleet. Consequently they took great interest in Marconi and his invention, and naval communications duly became one of the first major applications of wireless telegraphy. By late 1901, Marconi had even managed to set up a transatlantic radio link between Poldhu in Cornwall and St John's in Newfoundland. By the following year he was able to demonstrate 'ship to shore' radio contact over a range of about 1,000 miles (Roscoe et al., 2013), and the pace of advance was such that in January 1903, President Theodore Roosevelt and King Edward VII were able to exchange greetings sent between Wellfleet in Massachusetts and Poldhu in Cornwall (Roscoe et al., 2013).

By the time of World War I, valve based radio sets were being used in the trenches; in the peacetime between the wars most households had access to a radio receiver for news and entertainment; during World War II, portable radios were being used by agents of the SOE and the Resistance, and 'walkie-talkie' radios were being carried on the battlefield; in the 1950s people were watching television in their own homes; in the 1960s they were carrying transistor radios; by the 1970s they were watching colour television; the 1980s saw the mobile telephone networks roll out; in the 1990s we started sending text messages; in the 2000s we connected our telephones and personal computers to the internet by Wi-Fi, and so it seems to go on.

So far we have mentioned only the technological developments that have spun out from the discovery of electromagnetic waves, but it cannot go unsaid that their theoretical and practical benefits to science have been of just as great importance. Another discovery, this time by the American scientists Albert Michelson and Edward Morley (first in 1887) had shown that the speed of light in air does not depend on either the speed of the object that emits the light or the speed of the object that receives it. No matter how the experiment was repeated, or by whom, the result was always the same. Given that it was now pretty well established that light was an electromagnetic wave, this kind of behaviour was contrary to any other known type of wave. Not only that, it defied the sort of logic that scientists had been so comfortable with since the time of the great Sir Isaac Newton. There were also some theoretical problems with the forces between moving electrical charges, for it appeared that they could depart from Newton's laws. For over a decade scientists tussled with these uncomfortable 'electrodynamical' issues, until at last Albert Einstein provided an incontrovertible answer in his revolutionary paper (1905). The clue is in the title; he did not call it, as he might have done, 'On the Theory of Special Relativity', rather he called it 'On the Electrodynamics of Moving Bodies'. Relying heavily on Maxwell's equations, he put forward a radically different view of the world from Newton's. Difficult though it was to 'get one's head around', once special relativity was accepted it led to a whole new era of scientific thinking, the consequences of which made Einstein a household name. Einstein never forgot his debt of gratitude to James Clerk Maxwell, saying of him[2] that his discoveries were 'the most profound and the most fruitful that physics has experienced since the time of Newton'; 'One scientific epoch ended and another began with James Clerk Maxwell'; and 'I stood on the shoulders of Maxwell'. Coming from someone like Einstein himself, it is hard to doubt Maxwell's place in the history of scientific progress.

Through this seemingly never-ending evolution of scientific ideas and technology, electromagnetic waves have shrunk our world to the extent that we may now instantly know what is going on in any part of it; it is up to us, the people who benefit from it all, to see that this evolution leads to things that are for the better rather than the worse. What James Clerk Maxwell achieved, however, was not undertaken with the idea of bringing any immediate benefit. All one can say is that knowledge may be used for the general good, and at some stage knowledge that was originally pursued out of mere curiosity may be exploited to provide tangible benefits. As the theory of electromagnetic waves has clearly shown, the eventual benefit to mankind can be beyond the wildest dreams of the original discoverer. James Clerk Maxwell's great achievement was that he made all of this possible. How he did so lies at the heart of what physics is all about. For people such as him, lack of knowledge is no barrier, only an incentive.

APPENDIX 4

The Birth Dates of George and Dorothea's Children

The year of birth for Sir John Clerk, 5th Baronet of Penicuik, is given in some places as being 1736, but without firm evidence. His parents, George and Dorothea Clerk Maxwell, married irregularly in 1735, when George was only nineteen years old and Dorothea was not quite fifteen. The Baron, George's father, accepted the marriage but said that they were too young to live together, and sent George off to university in Leiden to finish his studies. He did not return until 1737. An obvious and significant question is therefore raised about the birthdate of the eldest child, could it be 1736, or would it have been 1738 or later? The following analysis gives a fairly conclusive answer, and also provides the dates of birth of the others, one of whom, whom we identify as James, does not even have a name against his birth record. As James Clerk HEICS, he is an important link to James Clerk Maxwell, for he was his paternal grandfather.

We know the eldest son was John, but this would not of itself rule out an earlier daughter. Mackay (1989) suggested that the first-born was a daughter, Agnes, but as we shall see, this must be ruled out. Secondly, as we know, the Baron informs us (Chapter 8) that after George and Dorothea's early marriage in July 1735 'his Wife and he being too young to live together, I sent him to Holand in January 1736', following which George did not return home until about July 1737 (BJC, p. 146). If the marriage had been consummated after their wedding but prior to George's departure, any child that resulted would have been born between April and October 1736, after which no further birth-dates would be possible before April 1738. Indeed, a daughter Janet was born, at Penicuik, on 7th July 1738.[1]

The next known record is for another daughter Agnes, born in September 1739,[2] which ties in with Mackay, except that he is clearly wrong in stating that she was the first-born. The next child was Joan, baptised on 16 March 1741,[3] which rules out, in all reasonable certainty, any other birth intervening between her and Agnes. It is likely that it was Joan who died the following November[4] because a Janet, an Agnes and a Dorothea all survived to marry.

281

Thereafter, a son was born in November 1742[5] and once again, by the dates, there is very little chance of any other child being born in between. While we do not know the name of this child, we can say that if there was no surviving issue in 1736, he would have had to have been the recognised first-born son, John. However, there is a mention in the Baron's memoirs (BJC, p. 204) that when John had the measles in about May 1746, he was at Penicuik, whereas at this time George's other children were at Dumfries (and one of them, William, died as a result of being inoculated against smallpox). The implication is that in 1746 John was away from the other children because he was old enough to be away at school, as was the custom for the sons of gentlemen. John therefore could not have been the son born in November 1742, for he would have been too young. He was consequently born by 1736, while the son born in 1742 was the second son, who was named George.

A third son named William was born at Dumfries and baptised on 14 June 1744,[6] and it is clear that he must have been the poor little fellow who died in 1746. There is evidence that a fourth son was baptised on 20 September 1745,[7] for although his first name does not appear in the transcribed record, the father is given as George 'Clark' (as is common in the transcription of such records) and the parish as Dumfries. This would fit in with the birth of James, later James Clerk HEICS, who is given by Foster (1884) as being the third surviving son, and once again we can be sure that no child has been missed from the sequence. Given that the Jacobite rising had been declared the month before and the rebels were already at Edinburgh, one can understand that naming a child James just at that very time could have been seen as an act of overt support for the Jacobite cause. Notwithstanding that the boy would actually have been named after his grandfather, Sir James Inglis of Cramond, fear of negative consequences could have been reason enough for not giving his name on the public record.

The last birth we have a date for is that of a fourth daughter, Dorothea, baptised on 27 February 1747.[8] The sons William, a second of that name, and Robert, are not recorded, possibly due to the family having by then moved to Edinburgh where such missing records are relatively common, especially in the Clerk family.

APPENDIX 5
Some Notable Scots of the Age of Enlightenment

Robert Adam, architect
William Aikman, artist
Joseph Black, chemist and discoverer of carbon dioxide and latent heat
James Boswell, diarist and biographer of Dr Johnson
Sir David Brewster, chemist and optical pioneer
Robert Brown, botanist and discoverer of Brownian motion
James Burnet, pioneer of linguistics
Robert Burns, poet
Thomas Cochran, admiral and founder of several foreign navies
Thomas Carlyle, philosopher and historian
James Craig, the designer of the New Town of Edinburgh
William Cullen, physician
Robert Fergusson, poet
James Gregory, physician
Gavin Hamilton, artist and neo-classicist
David Hume, philosopher
John Hunter, surgeon
William Hunter (brother of John), anatomist
James Hutton, pioneering geologist
John Macadam, engineer and road builder
Charles Macintosh, chemist and inventor of waterproof fabric
Henry Mackenzie, author of *Man of Feeling*
James Macpherson, poet and creator of the *Ossian* legend
Colin Maclaurin, mathematician
Alexander Monro I, surgeon
Roderick Murchison, pioneering geologist
William Nicol, geologist and inventor of the Nicol prism
John Playfair, mathematician
William Playfair (brother of John), pioneering statistician
Sir Henry Raeburn, artist

Allan Ramsay senior, poet
Allan Ramsay junior, artist
John Rennie, civil engineer
Sir Walter Scott, novelist
William Smellie, editor and publisher
Adam Smith, moral philosopher and economic theorist
Tobias Smollett, novelist
Robert Stevenson, lighthouse builder
James Stirling, mathematician
Rev. Robert Stirling, inventor of the Stirling engine
William Symington, steam engine pioneer
Thomas Telford, civil engineer
James Watt, steam power pioneer

Note: Even if they were not strictly part of the Enlightenment, James Clerk Maxwell and his friend William Thomson, Lord Kelvin, may be regarded as being products of that golden age.

Glossary

Advocate: a barrister

Annualrent: (one word) interest on a debt

Attainder: reserved for the most serious of capital crimes such as treason, the attainted person's property was confiscated by the Crown and could not be inherited by the heir

Assize: jury

Baillie: town councillor

Baron: title once given to the senior judges of the Exchequer Court

Baronet: a hereditary knight whose baronetcy is normally attached to lands. Baronetcies were granted by the Crown, usually as a reward for services rendered

Bow: distinctly narrow or crooked part of a street, e.g. West Bow and Nether Bow

Burgess: citizen accredited with certain rights, e.g. to carry out his trade

Causey: causie stane: a hard stone sett used for paving featuring in many streets of the New Town

Clandestine marriage: one not performed by a licensed minister of religion See also *irregular marriage*

Commissary: judge of the Commissary Court which had jurisdiction over wills and divorces

Close: built-over passageway. Many of the old closes have been preserved. Some of them opened out into *courts*, such as Milne's Court and James' Court

Court: open area enclosed by buildings

Covenanter: an adherent to the Scottish National Covenant of 1638, which demanded freedom from all forms of pressure to conform to episcopal beliefs and practices. See note 5 of Chapter 3 for details

Curator: legal guardian, e.g. of a minor. See also *tutor*

Dispone: to legally transfer; make a disposition

Dominie: schoolmaster, from the Latin *dominus*.

Drystane dyke: a wall made without mortar by using irregular flattish stones that lock together under their own weight
Dux: first in the class or school at the end of the school year
Dyke: wall
Eik: a supplement, as in the eik to a testament
Entail: a legally binding document, the main purpose of which is to dictate the line of succession of heritable property. See the section on the Middlebie Entail in Chapter 6.
Exchequer Court: the court that ruled over all matters related to government finances and revenue in Scotland
Feu: the right to hold property, as given by a 'superior' (Youngson, p. xxiv), ultimately one who holds land directly from the Crown; *adj.* feudal
Flit: move house
Gait: road or street, e.g. Cowgate, Canongate
Gardyloo: watch out for the water, from *gardez l'eau* (phonetically and euphemistically)
German: (as in brother-german) natural
Hospital: a place of refuge for the needy or infirm
Howff: habitual meeting place, frequently a drinking den
Irregular marriage: one meeting the legal standard, that a declaration of marriage had been made, but not meeting the requirements of a *regular marriage*.
Jus relictae: a widow's rights, e.g. a share of the joint movable property and a *terce*.
Laird: head or master, whether titled lord or otherwise
Land: large tenement with dwellings entered from a common stair, for example, Gladstone's Land in the Lawnmarket
Liferent: the use of something for the rest of one's life
Lord Advocate: the highest legal post in Scotland
Merk: (1) an ancient Scottish monetary unit equal to two thirds of *one pound Scots*; (2) a measure of land, e.g., the 'Twenty Merk Land of Middlebie'
Port: city gate
Pupil: a boy under the age of fourteen or a girl under twelve
Pound Scots: The original Scottish pound. On monetary union in 1707 it was worth a twelfth of a pound sterling, i.e. twenty pence in pre-decimal currency. See also *merk*
Pounds, Shillings and Pence: British currency up until 1971; the pound (£ or L) was divided into twenty shillings (s), each of which was worth twelve pence (d). Sums of money were represented as, for example, £1/2/3 or £1 2s 3d

Press: shelved cupboard
Regular marriage: one complying with rules of the established Church (c. f. *Irregular marriage*).
Retour: legally return (as heir)
Royal Mile: the long street running downhill from Edinburgh Castle to Holyrood Palace. It comprises four main sections, Castlehill, Lawnmarket, High Street and Canongate
Sasine: transfer of *feudal* property. Also the deed of transfer or the record of it. See also *seise*
Seise: to take possession of *feudal* property. See also *sasine*
Sheriff: county magistrate
Signature: the draft in English of a charter to be written out in Latin, on vellum, for signature by the monarch
Solicitor General: the highest legal post in Scotland after Lord Advocate
Tack: (1) lease; (2) share of product taken as rent
Tacksman: under the feudal system, a henchman granted a 'tack' from his superior in return for services, e.g., bringing in rents, and keeping order
Tailzie: alternative form of *entail*
Summons Declarator: a notification of a court action to obtain a legal ruling
Tenement: a holding; a building divided into separate homes
Terce: the right to a share (nominally a third) of a dwelling house and land
Tolbooth: town hall, often with a gaol
Town or Toun: anything from a city to a single house that was the chief building in a country area, for example, 'Waverley learned [...] that in Scotland a single house was called a TOWN' (Scott, Sir Walter, 1814, Chap. IX)
Turnpike stair: a spiral staircase, usually built onto the exterior of a building
Tutor: the *curator* of a *pupil*
Valve: vacuum tube
Writer: (1) general legal practitioner; (2) clerk or accountant with the HEICS
Writer to the Signet: a legal practitioner accredited to appear in the Court of Session, abbr. WS
Wynd: a narrow street with buildings on either side, for example, St Mary's Wynd

Notes

Preface

1. The title of Harman's three-volume work is *The Scientific Letters and Papers of James Clerk Maxwell*, which is easily confused with Niven's title, *The Scientific Papers of James Clerk Maxwell*.
2. The various Edinburgh Post Office directories for 1773 to 1912 are readily available online at www.nls.uk/family-history/directories/post-office/index.cfm?place=Edinburgh. We refer to them as 'EPOD' for short. The first directory is for the year 1773–74 and while no directories were published between 1776 and 1783, during the American War of Independence, the gaps are relatively few. Some directories were biannual rather than annual, nevertheless they all run from Whitsun to Whitsun, roughly speaking from May to May. For simplicity, however, we refer to a directory by the year in which it was published. Given that they date from an age when virtually everything was done by hand, the bulk of the information would have been compiled close to the start of the calendar year, and so this minor simplification is actually in accord with the currency of the information.

    Street numbering was not formalised until 1811, before which numbering was either lacking or at best ad hoc. After 1811, however, there is a high level of consistency with today's numbering, particularly in the Second New Town north of Queen Street. Not until 1833 did the directories include a section indexed by street name and house number.

1. Introduction

1. Repeating James Clerk Maxwell in full becomes tedious. His first biographer Lewis Campbell referred to him as 'Maxwell', which is still the general custom today. *Clerk Maxwell* seems never to have been in vogue. However, in referring to the boy or in a setting involving other Maxwells, 'James' seems more appropriate.
2. Hyphenated surnames were not common in Scotland at this time. Other examples are Hay Mackenzie and Scott Moncrieff.
3. Although Middlebie itself was a small town to the east of Dumfries, the estate consisted of various parcels of land and property scattered about Dumfries. While Dorothea was to become the proprietor of the estate, her guardian or husband would control it 'by right'.
4. Following the death in 1798 of his uncle Sir John Clerk, 5th Baronet, who died childless.

NOTES TO CHAPTER 1

5   The old county, also referred to as the Stewartry of Kirkcudbright, which now forms the central part of present day Dumfries and Galloway, with Dumfriesshire to the east, Wigtownshire to the west, the Solway Firth to the south and Ayrshire to the north.
6   SOPR: *Births*, 685/01 0560 0296 Edinburgh.
7   The house is now kept as a museum cared for by the James Clerk Maxwell Foundation, www.clerkmaxwellfoundation.org , registered Scottish Charity SC 015003.

2. James Clerk Maxwell's Life and Contribution to Science

1   'R.V.' Jones is better known for his World War II efforts as assistant director of Intelligence (Science) in which capacity he advised the prime minister and his chiefs of staff. He told his story in *Most Secret War* (London: Hamish Hamilton, 1978).
2   He was in fact Professor of Natural Philosophy at Marischal College.
3   (Campbell & Garnett, 1882), hereinafter referred to as C&G.
4   http://canmore.rcahms.gov.uk/en/site/212680/details/glenlair/. At one time, however, he had aspired to being a county sheriff, and had passed some time designing himself a house fit for one (C&G, p. 420).
5   From John Clerk Maxwell's diary for 1846 (C&G, p. 131), we find an example of just how well James could manage to endure physical pain, and without fear: 'Sa., Dec. 12. Jas. still affected by the tooth [...] got the tooth drawn [...] and Jas. never winced. This was in the days before anaesthesia, and it was an eye-tooth at that.
6   As part of the Middlebie estate, Nether Corsock as shown on Gillon's Plan of 1806 (DGA: GGD56/1) was already John Clerk Maxwell's by inheritance. Amongst the pages of Andrew Wedderburn Maxwell's memoranda (DGA: GGD56/4, 1893) is a note mentioning John Clerk Maxwell's purchase of Upper and Nether Glenlair from Sir John Muir Mackenzie in 1839. The Glenlair estate was therefore formed by these neighbouring lands being added to Nether Corsock at a cost of £10,530 9s 4d.
7   Literally a stupid person, but at that time implying someone who would later be described as being 'mentally challenged'.
8   *Gulliver's Travels* is now seen as an interesting comic adventure that has thrilled many children, but the original Swift version is an advanced work indeed for a boy of ten to twelve. Samuel Butler's *Hudibras* is a civil war epic. Thomas Hobbes was another seventeenth-century author; he published *Leviathan*, a weighty philosophical treatise indeed.
9   John Clerk Maxwell qualified in December 1811 while John Cay qualified the following March (Grant, 1944).
10 The lecturer and geologist William Nicol (1770–1851) was the inventor of the Nicol prism, probably the earliest device for transmitting only one polarisation of light. He lived at 4 Inverleith Terrace, just a short walk from the Edinburgh Academy.
11 Campbell mentions that class sizes were typically sixty or so. A total of ninety-one boys, including Maxwell, registered in Mr Carmichael's class for 1840–47. Seventy attended for year three, 1842–43 (Henderson & Grierson, 1914).
12 Campbell describes it as something that would come on under pressure, for example, when asked by the rector to give his prepared answers in class. To

overcome his problem, Maxwell created the idea of breaking the text down into smaller chunks, each of which he would write as entries into a table with the very same layout as the panes of the window in the rector's classroom. He would commit the text to memory in this form so that when he stood in front of the actual window he could recall all the words, pane by pane, without hesitation.

13 This medal, with the inscription 'Edinburgh Academical Club Medal', was specifically for geometry. The actual 'Mathematical Medal' was awarded to Maxwell in 1847.
14 Maxwell's parody of Burns' well-known song 'Comin' thro' the Rye' not only demonstrates both his poetical skill and sense of fun, but an ability to weave his words about a topic in his mind, in this case Newton's laws as they apply to the rigid molecules in a gas 'Flyin' through the air':
    IN MEMORY OF EDWARD WILSON,
    Who repented of what was in his mind to write after section.
    GIN a body meet a body
    Flyin' through the air,
    Gin a body hit a body,
    Will it fly? and where?
    Ilka impact has its measure,
    Ne'er a ane hae I,
    Yet a' the lads they measure me,
    Or, at least, they try. (C&G, p. 630)
15 The notebook is preserved in the James Clerk Maxwell Foundation's museum at 14 India Street.
16 Tait may have gone as far as attempting to make the 'signature' look as though it were Maxwell's own.
17 This note by Tait has been transcribed by E. F. Lewis, at www-history.mcs.st-and.ac.uk/Tait/Terrot_lecture.pdf. Charles Hughes Terrot FRSE (1790–1872) was a mathematician as well as a clergyman (as was Kelland).
18 Tait's slim manuscript headed 'Propositions on the Ellipse' bears no date. The decorative embellishments on the cover page suggest that it dates from his school era, raising the possibility that Kelland was giving him private tuition.
19 Founded in 1821 by Sir David Brewster under an earlier name, probably the Edinburgh Society of Arts mentioned by Campbell. The new technology of photography was then one of its interests. See www.rssa.org.uk/intro.shtml
20 On Arthur's Seat, a hill with notable rocky outcrops called Samson's ribs, that flanks the east side of Edinburgh.
21 See in Encyclopaedia Britannia, 1911. http://en.wikisource.org/wiki/1911_Encyclop%C3%A6dia_Britannica/Forbes,_James_David
22 James Forbes was first cousin to James Hay Mackenzie, who had a few years earlier married John Clerk Maxwell's niece, Janet Isabella Wedderburn (Jemima's older sister).
23 Reproduced in WDN1, pp. 1–3.
24 A phenomenon discovered in 1844 by Wilhelm Karl von Haidinger, whereby under suitable circumstances the eye perceives a yellow-tinged bar-shaped area that is perpendicular to the polarisation of the incoming light.
25 Provided that it is always held aligned with the earth's magnetic field, a steel blade tends to become magnetised by the action of friction on its surface during the process of sharpening. Alternatively, a steel bar can be magnetised by striking it

several times on one end, as recorded by Joseph Henry (www.princeton.edu/ssp/joseph-henry-project/permanent-magnet/).

26 There was indeed one at the south-east corner of his estate. See also note 25 above.
27 1788–1856, Scottish academic, Professor of Logic and Metaphysics at Edinburgh University from 1836 (Stephen, 1890).
28 Annealing is the process whereby locked-in stresses within a solid body are relieved by heating it to a significant degree and then letting it cool very slowly. Rapid cooling causes stresses that will remain until they are relieved by either a fracture or by annealing.

The Nicol prism (note 10 above) polarises light in the same way that Polaroid™ lenses do. Maxwell's experiment involved viewing a piece of glass placed between two polarisers, one of which is rotated at 90° to the other so that light is normally blocked from passing through. However, if any parts of the glass are stressed, they will rotate the polarisation of the light a little, allowing it to pass through the crossed polariser. This is an easy experiment to do nowadays with a pair of Polaroid lenses, which an old pair of sunglasses may readily provide.

This early work by Maxwell was highly significant because it was eventually taken up by Ernest G. Coker, who won the Royal Society of London's Rumford Medal who used it to investigate stresses in engineering structures.
29 Also WDN1, pp. 4–29.
30 Also WDN1, pp. 30–73.
31 At other times Maxwell could be lucid and explain things very clearly when that is what he set out to do.
32 Charles Mackenzie, the brother of James Hay Mackenzie, was one of the voices urging John Clerk Maxwell to send his son to Cambridge. He had been a second wrangler in the tripos, after which he took religious orders and subsequently became the Archdeacon of Pietermaritzburg and the first missionary bishop for central Africa.
33 Senior wrangler = first place, second wrangler = second place and so on. The top wranglers could compete in an additional examination for the Smith's prize. See also Forfar (1996). George Stokes (1819–1903) was later Lucasian Professor of Mathematics at Cambridge and a close friend of Maxwell's. Arthur Cayley (1821–1895) became a lawyer and mathematician. His name is remembered in the Cayley–Hamilton theorem. Philip Kelland had been Maxwell's Professor of Mathematics at Edinburgh. William Thomson, later Lord Kelvin, was another close friend (Craik, 2008; Harman, 1985).
34 Archibald 'Baldy' Carmichael, his old school master, seems to fit.
35 Meaning a cryptic pronouncement such as the oracles gave to enquirers.
36 William Swan FRSE, FRSSA (1818–1894) was a lecturer in mathematics and natural philosophy at Edinburgh University and later a professor at St Andrews. It was he who showed that a characteristic yellow band in the spectrum of sunlight was due to the presence of sodium in the Sun.
37 Also WDN1, pp. 80–114.
38 He did not actually invent the instrument; it had been suggested by William Cumming in 1846 and worked on by Charles Babbage in 1847; the first practical ophthalmoscope was made in 1851 by Hermann von Helmholtz, who became a keen follower of Maxwell's later work.
39 Also WDN1, pp. 76–79.
40 The terms refraction and reflection are quite different. Reflection refers to what

'bounces off' a surface, while refraction refers to the part that passes through to its other side. Whereas a reflected ray of light bounces off at the same angle, the refracted ray does not pass straight through, it is deflected.

41 A similar type of lens was discovered by Luneburg in 1944, and with modern technology it is now practical to make such things.
42 Also WDN1, pp. 271–285. Maxwell finished the paper at Aberdeen, on 12 January 1858.
43 Reproduced in 'James Clerk Maxwell at Aberdeen 1856–1860' (Jones, 1973, pp. 69–81).
44 By all appearances, from his niece Jemima Wedderburn's sketches, John Clerk Maxwell was of medium height and overweight. At the age of fifty-four he weighed very nearly 100kg (C&G, p. 10). He may well have suffered from congestive heart disease, for James mentioned that his father suffered from 'want of circulation', i.e. heart failure (Hilts, 1975, p. 59). He also had bronchial problems (C&G, p. 202).

He was a favourite uncle of Jemima's, who was to recall the last visit he paid to her and her husband Hugh Blackburn at Glasgow University (Fairley, 1988, p. 107): 'He was then in bad health and we had to send for a doctor in the middle of the night. It was a short visit but he feared he had been a trouble to us, which I assured him he had not. We had a very affectionate parting and I never saw him again.'
45 Only two doors down from where John and his mother built their town house in 1820. The two houses would have been more or less identical.
46 It was in fact a letter sent to Dr G. Wilson, who was writing a book on colour blindness. Wilson published Maxwell's letter in his book and it also appeared as Maxwell, 1856a.
47 Also WDN1, pp. 126–148.
48 See WDN1, p. 144. This contrivance was a simplification of an apparatus of Helmholtz, reported in the *Philosophical Magazine* of 1852. Both were based on Newton's experiment (Newton,1704) showing not only that white light could be split up into its spectrum by a prism, but the same spectrum could be recombined into white light by converging the diverging colours with a lens and then passing them through a second prism.
49 This idea completely contradicted the received wisdom of Sir David Brewster, which held that yellow, not green, was the third primary colour for light (C&G, pp. 215–216). He and Maxwell had first crossed swords on the subject at the 1850 meeting of the BA in Edinburgh. Brewster subsequently gave a paper on the subject at the 1855 meeting of the BA in Glasgow, at which James was present. In the expectation of demonstrating his colour top to Brewster later in the day at a prearranged private gathering, James politely kept silent about Brewster's error. Brewster may have had wind of James' planned demonstration that the three primaries for light were red, blue and green, for he did not turn up at the meeting.
50 WDN1, pp. 445–450.
51 WDN1, pp. 155–229. Despite the tardiness of the date of publication it does not seem that the paper was held back, for the papers both before and after are from a similar period.
52 Although he was sometimes referred to as Clerk Maxwell of Glenlair, Glenlair was only part of the enlarged Middlebie estate created by his father. It seems that the entail was never changed to call it Glenlair (see also note 71).
53 Jones (1973, Appendix, p. 69) gives the full text.

NOTES TO CHAPTER 2

54 WDN1, pp. 286–376 contains the full text of Maxwell (1859), and the later synopsis in 1862.
55 At the same meeting he also read 'On the Mixture of the Colours of the Spectrum', published in §15 of the report. The British Association for the Advancement of Science, usually referred to simply as the British Association, was formed in 1831 and held regular meetings in different cities throughout the UK. The abbreviation was 'B. Ass.', which so tickled Maxwell's sense of humour that he frequently referred to them as the *British Asses*!
56 Published in three parts, the first in January and the remaining two in July 1860 (WDN1, pp. 377–409).
57 A simple but effective animation of the concept is shown on http://en.wikipedia.org/wiki/Kinetic_theory
58 Daniel Dewar was a man of the cloth and a doctor of divinity, but his father had been no more than a blind fiddler. On the other hand, Daniel's wife, Susan Dewar, was a grand-daughter of George Gordon, Earl of Aberdeen, making her a cousin of the ex-prime minister George Hamilton Gordon (Flood et al., 2014). Maxwell said that she was 'a first-rate lady, very quiet and discreet, but has stuff in her to go through anything in the way of endurance'.
59 DGA: GGD56 (Unsorted Box), *Copy Summons of Declarator and Payment – The Right Honourable Sir George Clerk Baronet against James Clerk Maxwell Esquire of Middlebie*, 3/3/1858.
60 See Glossary.
61 Also a strict disciplinarian, he also gained himself 'the fiend' as a nickname with his students.
62 Maxwell wrote to Lewis Campbell from Glenlair in April 1848: 'We came here on Wednesday by Caledonian.' The Edinburgh to Glasgow train service began in 1842, and by 1848 one could get to Lockerbie on the main line from Glasgow to Carlisle. Dumfries, however, was not connected to it until 1863.
63 Charles Henry Fleeming Jenkin (1833–1885), first Professor of Engineering at Edinburgh University, had been in Tait's class at Edinburgh Academy and therefore knew Maxwell of old. He wrote up the results of the work on the standardisation of the Ohm (Jenkin et al., 1873).
64 WDN1, pp. 450–513.
65 WDN1, pp. 526–597.
66 WDN2, pp. 1–25.
67 WDN2, pp. 26–78.
68 See, for example, Earman & Norton (1998).
69 Riddell (1930, p. 36). Maxwell stayed in London during the winter of 1865–66 to continue giving working men's classes (C&G, p. 314; Flood et al., 2014, pp. 52, 65–66).
70 Previously, crossing the river near the house was by way of stepping stones, whereas bringing in any sort of carriage involved taking a long and circuitous route (Cay, 1887; DGA: GGD56/26, *Letter, W D Cay to Maj. A Wedderburn Maxwell*, 24/6/1898).
71 The nonsense of the estate being called Middlebie was historical. As explained in Chapter 8, George Clerk Maxwell of Middlebie was forced to sell off his property at Middlebie (and Dumcrieff) to pay off his debts. He could do so because Middlebie had not been part of the Middlebie entail since 1738. The so-called Middlebie estate was finally disentailed in 1922 by Brig. John Wedderburn-

Maxwell (NAS: CS46/1922/11/16, 11/1922). Until then, by the conditions of the entail, 'Maxwell of Middlebie' had to be the formal name legally adopted by each successor to the estate. Curiously, however, Maxwell's entry in Waterston & Shearer (2006) gives him as 'Maxwell of Glenlair' but this may well have come from Riddell's *Clerk Maxwell of Glenlair* (1930), a title referring to Maxwell's association with Glenlair rather than his formal designation.

72 For an excellent description see Cat (2012). See also CANMORE ID 212680 and www.buildingsatrisk.org.uk/ref_no/1038 . The plans on the RCAHMS website give the date 1868. James Clerk Maxwell told Lewis Campbell:
> My father was [...] always fond of inventing plans for country houses [...] as early as when he was twelve years old. He wanted to build his house on a scale suited to what he thought he would require [...] and had so built a small part of it [before] he died. We afterwards completed it, as far as possible according to his idea, but on a much smaller scale. (C&G, pp. 420–421)

73 Maxwell, 1868a, which was followed by Maxwell, 1869b. The former appears in WDN2, pp. 125–143.
74 WDN2, pp. 105–120.
75 He found the differential equations obeyed by governors and showed that such control mechanisms would be stable only under appropriate conditions. Otherwise they would either veer off in one direction or oscillate uncontrollably. Everyone is familiar with the public sound system amplifier that becomes unstable when the microphone gets too close to a loudspeaker; Maxwell showed us why even before such things existed!
76 WDN2, pp. 161–207.
77 The year as engraved on the medal, 'MDCCCLXXI'.
78 A short section of Trumpington Road, where numbers 6–14 are now the Royal Cambridge Hotel.
79 WDN2, pp. 257–266. Dimensional analysis is also given in his *Treatise on Electricity and Magnetism* (1873).
80 Also WDN2, p. 256. Maxwell investigated the same condition in 3D (Maxwell, 1873b; WDN2, pp. 297–300).
81 WDN2, pp. 538–40.
82 MacAlister, a fellow of St John's College Cambridge, later went into medicine and became chancellor of Glasgow University. He was created 1st Baronet of Tarbert.
83 The inverse square law also applies to gravity and to radiated heat, light and other electromagnetic waves. If Maxwell's experiment had shown that the inverse power was even slightly greater than two, then the total power radiated from a star would end up vanishing in distant space, while if it was less than two it would end up becoming infinite. Only a power of two exactly is consistent with the fundamental principles of physics.
84 WDN2, pp. 681–712.
85 See for example, www.oxforddnb.com/view/article/32639 (subscription or UK library registration required).
86 The mechanism has four diamond-shaped vanes that are mounted vertically on the arms of a little cross to make a sort of miniature horizontal windmill that is free to rotate within a sealed glass bulb. Each vane is blackened on one side but shiny on the other, and they are all mounted with the blackened faces orientated in the same way. Placed in bright sunlight, the vanes will rotate at speed, but the exact mechanism of its operation has always been a challenge to explain, for example, it

NOTES TO CHAPTER 2

will only operate when most of the air has been evacuated from the glass globe, but it will not work at all in a complete vacuum. In bright sunlight the black sides of the vanes get hotter than the shiny sides, and in so doing they affect the temperature of the gas molecules in their immediate vicinity. Maxwell showed that in a rarefied gas such temperature differences give rise to slight pressure differences that push each vane towards its shiny side, just as observed. While the radiometer appears to be little more than a curiosity, thinking about how it worked allowed Maxwell to put his advances in the molecular theory of gases to the test (Brush & Everitt, 1969).

87 RS: *Register of Sasines*, (6031) 1902.49, 15/2/1887.
88 www.royalsoced.org.uk/715_JamesClerkMaxwellStatue.html
89 http://alchemipedia.blogspot.co.uk/2009/12/100-greatest-britons-bbc-poll-2002.html
90 The seventeen scientists and engineers appearing in the BBC 2002 poll of the '100 greatest ever' Britons are given as: 2nd – Isambard Kingdom Brunel; 4th – Charles Darwin; 6th – Sir Isaac Newton; 20th – Sir Alexander Fleming; 21st – Alan Turing; 22nd – Michael Faraday; 25th – Prof. Stephen Hawking; 42nd – Sir Frank Whittle; 44th – John Logie Baird; 57th – Alexander Graham Bell; 65th– George Stephenson; 78th – Edward Jenner; 80th – Charles Babbage; 84th– James Watt; 91st – James Clerk Maxwell; 95th – Sir Barnes Wallis; 99th – Sir Tim Berners-Lee.
91 Dickson (DNB12); Knott (1911); O'Connor & Robertson (2003).

3. The Early Clerks of Penicuik

1 According to the National Museum of Scotland, the 'Penicuik Jewels' comprise a gold necklace, locket and pendant that were given to a servant of Mary's by the name of Egidia, or Giles, Mowbray, the great-grandparent of Mary Gray. See www.nms.ac.uk/highlights/ mary_queen_of_scots/the_penicuik_jewels.aspx , www.scotlandspeople.gov.uk/content/ help/index.aspx?1153 .
Other pieces of jewellery were reputedly made as bequests by Mary to her friends and supporters; Wilson (1891, p. 151) mentions a gold watch that was in the possession of Prof. Fraser Tytler of Woodhouselee, a descendant of Alexander Tytler (see note 4 of Chap. 5).
2 Near Arbroath, in what was then Forfarshire.
3 See Burke, 1834–38, vol. 4, p. 621, and Grant, 1944, p. 73.
4 Because of the Scottish context of this book, it will be simpler to write 'James VI' rather than 'James VI and I', and likewise 'James VII' rather than 'James VII and II'.
5 The Covenanters were stalwartly opposed to the principles and practices of episcopacy. The King, however, pressed in the other direction with measures designed to bring the Church of Scotland closer to the Anglican Church, of which he was the head. Wishing at all costs to prevent this, they banded together under the Scottish National Covenant, signed by the chief adherents to the cause in 1638. This was reiterated in 1643 by a treaty with the English Parliament in the form of the Solemn League and Covenant. There ensued a period of political and religious strife, rebellions and wars during which Covenanters were greatly persecuted. It did not end until James VII quit the throne. (Encyclopaedia Britannica, 2013; M'Crie, 1841).
6 Newbiggin, or Newbigging, was the original Penicuik House, which was 'close behind' the later structure built by James, 3rd Baronet, in 1761. A bigging was a house or home, and so, ironically, Newbiggin, 'the old house', had itself been 'the new house', suggesting that there had been an even earlier one.

7   NAS: GD1/1432/1.62, eighteenth century.
8   See also NAS: GD18/5218, *Letters to and from Sir John Clerk* [...] 1698–1708.
9   Later Robert Clerk of Colinton, he qualified as an advocate in 1725 and married Susan Douglas (Grant, 1944; Foster, 1884, p. 53; and Douglas, 2013).
10 CANMORE ID 51652.

4. The Baron Sir John Clerk, 2nd Baronet of Penicuik

1   Leiden is in Holland, about twenty miles north of Rotterdam.
2   He was Alexander Clerk (Worling, 2012), son of William Clerk (Grant, 1944, p. 37).
3   Presumably also on the grand tour, he was then Marquess of Tavistock and later 2nd Duke of Bedford.
4   The majority of it covered by his *own* borrowings, which by 1703 totalled £12,000 Scots (Hayton, 2002a). In order to pay back his debt to his father, his annuity of £1,000 Scots was held back.
5   While he was still a minor, John Clerk had been left an estate at 'Elvingstone' in East Lothian by his grandfather, Dr Henderson (BJC, p. 246n H). Today, there are two place names in the locality, Elphinstone and Elvingston, which are about four miles apart. The NAS catalogue records suggest the former, whereas 'Elvingstone' corresponds more closely with the latter. Dr Henderson also left him a substantial house at the head of Blackfriars Wynd in Edinburgh. He sold this only to purchase another at the opposite end of the Wynd, possibly Cardinal Beaton's old house down by the Cowgate. However, the original house seems to have been repurchased some time thereafter: '[about mid-October 1710] I lived in my own House at the head of Blackfrier Wynd [...] built, anno 1552, by Thomas Hendersone [...] my mother's Grandfather'.
    After the Baron's death in 1755, his son Sir James Clerk 'had a lodging in the head of Blackfriars Wynd either as proprietor or tenant' (BJC, p. 246n H).
6   For example, 'If I had consulted physitians they might have done me mischief' and 'this year I was trubled with a physitian' (BJC, marginal notes on pp. 146 & p. 229).
7   The original house is now part of the Scottish Parliament complex, www.scottish.parliament.uk/visitandlearn/15921.aspx
8   A further tradition given by Chambers (p. 309) is that the Treaty of Union was signed in a summerhouse in the gardens of Moray House, on the south side of the Canongate, a short distance from Queensberry House, or that it took place in a 'laigh shop', or cellar bar, further up the High Street. The only formal signing that took place, however, was when the Articles of Union were agreed in London on 22 July 1706.
9   Bertram (2012a, b); Clerk of Eldin & Laing (1855); Clerk of Eldin (1827).
10 BJC, pp. 161–162, 203. The Baron also tells us,
    Patrick [was] a fine Mathematician [...] [he] had a very great genius for the Ingeneering business [...] his Twin Brother, Henry, [was] a Liutenant in a 70 Gun ship called the Prince of Orange [...] [his] skill in navigation and the mathematical sciences, wou'd have rendered him a blessing. (BJC, pp. 158–159, 203)
11 The Baron tells us not a great deal about Mathew except that: 'In Aprile and May 1746 my son Mathew fell ill of a Feaver at Dalkieth [...] This boy is a fine scholar, and of great application to learning and business of all kinds' (BJC, p. 204 & marginal note).

NOTES TO CHAPTER 4

After his schooling he studied mathematics in hope of training as an officer in the army (NAS: GD18/5475, 22/8/1750), and the indications are that in 1751 he went to Woolwich Military College (NAS: GD18/4197, 11/12/1750), during which time his father was making enquiries about getting him a commission (NAS: GD18/4198, 1751). By the spring of 1757 the Seven Years War was well under way and Mathew, now a junior officer, was making preparations to go with his regiment to fight in America (NAS: GD18/4201, 2-5/1757), but by the end of the year he was writing home from New York to say things were not going so well 'we have been extremely unlucky in our expedition' (NAS: GD18/4202 20/12/1757). Only days thereafter he received his commission as a sub-engineer with the rank of lieutenant (NAS: GD18/2063A, 4/1/1758), and by April he is hopeful again, writing to his mother 'by the time you receive this we must have struck some blow in America' (NAS: GD18/4203 28/4/1758). In July came the battle of Fort Carillion at the Siege of Ticonderoga and in the days before the battle Mathew was involved in surveying and reporting on the French defences. It seems that, owing to his inexperience, he did not appreciate that the French had disguised them so that the area looked only lightly defended, whereas in reality the very contrary was true. The battle took place on 8 July, and Mathew was one of the many British soldiers that lost their lives in the ensuing defeat (NAS: GD18/1989, *Testament of Lieutenant Mathew Clerk*, 14 /7/1758).

Having perished soon after the battle and so being denied the opportunity of giving his own account, Mathew Clerk was made a scapegoat for the failure of the attack by his superior officer, Captain Abercromby, who maintained that Clerk had grossly underestimated the French defences. Captain Abercromby, on the other hand, had been rash in his own decision making and had not taken due heed of Clerk's lack of experience (NAS: GD18/4204 25/8/1758). Abercromby went on to become a general. A great deal of research has been done on who was more to blame; see for example Nester (2008) and Kingsley & Alexander (2008), while McCulloch (2008) gives a rebuttal of Kingsley & Alexander.

12 NAS: GD18/5321, 1751; GD18/4199, 6/1/1756; BJC, p. 146n2; NAS: GD18/4200, 19/2/1757.
13 NAS: GD18/5463, 26/11/1755.
14 It stood on Dock Street near its junction with Commercial Street, and would have been out of effective range of the castle guns. A fragment of it remains.
15 Fought near Stirling on 13 November 1715, with both sides claiming to have won.
16 He bought Cammo for about £2,800 by selling the lands of Elphinstone which Dr Henderson left him (BJC, p. 8). See also note 5 above.
17 The tower (CANMORE ID 50421) is assumed to have been built by the Bishops of Dunkeld (Scottish Castle Association, 2013) whereas for the house (CANMORE ID 74420) see Wood, 1794, pp. 44–51, etc.
18 A fever with alternating extremes of body temperature.
19 Loanhead was then within the precincts of his estate. It lies to the west of the steep valley of the River North Esk, about halfway between Edinburgh and Penicuik.
20 Father of the rather better-known classical architect Robert Adam (Macaulay, 2004; DSA, 2014; Gifford, 1989). He was also the architect who built the Scots Mining Company house for James Stirling (see Chapter 12) at Leadhills. The Clerk and Adam families were brought together by the marriage of John Clerk of Eldin, John Clerk's fifth son, to Susannah, William Adam's daughter (Bertram, 2012a, b).

21 In Annandale, about fifty miles south of Edinburgh on the road to Dumfries and Carlisle.
22 This section on Dumcrieff is based on Prevost, 1968. Charles had succeeded his father, the Baron's mentor of old, in 1711.
23 CANMORE ID 49725.
24 Once a very familiar feature in many parts of the Scottish countryside where a good supply of suitable stones was readily available. The first such dykes appeared in the area of Dumfries and Galloway about this time (Prevost, 1968).
25 BJC, p. 135. Archibald and Thomas Tod at Craigieburn were apparently unconnected with the Tods of Blackhall and Hayfield with whom the Baron had business connections. This Thomas Tod (1657–1742) was the minister at Durisdeer, west of Moffat.
26 Prevost (1968). While the Baron says they were at Dumcrieff, he was evidently not able to live in the unfinished house.
27 He was also in effect acting as the Chief Baron, for the incumbent Baron Lant was hardly ever there.
28 Algernon Seymour, later the Duke of Somerset.
29 The Board of Trustees for Fisheries, Manufactures and Improvements in Scotland was eventually set up to make grants for the encouragement of economic improvement throughout the land. After 1823, its funding was instead applied to the decorative arts and the encouragement of education in the fine arts, and from 1906 the Board of Trustees for the National Galleries of Scotland took its place (NAS, 2013; Select Committee on Arts and Manufactures, 1836, pp. 77–88). The Baron was also a member of the Honourable Society of Improvers in the Knowledge of Agriculture in Scotland, which was active from 1723 to 1745 (Maxwell of Arkland, 1743).
30 The King's Advocate for Scotland, he was the son of Sir David Forbes of Newhall who had married the Baron's aunt, Catherine Clerk (see page 62).
31 Charles Erskine, or Areskine, of Alva (1680–1763) was Solicitor General for Scotland (1725–37), Lord Advocate (1737–42), the judge Lord Tinwald from 1744 and Lord Justice Clerk from 1748. His wife Grizzel Grierson, was a descendant of Homer Maxwell of Speddoch (Chapter 6) and therefore distantly related to the Baron's niece and ward, Dorothea Clerk. By John Maxwell's entail of 1722, Grizzel Grierson's descendants were potential successors to Middlebie.
32 The Philosophical Society of Edinburgh was established in 1731 as a society for sharing medical knowledge. In 1739 it broadened its remit and became the Society for Improving Arts and Sciences, otherwise known as the Philosophical Society of Edinburgh, with interests covering medicine, literature, philosophy, mathematics and the sciences. The first co-vice-presidents were Baron Sir John Clerk and his cousin Dr John Clerk (C&G, p.165n4). In 1783, the Philosophical Society reincorporated itself under royal charter as the Royal Society of Edinburgh taking all the existing members over as its founder fellows (RSE, 2015; Campbell & Smellie, 1983; O'Connor & Robertson, 2004b). By then, however, the Clerk cousins were both dead.
33 The comets were in 1742 and 1744. The eclipses were: of the Sun, 1715 and 1737, of which he published an account (Clerk, 1737–38); of the moon, 1739; and Jupiter, 1751. He also attempted attempt to see Mercury during the comet of 1744. (BJC, pp. xxv, 86, 150, 164, 166–167, 227).

NOTES TO CHAPTER 5

5. William Clerk and Agnes Maxwell

1. NAS: GD18/5253, 1703–24.
2. NAS: GD18/5522, 15/9/1699 and GD18/5237, 1700–08.
3. NAS: GD18/2315, 1711.
4. Probably Alexander Tytler, writer in Edinburgh. His son William Tytler of Woodhouselea (b.1711), who was a Jacobite sympathiser, wrote an acclaimed defence of Mary, Queen of Scots. Newhall and Woodhouselee are both within a five-mile radius of Penicuik.
5. See note 3 of Chapter 3.
6. (1686–1758). His son Allan (1713–1784) was the well-known portrait painter.
7. The poet and topographer Dr Alexander Pennicuik (1652–1722) lived at Romanno Bridge, some distance from his old family seat of New Hall which had passed into the hands of the Forbes family. Despite the age difference, he and William Clerk were good friends and eventually Allan Ramsay joined their company. The howff at Cowie's Mill was near Macbiehill.
8. The inn at Carlops may be named after Allan Ramsay, but it was not established until the late eighteenth century (Hanson, 2009 & 2013). Until recently, however, there was an old inn called Habbie's Howe, CANMORE ID 50165, at Nine-Mile-Burn, about a mile north of Carlops, notwithstanding the fact that the actual Habbie's Howe was on the other side of the present road, close to Newhall.
9. In 1712 Queen Anne was on the throne and the Act of Settlement of 1701 meant that she would be the last Stuart monarch; instead she would be succeeded by her second cousin George, Elector of Hanover. He did indeed become George I in 1714, the eve of the Jacobite uprising. The Jacobites, however, maintained that Anne's brother, James Stuart, son of the James VII who had fled to France in 1688 was the true king. The Jacobites toasted James as 'the King over the water', but to those who wanted to move on, he was merely the Old Pretender (father of Bonnie Prince Charlie, the Young Pretender).
10. Unlikely to have been the Irishman Dr John Fergus, who was born *c.* 1700 and was Dublin based.
11. A pseudonym also used by Jonathan Swift and Richard Steele.
12. See the Glossary entry 'Royal Mile'.
13. Christian Ross (1683–1743) was the orphaned daughter of Robert Ross, possibly the brother of her guardian James Ross.
14. Ramsay's lines were apparently adapted from a passage in Gavin Douglas' translation of the *Aeneid* (Taylor, 2006, Book VI, Canto LXXXIII)
    There stands the traitor, who his country sold
    [. . .] *for a bribe of gold*
    It was this Douglas whose name Ramsay chose to take as his club alias. Burns employed Ramsay's wording in the form 'We're bought and sold for English gold' in his poem *Such a Parcel of Rogues in a Nation* that was written in much safer times.
15. He based his evidence on the love poem 'Harmonious Pipe' attributed to the Baron (BJC, p. xxii) and supposedly submitted to the Easy Club. If this had appeared at the Easy Club, it would have been more likely to ridicule him.
16. Henry Clerk, born in 1678, was 'bred to sea' and is therefore an unlikely candidate and their half-brothers would have been too young.

17 Hawkshaw, or Hackshaw, now lies beneath the Fruid reservoir, formed in 1963 by flooding the narrow glen that carries the Fruid Water.
18 SOPR: *Marriages*, 774/00 0010 0068.
19 NAS: GD18/5294, ?/12/1717.
20 BJC, p. 155. He was a cousin of Tobias Smollet (Herbert, 1870, pp. 9–11).
21 Middlebie is on the B725, two miles north-east of the A74(M) at Ecclefechan (Figure 2.1).
22 NAS: GD1/1432/1.7, 15/9/1718.
23 In late August 1718, Sir John Clerk had caught a serious infection on his back. Necrotised tissue had to be repeatedly excised, resulting in the wound getting so very large his life was feared for (BJC, pp. 97–98 and note).
24 NAS: GD1/1432/1.6, 22/1/1719.
25 Ramsay clearly influenced Jonathan Swift, whose *The Lady's Dressing Room* (1910) of 1730 is an even more satirical take on Ramsay's *The Morning Interview* of 1721. While the characters are Strephon and Celia rather than Damon and Celia, the parallels are abundantly clear.
26 NAS: GD1/1432/1.9, 5/8/1720.
27 FamilySearch.org, ID2:17NJNG4. This record does not come up in searching Scotland's People, but the information it contains does have the names of both parents accurately, and it ties in with the only other first-hand mentions of Dorothea's age or year of birth. Firstly, the Baron himself (BJC, p. 134) tells us, 'In March 1728 [when her mother died] [. . .] The girle was about 7 or 8 years of Age' and in DGA: GGD56/13, *Clerk Maxwell Family Tree*, a separate sheet showing the Maxwell of Middlebie line gives 1720.
28 DGA: GGD56/21, *List of Debts Paid by Mrs Clerk*, *c.* 1725. This shows that William rented a house from James Wright. Castlehill is the same small area of the town where James Ross of the Easy Club lived.
29 By this marriage Mary Clerk became the sister-in-law of David Moncrieff, great-great-grandfather of the Rev. William Scott Moncrieff (1804–1857) who was for some time the minister at St Mungo's in Penicuik (Seton, 1890; Wilson, 1891, p. 93). In 1841 he married Mary Irving (1811–1886), a second cousin of James Clerk Maxwell.
30 DGA: GGD56/21, *Accompt the Deceased Mr William Clark advocat*, 1723.
31 In his memoirs Sir John Clerk says forty, but the earlier notes give William's exact date of birth as 12 July 1681.
32 It appears that in the days before her father's death she may also have been unwell, for she received several of the doctor's prescriptions.
33 DGA: GGD56/21, *Accompt to the Funeral of Mr William Clerk Advocate to Wm Baillie*, 12/4/1723.
34 DGA: GGD/56/21, *A List of Debts Payed by my Wife for her Deceased Husband Mr William Clerk*, *c.* 1725.
35 See, for example, his complaints about debtors in 'To John Wardlaw' and 'To James Clerk of Pennycuik' (Ramsay & Robertson, *c.* 1886, pp. 249–255).
36 There is no record of a testament on Scotland's People. It seems strange that Agnes was left to clear off William Clerk's debts, but as revealed by DGA: GGD56/21, *Vouchers*, 1723–24 read along with DGA: GGD56/21, *A List of Debts* [. . .] *c.* 1725, this she did over a period of time between June 1723 and September 1724.

6. The Maxwells

1. Named similarly to other stretches on the River Tweed near Kelso, such as Doctor's Well, Sprouston Dub and Carham Weil.
2. DGA: GGD56/8, loose sheet.
3. On OS maps, the main site is shown as being between Lochmaben and Lockerbie, on the east bank of the Dryfe water, about a mile upstream of its confluence with the Annan. For an account of the battle of Dryfe Sands see, for example, Pitcairn (1833, pp. 28–53).
4. The sources of this account of the battle are Johnstone (1889, pp. 122–124), Lawson (1849, pp. 79–84) and McDowall (1873, pp. 270–281).
5. His conduct would seem to border on the insane, and it may be relevant that his grandmother had a history of mental illness.
6. Pitcairn (1833, pp. 28–53) gives an account of the trial.
7. Johnstone (1889, pp. 153–155), Lawson (1849, pp. 84–87) and McDowall (1873, pp. 282–293). The instrument of his death was the 'Maiden', an early form of guillotine. He and the Earl of Morton therefore suffered the same fate.
8. According to Paul (1909, p. 483), 'As a Catholic he [the 8th Lord Maxwell] played a somewhat disturbing part in public affairs', while according to Mackenzie (1841, vol. 1, p. 21) the ninth Lord died on the scaffold 'in the profession of the Roman Catholic faith'. The tenth Lord, the Earl of Nithsdale, was described as being 'a papist' in 1678 (Mackenzie, 1841, vol. 1, p. 209). For refusing to take the oath under the 1681 Act Anent Religion and the Test, he forfeited the Stewardship of Kirkcudbright (Mackenzie, 1841, vol. 1, p. 243). By 1715 the Earls of Nithsdale, and in general their people, were still staunchly Roman Catholic:

    many Roman Catholic families [...] religion to which they were sincerely attached [...] Amongst the[ir] number was [William] the [5th and last] Earl of Nithsdale, combining in his person the representation of the noble families of Herries and Maxwell. (Mackenzie, 1841, vol. 1, p. 364)

    and, of the same Earl of Nithsdale, Scott (1913) said, 'His mother [...] educated him in sentiments of devotion to the Catholic faith and of loyalty to the House of Stuart, for which his family was famous.'

    In February 1716 he escaped execution for his part in the rebellion by famously being spirited out of the Tower of London disguised as his wife's maidservant. They lived in exile and penury in Rome, adherents of the Roman Catholic faith to the end of their days. See also Mackenzie (1841, vol. 2, pp. 209–210, 243, 364).
9. Curiously, so did James Johnstone, 1st Earl of Hartfell, whose father Sir James Johnstone of Dunskellie had been murdered by Nithsdale's brother, John, 9th Lord Maxwell!
10. John and Thomas Blake, or Black, were imprisoned at Dumfries for the murder of John Maxwell of Middlebie in 1639. They were held at the order of Middlebie's half-brother, Lord Robert Maxwell, 1st Earl of Nithsdale, and kept in irons for over two years without any sort of trial. It would seem that the Earl had no more than a suspicion that the Blakes had a hand in his half-brother's death, and his actions speak volumes about the degree of impunity with which the March Lords acted.

    NB: In Burke (1894, vol. 2, pp. 1365–1366) Thomas Blake is referred to as Robert Blake.
11. DGA: GGD56/13.22, *Inventory of Writs for Middlebie* (no. 25), 1738.

12 Burke (1894, pp. 1365–1366) says she was from a Northumberland family, but the principal seat of the Dacres was in Cumberland, around Naworth and Kirklinton, just east of Carlisle. The Dacres of Naworth were already connected with the Maxwells; Hugh, 4th Lord Dacre, married Elizabeth, daughter of John, 3rd Lord Maxwell c. 1350.
13 DGA: RGD56/13; GGD56/8, loose sheet.
14 NAS: GD1/1432/1.2, 25/5/1714.
15 DGA: GGD56/21, *Discharge Mrs Maxwell to Mrs Leblanc* 1728, 1/2/1728.
16 SWT: CC5/6/9, *John Maxwell of Middlebie*, 22/1/1729 & 11/2/1729. The testament, written a year after his death, leaves the date blank, whereas the 'eik', written about three weeks after the testament, indicates only February 1728. Since the discharge in note 15 above given by Mrs Maxwell (Grace Smith) is dated 1 February 1728, January seems to be far more likely.
17 NAS: GD18/2575, 1702 & 1715. The names McLellan and McMillan are so similar there could be some confusion here, with the potential that both this Robert 4(ii) and his uncle Robert 3(ii) are linked to the same family, whichever the name.
18 NAS: GD18/2593, ?/6/1730.
19 DGA: GGD56/31/13, 7/4/1707 refers to a bond of provision that John Maxwell 3(i) made for 'his eldest daughter, Anna, by Dorothea Ballantine his spouse'.
20 FamilySearch.org, reference 2:1615B50 (accessible through NLS digital collections, registration required).
21 FamilySearch.org, ref. 2:17NJNG4.
22 FamilySearch.org, ref. pal:/MM9.1.1/XYWV-BHW.
23 DGA: GGD56/36, ?/6/1722.
24 DGA: GGD56/36, *Copy of the Middlebie Entail*, pp. 29–30, 9/6/1722.
25 SCAN: Person Code NA24209, *Maxwell, of Middlebie Dumfriesshire*, 17th–19th C. See the record narrative under 'activity'. See also Burke (1894, pp. 1365–1366).
26 McDowall (1873, p. 208), DGA: RGD56/13, *John Clerk Maxwell's Draft Family Tree*, 19th C.
27 As implied by Aitken (1887–88, p. 122).
28 Bryson & Veitch (1825) and as reviewed in *Literary Gazette* (1825, p. 419).

7. Agnes Maxwell and James Le Blanc

1 Turnbull (2001, p. 190); NAS: PA7/19.110; 14/8/1705; and PA7/19.149, 14/9/1705 all give it as the Duke of Atholl's Regiment, whereas NAS: GD18/2602 gives it as the Earl of Tullibardine's Regiment. However, Lord John Murray became Earl of Tullibardine in 1696, and in 1703 the Duke of Atholl. All are therefore correct. The regiment was disbanded in 1697.
2 NAS: GD112/15/118.25, 1707.
3 NAS: GD112/64/18.14, 1715 and GD248/107/17.10, 1704.
4 NAS: PA7/19.110, 1705.
5 NAS: PA7/19.149, 1705 and GD124/10/444.13–15, 1705.
6 NAS: GD18/2602, 13/6/1735 refers to Le Blanc's marriage to Elizabeth Houston; RPS: 1706/10/457 and NAS: PA6/34, 25/3/1707 refer to his naturalisation.
7 NAS: GD18/5227, 1699 and GD18/5367, 1728–34.
8 John Houston of Webster Southbarr was in 1678 the sheriff depute of Renfrew (RPS: 1678/6/22).

NOTES TO CHAPTER 7

9 NAS: GD18/5295, 1716–25.
10 NAS: GD3/14/3/1/15, 24/12/1720 refers to him as being paymaster, while NAS: GD1/49/83, 23/11/1723 and GD1/49/84, 2/12/1723 refer to a debt of 3,000 merks owed to Le Blanc mandating a sasine, presumably to take land as security.
11 NAS: GD18/4152, 24/9/1724.
12 NAS: GD18/5295, 26/12/1724.
13 NAS: GD18/2584, 1727–31.
14 DGA: GGD56/21, ?/3/1720–?/9/1724.
15 As explained in Chapter 6, her family were mostly Roman Catholics but may have dissembled adherence to the Protestant faith in order to survive in difficult times. At any rate, her husband was a Protestant and they had a Protestant wedding, which took place on 26 August (SOPR: *Marriages*, 685/010460/0322 Edinburgh). Note that in the transcription of the record, the new month appears to have been overlooked and July is given in error.
16 DGA: GGD56/21, *Medical Bill*, 1727. John Knox was the surgeon to the castle garrison, while David was his son. Major Le Blanc was buried in Greyfriars Kirkyard (DGA: GGD56/ [Unsorted Box], *Accompt Madam Leblanc for Funeral Charges* [. . .], 29/7/1727).
17 DGA: GGD56/21, *List of Debts Paid by Mrs Clerk*, 1726.
18 NAS: GD1/49/83. *Extract Heritable Bond by James Clephan of Powguild and Janet Beatson, his spouse, to Major James Leblane* [sic] *of the Castle of Edinburgh, for 3,000 merks*, 13/11/1723.
19 DGA: GGD56/21, *Regrat Disposition of Major James Le Blanc to Agnes Maxwell*, 14/7 & 1/8/1727.
20 Now called Woodside, a farmstead about one and a half miles north of Armadale in West Lothian. There is no obvious connection with James Young, the pioneer of shale oil extraction, who based himself in the same general area.
21 NAS: GD18/2587, *c.* 1728.
22 RPS: 1700/10/73.
23 NAS: GD1/1432/1.43, *Personal Spiritual Covenant with God*, early 18th C. The Baron described her as being 'a very virtuous woman' (BJC, p. 114).
24 NAS: GD18/2602, 13/6/1735.

8. The First Clerk Maxwells

1 DGA: GGD56/(Unsorted Box), *Life of George Clerk Maxwell, c.* 1788. This detail was omitted from the published version (Clerk , 1788, pp. 51–56). Jock's Lodge is on London Road to the east of Edinburgh (Figure 13.1).
2 Described by the Rev. Mr William Scott in Statistical Accounts (1791–99, vol. 12, p. 24).
3 NAS: GD1/1432/1.14, 2/5/1732.
4 This school is unconnected with the later Lowther College in Lancashire and thereafter in Wales. It was in the village of Lowther, near Penrith in Cumbria and less than a half mile west of the present M6 motorway, about three miles south of the Penrith Junction. The school was founded by Sir John Lowther Bt, afterwards Viscount Lonsdale, who was born in the parish of Lowther, then in county Westmoreland. Based on his own experience of school he formed the opinion:
> if a parent send his son to the university to be educated as a gentleman, he should by all means engage a governour to attend him, who should be so

much of a scholar, as to improve his school-learning, yet so complacent and polished in his manners, that his pupil might respect and love him, and delight in his company. (Lowther, 1808)

Having rebuilt the church at Lowther, in 1698 he built and endowed a school, Lowther College, in fulfilment of these philanthropic notions. In George Clerk's time the master was a Mr Wilkinson, who the Baron thought to 'answere the character I had got of him, for he was indeed a learned, honest, diligent, careful man' (BJC, p. 139).

At the behest of the then Viscount Lonsdale, the school closed when Mr Wilkinson retired in 1740, in order to allow the charitable funding to be directed towards manufacturing, making use of the local wools, rather than for the benefit of the education of young gentlemen. George would no doubt have approved, for it was in 1739 that he was also setting up a similar scheme in Dumfries for the manufacture of linen (Curwen, 1932).

5 NAS: GD1/1432/1.16, 25/2/1731.
6 NAS: GD1/1432/1.14, 2/5/1732.
7 NAS: GD1/1432/1.15, 18/4/1733 and GD1/1432/1.17, 19/4/1733.
8 NAS: GD18/5396, 8/11/1732.
9 NAS: GD18/2338, 1735.
10 For example, Henry Walker (d. 1730) of Blackfriars Wynd in Edinburgh was accused of conducting irregular marriages (Scott, 1915, vol. 1, p. 278). He could well have been the very man as the Baron's town residence was in that same street (note 5 of Chapter 4).
11 SOPR: *Marriages*, 697/00 0030 0093 Penicuik.
12 'The betrothal of George Clerk and Dorothea Clerk Maxwell is said by the Baron [...] to have been in accordance with her mother's dying wish. (Dorothea was seven years old when Agnes Maxwell died!)' (C&G, p. 19). The all-important exclamation mark is Campbell's.
13 NAS: GD18/1973, 11/10/1735.
14 NAS: GD18/1975, 28/12/1737.
15 DGA: GGD56/5, *Extract Interlocutor in the process of Sale* [...], 9/2/1738.
16 DGA: GGD56/27, *An Act to Enable Dorothea Clerk to Sell Lands* [...], 1738 and DGA: GGD56/36, *Copy of the Middlebie Entail (9/6/1722)*, 9/10/1735.
17 NAS: GD1/1432/1.2, 14/10/1735.
18 NAS: GD18/5396, 25/9/1736 and 22/4/1737.
19 NAS: GD1/1432/1.44, 6/10/1736.
20 Shalloch was to come into the Clerk family at a later date, for it was given to John Clerk Maxwell in 1826 by his brother George, 6th Baronet, as part of his wedding settlement (see later). Dorothea must have inherited the property sometime after 1737, for she got little out of her uncle's will when he died in that year.
21 NAS: GD1/1432/1.12, 14/12/1737.
22 NAS: GD1/1432/1.23, 4/4/1738.
23 NAS: SIG1/32/18, *Signature of the lands of Drumcrief etc* [...], 13/2/1738.
24 NAS: GD18/2606.3, 24/2/1738.
25 See documents cited in *Inrollment of George Clerk Maxwell* in DGA: GGD56/ (Unsorted Box), ?/5/1741. The relevant item is *General Retour in Favour of Dorothea Maxwell alias Clark as Heir to the Said John Maxwell*, dated 6 February 1738. The fact that Dorothea was returned as heir meant that it was accepted beyond doubt that she was not a Roman Catholic.

NOTES TO CHAPTER 8

26 NAS: GD18/1880, 15/8/1738.
27 NAS: SIG1/35/21, *Signature of the Lands of Middlebie*, 6/8/1765.
28 DGA: GGD56/ (Unsorted Box), *Inrollment of George Clerk Maxwell*, ?/5/1741.
29 Miss Isabella Clerk was the seventh child of Sir George Clerk, 6th Baronet of Penicuik, and therefore James Clerk Maxwell's first cousin. Being born after 1828, she was probably of a similar age (Foster, 1884). She lived with her older brother, Major-General Henry Clerk, at Hobart Place, Eaton Square, London (*London Gazette*, 25/2/1880, p. 1537).
30 He was involved in projects such as: the repair of a road at the Bield near Tweedsmuir to the north of Moffat (NAS: GD1/1432/1.25, 18/4/1749); a scheme to realign the road at West Linton (NAS: GD18/5396, 13/9/1749); roads to the Hartfell Spa and the Moffat Well (Prevost, 1968, pp. 204–205); and a new turnpike built alongside Annan Water, which required the building of two bridges in 1763 (Prevost, 1968, pp. 204–205).
31 NAS: GD18/5396, 2/10/1738.
32 NAS: GD1/1432/1.18, 25/10/1738.
33 DGA: GGD56/ (Unsorted Box), *Notebook of George Clerk Maxwell*, 1/11/1774.
34 NAS: GD18/5396, 13/3-10/9/1739 & 12/11/1739.
35 The Baron had been a trustee of their Board since 1727, which no doubt helped. John Cockburn of Ormiston (1685–1758) had already established such a school at Ormiston in East Lothian in the same way (NAS: GD18/5905, 1739; *Gazetteer for Scotland*, 2013a). George Clerk Maxwell's spinning school in Dumfries is cited in DGA: GD18/5903, 1738–39, which mentions the Board's approval for setting up the school; DGA: GD18/5916, 28/11/1748, which concerns the school itself and purchase of linen yarn; and DGA: GD18/5918, 1748–49, which refers to letters from the spinning mistress, Elizabeth Hill, about school matters and her salary. In 1751 the school was still going with forty pupils (McDowall, 1873, Chap. 42).
36 NAS: GD1/1432/1.46 25/9/1739 and GD1/1432/1.47, 28/9/1739.
37 NAS: GD18/2232, 1740; Prevost (1969).
38 NAS: GD18/5396.43, 29/3/1742 and GD18/5396.48, 14/4/1742.
39 His post was Clerk of Discoveries (see, e.g., NAS: E749/1, 1748–49; E769/6 1749–52). In the earliest available directory of 1773, his address is given as James' Court off the Lawnmarket. The records of the marriages of his daughters Agnes and Dorothea suggest the Clerks had been there or nearby since 1758, as they are given as being in the Tolbooth Parish, which covered the Castlehill area. James' Court dates from *c.* 1725. David Hume also lived there until the early 1770s, after which he let his apartments to James Boswell. George Clerk Maxwell must therefore have known Boswell and Hume fairly well, and indeed Boswell records having been at a dinner in 1780 at which George was amongst the company (Boswell, 2013, p. 397).
40 DGA: GGD/13.22, *c.* 1748.
41 He was buried at Penicuik Churchyard (St Mungo's) on 24 February but the actual date of death is unstated in SOPR.
42 The East India Company had started out with a charter giving them the monopoly of trade with India and soon formed settlements there on the East Coast. Before long they virtually ruled a large portion of the country, had an army on land and merchant fleet at sea. By the monopoly of their exports back to Great Britain they created fantastic wealth.

43 Foster (1884) confuses this George Clerk Craigie, advocate, with Captain George Clerk, son of Agnes Clerk and John Craigie of Glendoik, who died in the US war of Independence in 1781.
44 http://en.wikipedia.org/wiki/Great_Siege_of_Gibraltar
45 SOPR: *Deaths*, 685/010970/0266 Edinburgh, 27/11/1781.
46 NAS: GD18/2327, 17/3/1742.
47 NAS: GD1/1432/1.59, 1763.
48 General George Wade (1673–1748) is better known as the builder of the first system of military roads and barracks in the Scottish Highlands (1724–36). His bridge over the River Tay at Aberfeldy still stands. The Duke of Cumberland (1721–1765) was William, third son of George II, who became notorious for his merciless pursuit of the Jacobite rebels after Culloden, earning himself the soubriquet 'the Butcher'.
49 NAS: GD18/3249, 1745.
50 *The Land of the Mountain and the Flood* is the very apt title of the 1887 overture by Hamish McCunn. For an idea of the old Highland way of life, see the works of John Prebble. *Lion in the North* (1971) is a concise history of Scotland, while *Culloden* (1967) gives a comprehensive account of the '45 and its immediate aftermath. *Glencoe* (1968) and *The Highland Clearances* (1982) are also rich in their discussion of two other highly dismal episodes of Scottish history.
51 It was in fact the second such Commission, the previous one having been set up in 1716 in the wake of the first Jacobite rebellion. Forts William, Augustus and George, and Wade's military roads had all been built as a result.
52 Son of the John Baxter (d. 1770) who had been a builder to the Baron (Chapter 4). While it was the father who built the new Penicuik House to Sir James Clerk's design, the son, having studied in Rome as one of Sir James' protégés, did some of the finishing details (DSA, 2014).
53 NAS: E728/33, 1774–79.
54 NAS: E730/21, 1771–79.
55 NAS: GD1/1432/1.55, 2/9/1754 and Smiles (1904, p. 203).
56 Fresh income had become available because the forfeited estates were now being returned to the heirs of the attainted owners. Those who benefited by the return of an estate had to pay money to the Crown in order to clear off the debts lying on the estate at the time of the forfeiture. Baron Sir John Clerk noted that almost all of the estates were far in debt for, if it had been otherwise, the proprietors would have thought twice about coming out in the rebellion (BJC, pp. 220–221).
57 NAS: GD18/2615, 1768–1778.
58 John Stuart (1713–1792), Prime Minister 1762–63, later 3rd Earl of Bute.
59 NAS: GD1/1432/1.28, 15/6/1762.
60 NAS: GD1/1432/1.29, 7/8/1762 and GD1/1432/1.56, 1/11/ 1762.
61 NAS: GD1/1432/1.30, 14/12/1762.
62 Charles Duke of Queensberry had become one of a growing breed of lords to leave his factors to attend to business at home while he himself was in London or at his Amesbury estate in Wiltshire. Amesbury is a few miles north of Salisbury and over 360 miles distant from Drumlanrig. The Duke's mansion there was built by Inigo Jones (Virtue & Co., 1868).
63 William Muir of Caldwell had been MP for Renfrewshire and was appointed Baron of the Exchequer in 1741 (Hamilton, 1894).
64 NAS: GD1/1432/1.31, 10/1/1763.

NOTES TO CHAPTER 8

65 NAS: GD1/1432/1.32 and GD1/1432/1.33, ?/3/1763.
66 NAS: GD1/1432/1.36, 17/3/1763.
67 NAS: GD1/1432/1.37, 22/3/1763 and GD1/1432/1.34, 24/3/1763.
68 William Craik (1703–1798). Arbigland is on the Solway coast due south of Dumfries. John Paul, father of the US naval hero John Paul Jones had a gardener's cottage on that estate; it is now preserved as a museum.
69 The date given for the visit to London in this article is 1761, but since George was not appointed until 1763, it is likely to have been a misreading for 1764. In the same article, it is said that Craik could have been a Commissioner for the Board of Customs, but he would not go to Edinburgh to take up the post. That being the case, George must therefore have had to be regularly in Edinburgh.
70 Frandzen, 2014; Kay & Paton, 1877, vol. 1, no. CLIII, pp. 384–385.
71 Rae, 1895, Chap. XXI, pp. 325–338. The Grassmarket is an open low-lying area at the west end of the Cowgate to the south of Castlehill. In years gone by it was the livestock market, from the French 'grasse', meaning 'fat', as in fat-stock.
72 The customs house was then in the town itself, not at the later site on Commercial Street in Leith. John Ainslie's 1780 map of the city shows it to be at 'K' (see legend), which is off the High Street in the exchange building, now the City Council chambers.
73 Not to be confused with the Isle of Man, mentioned previously. The Isle of May is yet another island. This one lies off the east coast of Scotland at the entrance to the Firth of Forth.
74 The ship had been loaned to the colonial Continental Navy by the French, hence the possibility of confusion.
75 NAS: GD1/1432/1.27 12/9/1753.
76 CANMORE ID 48491.
77 Ramsay, 1888, vol. 2, pp. 327–335. John Ramsay (Horn, 2004) wrote a letter to George Clerk Maxwell in which he praised the new spa and referred to Williamson as a 'so-called *daft man*' (NAS: GD1/1432/1.51, 18/4/1750). The mine workings may have been a copper mine lying on the same contour only a few hundred yards west (Pococke, 1887; CANMORE ID 48490).
78 NAS: GD1/1432/1.29, GD1/1432/1.56 and GD1/1432/1.30, 7/8/1762–14/12/1762.
79 NAS: GD18/1123, GD18/1164-66, GD18/1168-70, GD18/1173 and GD18/1179, 1755–82.
80 NAS: GD18/1164, 1755.
81 On the 1864 6" OS map recommended by the NLS for looking for old workings, the nearest old mine lies about a mile north of Glendorch at Glendouran (CANMORE ID 258031), but there is no indication at Glendorch itself. The mine is an isolated spot high up in the glen.
82 A wealthy father would generally leave individual bequests, in the form of bonds of provision, for those children who were not in the fortunate position of being his heir. For example, the Baron mentions the individual bequests left by his own father (BJC, pp. 110–111), in particular the 15,000 merks (about £750 sterling, a considerable amount of money roughly equivalent to £60,000 today) left to his half-brother James. We may consider that George would have benefited from at least a similar sum on the death of the Baron.
83 NAS: GD18/1165, 10/9/1756.
84 NAS: GD1/1432/1.26, 26/5/1757.

85 DGA: GGD56/18/11, *Lease or Tack by the Earl of Hopetoun to George Clerk Maxwell & his Partners of Three Fourth parts of the Mines in South Shortcleugh. 11 years from 1st Jan'y 1758*, 1758–69. The opening words were later overwritten with 'Renounced'.
86 It was near Blackcraig farm (Donnachie, 1971, p. 121) just to the north of the present A75.
87 Patrick Heron, landowner and 'joint-adventurer' with the Craigtown Mining Company, was struck by a calamity in 1773 when the bank Douglas, Heron & Company was caught out with too many loans on their books. They went bankrupt with a loss of over £663,000 to its 225 shareholders, the Duke of Queensberry amongst them (The Archive Hub, 2014; RobertBurns.org, 2014). Although he was only a co-adventurer with the Craigtown Mining Company it is feasible that some of the partners felt the shock of this collapse. Though Heron himself managed to recover from the collapse of his bank, it was a great scandal.
88 DGA: GGD56/1811, 1758–69.
89 NAS: RHP3849, *Sketch Plan of Daltamie and Blackcraig*, 1768 (Donnachie, 1971, p. 121).
90 A distant cousin of Patrick Heron.
91 NAS: GD18/1173, 1770.
92 DGA: GGD56/18.12, 8–11/5/1770.
93 NAS: GD18/2254, 1766–77.
94 NAS: GD18/1126, 21/11/1778.
95 The mines in the area enjoyed a revival between 1853 and 1880 as Blackcraig East and Blackcraig West, the workings that were eventually recorded by RCAHMS. They produced an average of 130 tons of lead a year between them, and so it was still not a huge yield.
96 NAS: GD1/1432/1.34, 24/3/1763.
97 NAS: GD1/1432/1.35, 5/1/1763.
98 An offshoot of the Select Society (Emerson, 2004) of which he had become a member around 1755.
99 The Society of Antiquaries of Scotland (2014). Amongst their proceedings is the article on the Trustees Drawing School referred to above (Laing, 1869).
100 CANMORE ID 66815.
101 Sanquhar was not close to home. It lies on the Nith some twenty-five miles up river from Dumfries, and it is some thirty miles distant from Moffat by a difficult route.
102 It seems unlikely, therefore, that he was one of the unfortunate people to have made substantial losses as a result of the 1772 banking crisis, or, despite his close business association with Patrick Heron, the ensuing collapse of the Douglas and Heron Ayr Bank in 1773.
103 NAS: CH2/537/8, *Dumfries Kirk Session Records*, pp. 20 and 79.
104 All of Dumcrieff and the disentailed lands of Middlebie that had been repurchased by his father. That he needed Dorothea's consent is an indication that there was some further complexity to it. Alexander Ogilvie Farquharson of Haughton was an Aberdonian who had a significant reputation as an accountant (Brown, 2004). He may well have been known to George beforehand, for one of his tasks had been reporting on the finances of some of the forfeited estates.
105 DGA: GGD56/29, *Roup of Dumcrieff and Middlebie*, 29/12/1787. The date on this document is 1787 because the document quoted as evidence was drawn up for a

NOTES TO CHAPTER 8

separate process in the Court of Session that took place some five years later. It refers to the roup, for which it gives 'the tenors'. James Boswell and his friend Charles Hay were the advocates in the case.

106 DGA: GGD56/29, 1787.
107 The Rev. John Walker, minister of Moffat, acted as an intermediary on Lt Col Johnson's behalf at the roup presumably because the latter had not yet returned home from service. The ownership of Dumcrieff after Col Johnston is given by Prevost (1968, p. 210). Dumcrieff was at one time rented by John Macadam (1756–1836), the road-building pioneer, whose original sandstone roller is said to remain there: www.dumcrieff.com/content.php?id=175g=0/History
108 With thanks to Catherine Gibb at Dumfries and Galloway Archives for extracting this information out of the articles of roup.
109 See Figure 1.1. It was progressively drained between 1759 and 1763. (Royal Scottish Geographical Society, 2014)
110 DGA: GGD56/18/1, *Statement of Lady Dorothea's and James Clerk's Affairs*, 1794.
111 DGA: GGD56/18/14, *Statement of Account*, 1796.

9. The Successors of George and Dorothea

1 NAS: GD18/5487, *Letter from John Clerk* [at Chatham] *to his father* [...], 17/9/1757.
2 NAS: GD18/5494, 1768–70. The artist Ann Forbes (Sloan, 1997) was a grand-daughter of William Aikman the painter, a cousin of John Clerk's own grandfather, the Baron.
3 NAS: GD18/5506, 1775–76. The spelling *Enterprize* was contemporary. NB: this John Clerk was actually John Clerk of Eldin's nephew rather than his son, who was an advocate.
4 NAS: GD18/4207, 27/11/1776.
5 She is also mistakenly referred to as Mary Appleby Dacre. Later in life, for reasons explained elsewhere, she styled her first name as Rose Mary or Rosemary (Prevost, 1970, pp. 162, ff. 176 and 178–80).
6 NAS: GD1/1432/1.39, 15/10/1778.
7 NAS: GD18/2125A, *c.* 1771–72.
8 NAS: GD1/1432/1.41, 22/10/1778; GD1/1432/1.60, 27/10/1778; and GD1/1432/1.42, 3/11/1778. James Veitch. Lord Eliock was a Scottish judge and fellow commissioner of George Clerk Maxwell in the 'Forfeited Estates'. Like his friends the Clerks, he was also a friend of the Duke of Queensberry, who was then seriously ill in London. It appears that Veitch had seen James Clerk HEICS, but not 'the Captain', i.e. Lieutenant John Clerk RN, who had been moved to Chatham by Lord Sandwich. Although then a lieutenant by rank, John Clerk had been in command of a ship, for in NAS: GD1/1432/1.39, 15/10/1778 he also is referred to as 'Captain'.
9 The battle of Cape St Vincent had taken place on 16 January of that year. HMS *Alfred* may have been with the fleet but she does not seem to have taken part.
10 NAS: GD18/4214, 10/3/1780.
11 NAS: GD18/281, 6/8/1782.
12 Raeburn got sixty guineas (£63) for it (Greig, 1911, p. li).
13 Now the Royal Highland and Agricultural Society.

14 Due to the American War of Independence.
15 Number 37 Princes Street would be somewhere around where number 70 is today. In 1785, Miss Clerk was still listed at Crichton St, but Lady Clerk was no longer there, ergo, it seems she would then be at 37 Princes St with Sir John.
16 She was baptised on 3 November 1745 and died 'in her 89th year' on 1 November 1834 (Prevost, 1970, pp. 162, 172). Given the circumstances of her birth, she would have been baptised without delay, and so she would have been almost exactly 89 years of age at her death.
17 See http://list.english-heritage.org.uk/resultsingle.aspx?uid=1335577
18 Bishop from 1734 to his death in 1747.
19 Just six miles to the south of Carlisle, near Dalston, http://list.english-heritage.org.uk/resultsingle.aspx?uid=1087473
20 The white cockade was the Jacobite emblem chosen by Prince Charles Edward Stuart for his campaign of the '45. It was made from some white silken material fashioned to represent a rose.
21 See Chapter 8 for George Clerk Maxwell's part in the pursuit of the retreating rebels through Westmorland and Cumberland to Carlisle, thirty-two years after which he became Mary Dacre's father-in-law.
22 NAS: GD18/3263 and GD18/3266, 1746.
23 The Baron does not refer to this Henry Clerk in his memoirs, but since his own brother and son of that name were already dead, he was probably a nephew. The name Henry seems only to have been introduced to the Clerk family in the Baron's own generation, in honour of his maternal grandfather, Henry Henderson. If this Henry Clerk got his name in the same way, he could not have been more distantly related than a cousin, or more likely, a nephew.
24 He is believed to have been Captain Ranald Macdonald, third son of Ranald Macdonald 3rd of Kinlochmoidart. He received a pardon for his part in the rebellion and it may be that his gallant behaviour towards Joseph Dacre's wife and newborn child had something to do with it, for Joseph was one of the judges.
25 With the assistance of David Stewart of Garth, who founded the Celtic Society of Edinburgh in 1820, with Scott as one of the original members.
26 This is the subject of a curious but once well-known story of the supernatural (Grant, c. 1887, vol. 5, p. 192).
27 Ramsay (1861, p. 88). Ramsay erroneously states that Mary Dacre was born in Newcastle and that her father was a Jacobite sympathiser; Prevost's evidence is quite to the contrary.
28 DGA: GGD56/27/8, 1775.
29 NAS: GD1/1432/1.40, 27/4/1776.
30 NAS: GD18/5512, 12/2/1778.
31 NAS: GD1/1432/1.39, 15/10/1778.
32 NAS: GD1/1432/1.39, 15/10/1778.
33 NAS: GD1/1432/1.41, 22/10/1778.
34 NAS: GD18/4228, 1782.
35 The limited information available at www.eicships.info/voyages/lost_m.htm and www.eicships.info/ships/shiplistM.asp give the 'Major' as having been in service during 1781–82, but lost in 1784. Although the East India Company monopoly formally ended in 1783, it is not clear whether the ship could have made its last voyage in private hands. It is more likely that the ship made one successful voyage during '81–82, but as it did not return from its second voyage, it went unrecorded.

36 In 1784, we have James Clerk in the Bay of Bengal, whereas he is listed in that year's directory as 'Clark James Esq., at Lady Maxwell's'. In turn, Lady Maxwell's entry is given as 'Maxwell Clerk Lady, No 30, Princes street', that is to say, Dorothea, James' mother. However, James' brother John, who was also frequently away at sea, provided a forwarding address for the directory during the years 1774 and 1775, namely 'at Mrs Shaw's, James' Court'. Curiously, in the 1786 directory, there is no listing for James Clerk, but he is back in 1788, again at his mother's.
37 DGA: GGD56/18/17, 21/05/1785.
38 SOPR: *Marriages*, 685/01 0510 0345 Edinburgh; Grant (1922, p. 145).
39 DGA: GGD56/13, c. 1792–93 contains three extract baptismal certificates. The outer paper was originally sealed with red wax to form an envelope addressed to 'James Clark, Princes Street, Edinburgh'. Written on the reverse side is 'Certificates of the Births of my Children'. This is in James Clerk's own hand, for it is signed 'Ja: Clerk'.
    The first certificate informs us that a son, named as George, was born on 19 November 1787.
    The second is for Isabella Clerk, baptised on 20 May 1789.
    The third, dated 14 August 1792, is for a second son, James. It is marked 'died an infant' in a later hand.
    Unfortunately the baptismal certificate of John Clerk, later Clerk Maxwell, is not to be found with the others. However, the official record (SOPR: *Births*, 685/01 0380 0216 Edinburgh) gives his baptism as being on 10 November 1790.
40 NAS: GD18/1887, 4/7/1788 and DGA: GGD56/4, n.d.
41 The reason usually quoted is that the entail of Middlebie required the inheritor to take the name of Maxwell, but that on its own would not have prevented it, in fact it was forbidden by a like condition of the Penicuik baronetcy, which is actually how it is worded in the marriage contract itself.
42 DGA: GGD56/18/1, *Statement of Lady Dorothea CM's Account from 1788 to Aug. 93*, 1793.
43 NAS: GD18/1537, 3/5/1793.
44 CANMORE ID 150395 and OS 6" map series, *Edinburghshire, Sheet 7*, 1854.
45 DGA: GGD 56/11, *Payments made to Mrs Janet Clerk & Rents on the House and Lands of Over Lasswade on Behoof of Mrs Janet Clerk*, 1798–1812. The names Upper and Over Lasswade are mentioned together within these papers.
46 Edinburgh University Library, 2014; CANMORE ID 150395; OS 6" map series, *Edinburghshire, Sheet 7*, 1854.
47 DGA: GGD56/4, *Memoranda by Andrew Wedderburn Maxwell on the Descent of Middlebie*, c. 1893.
48 DGA: GGD56/18/8; it included James Clerk's whisky, valued in DGA: GD56/18/13.
49 DGA: GGD56/4, *Memoranda on the Descent of Middlebie*, c. 1893; DGA: GGD56/11, *Various Allowances to be made to Mrs Janet Clerk & Payments made to Mrs Janet Clerk*, 21/5/1798; and DGA: GGD56/27/4, *Commission for Proclaiming Breve of Service*, 1803.
50 By the entail of 1722, she would have succeeded to Middlebie if she had outlived her brothers and neither of them had produced an heir, but that did not happen.
51 DGA: GGD56/4, *Descent of Middlebie*, c. 1893.
52 The 'Daisy of Pentland' (Fairley, 1988, p. 97).
53 DGA: GGD56/13, *Disposition by Sarah Irving in Favour of Janet Irving*, 25/10/1783.
54 Elizabeth Cleghorn, Sir James' widow had died in 1783, and Janet Inglis had died long before in 1760.

55 DGA: GGD56/11, *Various Allowances to be made to Mrs Janet Clerk* [...], 21/5/1798.
56 The first directory entry that unambiguously identifies our Mrs Clerk is for 1805. The entry of her sister-in-law, Mary Clerk (Dacre), in the same directory is given two lines above as 'Lady Dowager of Pennycuik'. There was no actual Lady Clerk of Penicuik at that time because the baronet was now Janet's eldest son George, who was then only eighteen years of age. Janet had missed out on having the title Lady Clerk because her husband James had died while his elder brother John, the 5th Baronet, was still alive, leaving the title to pass on down to the next generation. It is understandable that, as the mother of the young baronet, Janet would have wished her station in life to be recognised and so she was allowed the privilege of appending 'of Pennycuik' to her name.

From 1805, we may trace the same Mrs Clerk (or Clark) at that address in the directories back to the year 1800. In 1799, however, the only address given for a Mrs Clerk at George Square is at number 9, which was on the north side, while in the previous directory, which was for 1797, we have simply 'George's square, south side'. There is no way of knowing for certain whether or not these relate to our Mrs Clerk. However, at the very least, we can say that Mrs Clerk (Janet Irving) lived at George Square from at least 1800, and indeed the directories confirm that she was there until 1808.
57 Bristo Street was the continuation of present day Bristo Place. Now obliterated by the University Student Centre, it was once part of the main road south.
58 In 1809 the house was unnumbered, but in 1810 it was listed as '12, west Heriot row'. In the year following, however, the general renumbering exercise made it '31, Heriot-row'.
59 James Wedderburn purchased 31 Heriot Row in 1821. RS: *Register of Sasines*, PR 1902.277 (688), 24/5/1851.
60 RS: *Register of Sasines*, PR 629.142 (13,869), 30/1/1809.
61 RS: *Register of Sasines*, PR 1902.277 (688), 24/5/1851.
62 Maria was the grand-daughter of Edmund Law DD, Bishop of Carlisle between 1768 and 1787. Curiously, the maternal grandfather of George's aunt, the Dowager Lady Clerk, aka Mary Dacre, whose grandfather had been the Bishop of Carlisle between 1734 and 1747 (Chapter 9). Lady Clerk, was still very much alive at the time of George's marriage.
63 DGA: GGD56/13, *Inventory of the Personal Estate of Mrs Janet Clerk*, 1822.
64 DGA: GGD56/13, *Extract of Birth Register for Isabella Clerk,* 20/5/1789.
65 The narrative here broadly follows the information given in Wedderburn (1898) supplemented by other references as cited.
66 Blackness is a promontory on the south shore of the Firth of Forth about twelve miles west of Edinburgh.
67 In more recent times they have provided the basis for the novel *Joseph Knight* (Robertson, 2003).
68 He had 565 slaves on two estates in Jamaica, Wallens and Rose Hall, both in St Thomas-ye-Vale parish (UCL Department of History, 2014b).
69 He held several important public offices in Jamaica including Island Secretary and Governor's Secretary, from which he derived a substantial income. His house, Bullock's Lodge, overlooks the harbour at Port Henderson near Kingston Jamaica (Cundall, 1904, p. 26).
70 Brown & Lawson (1990, p. 362) give the years of his tenure as 1810–19.

NOTES TO CHAPTER 9

71 Walter Scott was born in August 1771 and after qualifying as an advocate alongside his friend William Clerk, son of John Clerk of Eldin, he was appointed Sheriff of Selkirkshire on 16 December 1799 (Grant, 1944). He and James Wedderburn were friends, for in a letter to Daniel Terry of 10 November 1822, Scott mentioned having 'a depression of spirits following the loss of friends (to whom I am now to add poor Wedderburne)' (Lockhart, 1837, vol. 5, p. 183).

72 Janet Isabella Wedderburn was referred to by the family simply as Isabella. Her husband, James Hay Mackenzie WS (1809–1865), was the third surviving son of Colin Mackenzie of Portmore (1770–1830) and a nephew of Sir William Forbes the banker. He was Deputy-Keeper of the Great Seal, 1858–65, in which role he was succeeded by his son, Colin, who was James Clerk Maxwell's cousin and legal adviser. After their marriage, the couple lived at 31 Heriot Row with Mrs Wedderburn, but by 1841 they not only had a house of their own at Randolph Cliff, just east of the Dean Bridge, but also a newly built seaside retreat called Marine Villa or Silverknowe (sic). About the winter of 1849–50, Isabella took ill, probably with consumption, and died in 1852 leaving six children, Colin being the eldest (Fairley, 1988, p. 157). Having remarried in 1861 to Selina Jane Norton, James died in 1865. All three, James, Isabella and Selina, are buried side by side in the churchyard of St John's Episcopal Church off Lothian Road, not far from the grave of Peter Guthrie Tait.

73 NAS: GD495/48/1/60-61, 20–23/6/1819.

74 She was to be provided for after his death with an annuity of £350 from £9,000 to be made up of £6,000 from his estate, £2,000 from her elder brother Sir George Clerk, and £1,000 from the estate of her mother, Janet Irving (DGA: GGD56/7, *Contract of Marriage between James Wedderburn Esquire and Miss Isabella Clerk*, 1813.

75 RS: *Register of Sasines*, PR 1902.277 (688), 24/5/1851.

76 SC: *Wedderburn, Isabella,* 685/01 158/00 009 Edinburgh, 6/6/1841.

77 Grand-daughters of James Wedderburn senior's brother Andrew Colvile, formerly Wedderburn.

10. Sir James Clerk, 3rd Baronet of Penicuik

1 Alexander Gordon (*c.* 1692–1754) was a celebrated Scottish antiquarian whose *magnum opus* was *Itinerarium Septentrionale* (Gordon, 1726).

> His great patron was Sir John Clerk of Penicuick [...] he was a frequent guest at Old Penicuick House, where he had access to a splendid museum of antiquities, and was accompanied by Clerk in his Northumbrian explorations. (Goodwin, 1890)

2 NAS: GD18/5357, 5/9/1727. That William Simpson was a schoolmaster at Dalkeith is known from Scott (1925a, vol 3, p. 47) under the entry for 'Claud Hamilton'.

3 Rock, 2013b. The school was the forerunner of the Trustees Drawing Academy set up by the Board of Manufactures (Chapter 8; Laing, 1869). It was established by Richard Cooper in October 1729 under the formal name of the Edinburgh Academy of St Luke. Through the practice of anatomical drawing, it had had a close connection with the classes of Alexander Munro Primus at Edinburgh University.

4   NAS: GD18/2321, 25/11/1732. The appeal was made by 'James Clerk, son of Sir John Clerk'. There could therefore be scope for confusion here between James Clerk, son of the 1st Baronet by his second marriage, and James Clerk, the Baron's son and subject of this chapter. The former, having been first a merchant and then a customs officer, died in 1830 and so the reference must be to our present subject.
5   NAS: GD18/5340.6, 8/6/1736 and GD18/5396.2, 25/9/1736.
6   NAS: GD18/5340.8, 21/8/1738 and GD18/5340.9, 27/11/1738–23/1/1739.
7   NAS: GD18/2334, ?/2/1748.
8   NAS: GD18/5463, 10/11/1755.
9   CANMORE ID 254568 and Wilson (1886, p. 144–145).
10 John Cleghorn was minister in the Church of Scotland first at Burntisland then Wemyss in Fife. While his entry in Scott (1925b, vol. 5, pp. 82 and 121) contains information concerning a dispute with the Church, note that it refers not to him but to the subsequent presentee.
11 NAS: GD18/1986, GD18/1987 and GD18/1988, 11/12/1756–25/2/1758.
12 CANMORE ID 51637 and 51638.
13 Later a general, he was the second son of the Baron's cousin, Dr John Clerk, MD (Burke, 1834–38, vol. 3, p. 188).
14 As a consequence of his younger brother John's marriage to Susannah Adam.
15 NAS: GD18/1408, 12/5/1762.
16 NAS: GD18/1700, *Minute of Sale*, 4/11/1761 gives the purchaser as having been 'Robert Clerk, merchant in Edinburgh'. Robert Clerk (1722-1814) and his father Hugh, were both merchants in Edinburgh (Boog Watson, 1930). This Hugh Clerk was the son of Robert Clerk the surgeon, uncle of the Baron (Foster, 1884, p. 53; BJC, p. 248).
17 NAS: GD18/5015, 3/1/1765.
18 NAS: GD18/4679, 19/4/1768 and GD18/4680, 16/5/1770.
19 NAS: GD18/4682, 12/1/1771.
20 Curiously, the Gaelic poet Dugald Buchanan (Cooper, 1886) wrote to Sir James asking him to impress upon the Commissioners of the Forfeited Estates (of which his brother George was one) the need for a Gaelic dictionary to preserve the ancient Gaelic language (NAS: GD18/4529, 31/1/1767). Together with his interest in Macpherson's *Ossian*, it suggests that the 3rd Baronet was known for having more than just a passing interest in the Gaelic language and culture.
21 Part 1 of Gordon (1726) contains a plate of Arthur's O'on (CANMORE ID 46950), a beehive-shaped Roman monument at Stenhousemuir that had been much admired by the Baron, but was scandalously demolished in 1743. The monument was said to be similar in shape to an old brickmaking oven, of which 'o'on' appears a likely corruption. While 'oven' clearly refers to the shape of the monument, any connection with Arthur is obscure.
22 Skempton, 2002; NAS: GD18/5832, 1765.
23 NAS: GD18/1247, 29/3/1769 and CANMORE ID 213296.
24 CANMORE ID 120626.
25 NAS: GD18/1116, 26/3/1771; GD18/4208, 11/7/78; GD18/1119, 20/2/1772.
26 NAS: GD18/5072, 22/5/1781.
27 SOPR: *Deaths*, 697/00 0030 0292 Penicuik.

11. From Weir to Irving

1. This was Kirkton in South Lanark, a different Kirkton from the one at Kilmahoe, Dumfriesshire, that was part of the Middlebie entail.
2. One could be forgiven for failing to comprehend that the river at Newton and the mighty River Clyde that flows through Glasgow are actually the same river. It is fed by tributaries that flow down from the Lowther Hills in the southern uplands of Scotland; one of these, the Elvan Water, meets the Clyde at Elvanfoot just opposite Newton. Before reaching Glasgow, the river flows through the county town of Lanark and thence it passes between the industrial towns of Motherwell and Wishaw.
3. DGA: RGD56/13, early nineteenth century.
4. RCAHMS Mapping (registration required); sites around Elvanfoot and Newton.
5. The battle of Flodden Field on 9 September 1513 was a heavy defeat for the Scots. King James IV was slaughtered along with perhaps a third of his army and many of his nobility. The grievous loss is now associated with the haunting lament by Jean Elliot (1727–1805), 'The Floo'ers o' the Forest'.
6. But somewhat confusingly, there was also a separate family by the name of Blackwood from Ayr, much further to the west. While no specific evidence of a connection has been found, in a later part of our story there is a marriage between a Blackwood and a Weir, and so it seems likely.
7. DGA: RGD56/13, *Clerk Maxwell Family Tree*, early nineteenth century; Irving (1907, p. 3).
8. SWT: *Margaret Weir*, CC8/8/87, Edinburgh Commissary Court, 18/6/1718. Margaret Weir had died in 1695, but this testament dative was in respect of monies owing to her estate.
9. SWT: *Johne Balmain*, CC8/8/76, Edinburgh Commissary Court, 25/2/1679; Grant (1898); Armet (1951).
10. Rather than president of the court, the Lord Chancellor was actually the head of the Privy Council, and therefore he was effectively the King's vice-regent in Scotland, but such were the times, he was deliberately interfering in the court's business.
11. This William Weir also went by the surname of his stepfather, that is to say, Lawrie or Lowrie. For whatever reason, he was also known as Robert Blackwood. It is not impossible that his mother remarried twice, but why the change of given name?
12. We have an account of its lengthy progress from Fountainhall (1848, pp. 465, 529, 559, 568, 600, 709) and, for example, within RPS: 1685/4/1–1685/4/76, 23/4/1685–16/6/1985.
13. Her name is mentioned in NAS: GD120/178, 1687.
14. Irving (1907, pp. 4–5) states that George inherited directly from his father, but in SPWT: *Testament of George Irvine of Newtoun WS*, CC8/8/106, 11/8/1742, we find George as executor but 'Robert Irvine of Newton' as cautioner. When George was served as heir to Newton in 1748, it must have been as heir to his brother.
15. DGA: GGD56/13, *Disposition by Sarah Irving*, 25/10/1783.

12. The Enlightenment of Edinburgh

1. Canonmills is about a quarter of a mile north of the line of Heriot Row, while Bearsford Park was on the south side of that line.

2   SOPR: *Deaths*, 685/010 0970 0289 Edinburgh.
3   Irving (1907) gives her first name as Janet, but in their marriage articles (DGA: GGD56/ (Unsorted Box), 1788) she is Isobel or Isobell, and on John Clerk Maxwell's family tree (DGA: RGD56/13) she is given under the alternative form, Isabella.
4   He was one of the Colquhouns of Camstraden (Fraser, 1869, p. 194) near Luss on Loch Lomond, home of Clan Colquhoun. John Clerk Maxwell's family tree (DGA: RGD56/13, early 19th C.), has his father as being a Glasgow merchant, but he mistakenly wrote Garscaden, a different branch of the Colquhoun family.
5   Coincidentally the same name as the Baron John Clerk's wife. She was the daughter of Thomas Inglis of Edinburgh, described variously as a merchant or as a pewterer, hence he was likely to be both. The Baron's wife was the daughter of James Inglis of Cramond, and no relationship between the two Inglises has been found.
6   An ordinary Scottish judge's formal title was actually 'Senator of the College of Justice'. There had been a previous Lord Newton, Charles Hay, but they were not related.
7   Irving (1907) gives 1773–1850.
8   'I wrote to Irving before leaving Kelso. Poor fellow, I am sure his sister's death must have hurt him much; though he makes no noise about feelings, yet still streams always run deepest.' Letter to William Clerk, 26/8/1791 (Lockhart, 1837, vol. 1, p. 110).
9   No place name corresponding to 'Partraith' has been found, but it may be connected with the Portrail Water, one of the main tributaries of the Daer Water that flows into the Clyde at Newton.
10 According to Cockburn (1874, vol. 1, pp. 26–27, my emphasis), 'No man ever rose so much above expectation after being made a judge.' He did not have sufficient force of personality to hold sway amongst other judges in great cases, but:
    in civil causes, deciding by written leisurely judgments, he was perfect [...] great knowledge of law, general intelligence, *especially in science*, a laborious patient manner, admirable listening, only broken by short judicious interrogations, perfect serenity, complete candour, and a devotion to his business.
11 In a breach of the conventions that have been observed so far, Alexander chose to give his son George a middle name. Vere is an early spelling of Weir, recalling the origins of Newton.
12 It is also evident in reading Lockhart's *Life of Scott* that the Irving and Scott families lived fairly close to one another.
13 To be fair, the engines were relatively small and capable of propelling the ship, via a screw, at something less than four knots, which would have been quite ineffective in pack ice.
14 NAS: CH3/1574, *Records of Falkirk: Camelon Irving Memorial Church*, 1890– .
15 After whom we have already presumed Archibald Stirling Irving, son of John and Agnes Irving, was named.
16 Grant (1884, pp. 365–366). The Pandects comprise key excerpts from Roman law.
17 Allen (2015) and Fletcher & Brown (1970, Chap. 5). Curiously, Dr Alston's immediate predecessor at the Physick Garden, Dr William Arthur (Fletcher & Brown, 1970, Chap. 3) was the second husband of Baron Sir John Clerk's sister Barbara. Dr Arthur fled his post after being involved in a Jacobite plot in 1715.

NOTES TO CHAPTER 13

13. Dr Thomas Weir and Anderson's Patent Pills

1 Wootton (1910), however, seems to have had access to the information they contained, at some point.
2 Sometimes referred to as Jane, as already mentioned she was the sister of George's wife Mary or Molly Chancellor.
3 See for example (subscription required): http://search.ancestry.co.uk/cgi-bin/sse.dll?db= NEWS-UK-MIDL-ED_AD&rank=1&new=1&so=3& MSAV =1&msT=1&gss=ms_db&gsfn=Thomas&gsln=Irving&msydy=1797& rg_810002A1__date=1797&uidh=000&rs_810002A1__date=1
4 It appears that many of the ages given were mere estimates, for they often seem to be multiples of five.
5 At this time officers were promoted on the basis of seniority and this continued even after they retired. He was made Lt General in 1854 (*London Gazette*, 1854) and he eventually ended up as a full general, as did his brother George.
6 James and John Gillespie of Spylaw, in Colinton just outside Edinburgh, became famous as tobacconists and philanthropists. John ran their shop in the High Street while James ran their snuff mill on the Water of Leith, just at the back of their house. Despite his wealth, James was a very kindly and modest man. The one extravagance he allowed himself was a rather plain looking carriage, at which Henry Erskine, author of 'Man of Feeling' poked fun: '*Wha wad ha'e thocht it, / that noses had bocht it?*' When James died, he bequeathed his fortune to the setting up of a school and hospital on Bruntsfield Links (Kay & Paton, 1877, vol. 2, no. CCXLIV, pp. 218–222). A modern school in the area still bears his name. George Cotton took on the tobacco shop about 1806 (EPOD). Cotton's tobacco blends came to be sold worldwide and, at the time of writing, Chief Catoonah Tobacconists of Bridgeport, Connecticut, still supplies a 'Princes Street Mixture'.

14. The Cays and the Hodshons

1 The hoastmen were a chartered society of coal traders and porters (Mackenzie, 1827).
2 Burke says 'impoverished', but by 1704 her eldest surviving son John Cay was well-off, having succeeded to his brother Jabez' fortune and moiety of North Charlton.
3 The burning of an effigy on 5 November each year had been encouraged as a celebration of the anniversary of the foiling of the 'Popish Plot', or 'Gunpowder Plot', of 1605 which had attempted to blow up James VI while he attended parliament. Although at first the effigy had been a 'Guy', so called after the best known of the conspirators, Guy Fawkes, in Jabez Cay's time it had generally been replaced by one representing the Pope. This was done as a protest against the Roman Catholic leaning monarchy, in particular the Duke of York himself. The practice was consequently banned, but to little effect it would seem, for the '5th of November' is a tradition carried on with enthusiasm even to this very day. Happily, the anti-Catholic symbolism that was once attached to it has long since been forgotten.
4 Richard Gilpin, her father, was not only a physician but also a leading Puritan. See Cay (1696–1702). Burke mistakenly states that Jabez Cay was unmarried.

5   According to Horsley (1732, p. 211), Jabez Cay got the Benwell Stone from Mrs Mary Shafto. She was the wife of Robert Shafto of Benwell near Newcastle, and great-grandmother of 'Bonny Bobby' Shafto (c. 1732–1797), MP for Durham County and Downton. Baron John Clerk would have been most interested, for he visited Benwell on his 1714 excursion to Newcastle and Carlisle, and at Benwell Tower he was 'kindly entertained by the Master of the House, one Mr. Shaftoe', grandson of the aforementioned Mary Shafto.
6   Cay (1696–1702). Note that while the title of the web page referred to is 'Jabez Cay, d. 1702', by the new calendar his death was in 1703.
7   Two deep pools near Darlington, County Durham. They were formed in ancient times by land subsidence.
8   If so, this would have been in addition to his interests as a merchant, salt manufacturer and part coal mine owner.
9   Each county has a lord lieutenant who acts as the monarch's representative, e.g. on ceremonial occasions, and more than one deputy lieutenant could be appointed.
10 Hence the old expression for something pointless, 'taking coals to Newcastle'.
11 John Cay (1700–1764) was the third son of Jabez Cay's brother John. He studied at Edinburgh University after which, in 1724, he became a barrister at Gray's Inn in London. He was subsequently appointed a steward and then Judge of the Marshalsea, a reviled debtors' prison. He was a fellow of the Society of Antiquaries of London and amongst his written works were *The Statutes at Large* and *An Abridgement of the Statutes*. His son, Henry Boult Cay (1731–1795), was a fellow of Clare College, Cambridge, where he was a second wrangler, but later he followed in his father's footsteps in the law (Goodwin, 1887; Cay, et al., c. 1947).
12 Darlington (1955) described the Marshalsea thus:
> In 1718 an anonymous rhymester called the Marshalsea an 'earthly Hell', a description which was more than justified by the findings of the Committee, appointed by Parliament in 1729, to inquire into conditions in the gaols. The prisoners were tortured with irons, beating, and by being locked up with human carcases. They were confined in so small a space that many were stifled to death in the heat of summer, and those who survived ill-treatment often died of starvation, since the keepers took most of what was given in charity. The report states that in the warmer weather 8 or 10 died every 24 hours.
13 NAK: NRO 324/F.2/28, 1760–1769. While Ord is on record as having obtained his estate of Whitfield in 1750 from its owners who were heavily in debt, it is impossible to know whether he was breathing down John Cay's neck or simply trying to help.
14 Mary entered the Mary Ward Institute of the Bar Convent in York where she became the Mother Superior. Elizabeth also entered Bar Convent, whereas Catherine entered the Convent of Poor Clare at Gravelines in northern France, also associated with Mary Ward (Bowden & Kelly, 2013).
15 Burke (1835–38, pp. 384–385). Frances Hodshon's gravestone and burial record (Grant, 1908) states that her father's name was Richard rather than Ralph, but the error is in the inscriptions, which were not begun until sixty years after Frances' death (Cay Family, 1910, no. 45) and so a mistake such as this is not surprising. A letter given to Frances Hodshon by her parents on her marriage to John Cay (Cay Family, 1910, no. 69) is given as being from 'Ralph and Mrs Hodshon'. Brown (1910) and Cay et al. (c. 1947) concur.

16 'The working of the [Elswick] colliery and the losses to the partners that were then being incurred. These partners were Mr. Leo. Shaftoe, 5 shares; Mr. Ledger, 6; *Mr. Cay*, 3; *Mr. Hodghson*, 5; and Mr. Bellasise, 1 share' (Leighton, 1906–07, p. 123).
17 Because anti-Catholic legislation was much the same in England as in Scotland, there would have been serious discrimination when it came to socialising, business, property ownership, legal rights and so on. Moreover, the Roman Catholic Church would not have accepted the marriage as being valid. Not only could Frances have been excommunicated, she would also have had difficulties of conscience about her own faith and, to make things worse, regarding her children being brought up as Protestants. At least John Cay must have been an Anglican rather than a Puritan, for surely the latter would have made a marriage nigh-on impossible. Later in life, the couple's solution to the problem was simply to eschew all discussion of religion, and those friends that they could trust to remain friends would have accepted that strategy. James Clerk Maxwell responded to Francis Galton's questionnaire concerning 'English Men of Science' (Hilts, 1975, p. 59) thus: 'Religion was a forbidden subject in her father's family as his mother [Frances Hodshon] was a Roman Catholic and all discussion was avoided religiously.'
18 Catherine Hodshon married Thomas Selby of Biddlestone (Cay et al., *c.* 1947, p. 11). We may deduce from Burke (1834–38, vol. 2, pp. 705–706) that this took place sometime about 1780, long before Robert Hodshon Cay married, for her sixth son was born about 1788. Despite the Selbys being Roman Catholic, Lintz became part of Selby's estates on Ralph Hodshon's death. The law in England was much the same as the law in Scotland inasmuch as both prohibited the inheritance of property by Roman Catholics unless they renounced popery by means of the prescribed oath, but as in the case of some of the Maxwell family, many did what the law required of them while secretly maintaining their true faith.
19 A Navy Pay Office certificate states that Robert enlisted on 26 March 1804 when he would have been about ten and a half years old. Boy volunteers could join the Royal Navy at eleven. Did he lie about his age to enlist? The certificate also states that he died on 1 September of the same year, 1804, whereas the family gravestone at Restalrig Churchyard says the year was 1805. It appears to have been one of two minor errors made when the gravestone was commissioned (see note 15 above).
20 Dr Daniel Sandford (1766–1830), Bishop from 1806, had been the prime mover in founding an Episcopal congregation in the New Town. His wife was Helen Douglas (1770–1837), a relation of the Duke of Queensberry (Chapter 4). They lived for a few years at 6 W. Heriot Row (now numbered as 25). He was a founder of St John's Church, the present-day landmark building dating from 1818, which stands at the south-east corner of Princes Street and Lothian Road. Its congregation included many people connected with our story (Harris, 2011). It was where John Clerk Maxwell and Frances Cay were married, and where Jane Cay took her nephew James Clerk Maxwell to Sunday afternoon service when he was living in Edinburgh.
21 DGA: GGD56/13. *Letter, John Cay to his sister Frances*, 9/10/1839.
22 The provenance of the Cay Calotypes suggests that they left after 1843, whereas Robert's obituary (*The Argus*, 16/8/1888) states that he arrived in 1842.
23 NAS: GD1/1372/3, 1868–83; NAS: GD1/1372/17, 1869–73; Cay et al. (*c.*1947, p. 35).
24 Later known as 'The Cedars' at 199 St John's Rd, it was until about 1970 the surgery of a Dr Cormack. The cedar trees still stand, but unfortunately the building itself was demolished about 2000.

BRILLIANT LIVES

25 This house was owned in 1864 by a mill owner by the name of Tod. It became the scene of a murder for which George Bryce was the last person to be publicly executed in Edinburgh.
26 Dyce may have been impressed by Mrs Cay's (Elizabeth Liddell) pastel portraits of her family (Chapter 2), for he took up the same medium for some of his early portraits (Dyce & Dyce, aft.1864).
27 The RSE biography of past fellows (Waterston & Shearer, 2006) erroneously has the painter William Dyce as the cousin of the eminent literary critic and collector the Rev. Alexander Dyce (Bullen, 1888). This Alexander Dyce was the son of Major-General Alexander Dyce of Rosebank FRSE, but he and Dr William Dyce were from separate branches of the Dyce family.
28 Prior to its amalgamation with the University of Aberdeen.
29 'Bacchus Nursed by the Nymphs [...]'
30 From 1831 they lived at 8 Mansfield Place, which is the address Dyce gave when he was exhibiting at the Royal Academy in 1832, along with 19 Berners Street as his London address (Royal Academy of Arts, 1832, p. 57).
31 By a letter addressed to A. W. Callcot RA (Dyce & Dyce, aft.1864), Dyce appears to have first taken a flat at 8 London Street, just round the corner from his sister Mrs Ross, before taking up residence in 1833 at 128 George St (now the Alexander Graham Bell pub). Dyce then moved in 1836 to an even grander residence at 43 Moray Place, where he stayed for his last two years in Edinburgh. He was in good company, for at number 46 was his friend and fellow artist, James Skene of Rubislaw. Living in the fifty or so houses in Moray Place, there were at least four lords, three knights, ten other landed gentry, twelve lawyers, a general, a professor and a minister. Coincidentally, 128 George Street is next door to the house where James and Isabella Wedderburn lived until 1821 (Chapter 9).
32 James Clerk Maxwell was elected when he was about two months short of his twenty-fifth birthday, but Dyce was elected fully five months before his (Waterston & Shearer, 2006).
33 Hewitsons LLP, private communication, a letter from Hewitsons concerning a bequest made by Mrs Wallis, 23/7/2009.
34 In a commentary note added below Dyce's letter of 22/10/1834 to Robert Dundas Cay (Dyce & Dyce, aft. 1864).
35 It is stated in Pointon (1979, p. 30) that the Miss Cay here referred to was Robert's younger sister Tibby (diminutive of Isabella). This is a misreading, for there was no such younger sister and, in fact, Dyce does not mention the sister's name at all. The name Tibby is used by Dyce only in the context of his congratulations to the happy couple, Robert and Isabella Dyce. The other Miss Cay, if a sister, could only have been Jane (Cay et al., c. 1947).
36 Select Committee on Arts and Manufactures (1836, pp. 77–88). Sir Thomas Dick Lauder of Fountainhall FRSE FSA (Scot) was probably also a trustee at the time, for he became the Board's secretary after Skene. Lord Meadowbank was also involved and may have been its president.
37 Gladstone was of a similar age to Dyce and, though born in Liverpool, was of Scottish descent. His father had an estate in Forfarshire, south of Dyce's Aberdeen. He was not Prime Minister in Dyce's lifetime, for his first term of office did not come until 1868.
38 SOPR: *Marriages*, 168/0A 0290 0172 Aberdeen, 29/10/1835. For the Rev. John Murray, see Scott (1926, vol. 6, p. 4).

39 SOPR: *Births*, 168/0A 0100 0266 Aberdeen, 26/5/1801–26/4/1804 and Bertie (2000).
40 SOPR: *Births*, 168/0A 0100 0266 Aberdeen, 19/9/1806 and Scott (1926, vol. 6, p. 4).
41 The even numbers up to twenty in this street were demolished sometime after 1870 to make way for the Caledonian Railway Station.
42 According to Norton-Kyshe (1898, p. 325) they sailed together but this seems to be no more than an erroneous personal recollection many years after the event, for Campbell also says they went separately (C&G, p. 14).
43 SOPR, *Births*, 685/02 0360 0183 Edinburgh, 26/6/1844.
44 In a letter from James to his father (C&G, p. 59), he mentions what he has heard of Robert's progress across the globe. It clearly indicates he took the overland route: '"Old 31", *28th March 1844* [...] There have been letters from uncle Robert, dated Gibber Altar, but I have not seen or Heard what is in them farther than that he was to be *at Suez* on Monday last.' (my italics).
45 SSR: *Deaths*, 685/04 0921, Edinburgh, 1860.
46 Alexander's death was too close to Robert's departure to have directly prompted it, but Robert probably had sufficient forewarning to give him cause to reappraise his entire situation. Perhaps, too, there had been calls for him to return whatever the personal consequences.
47 EPOD (1856–59). James Clerk Maxwell visited his aunt and uncle there in October 1857 (C&G, p. 284).
48 James and William Thomson were the two eldest sons of the Glasgow University mathematics professor James Thomson (Smith, 2004).
49 Although he did not graduate from Edinburgh University, he won the Stratton Medal. His works include the design of the stone bridge over the Urr at Glenlair (1861), Southern Breakwater Aberdeen (1874) and Lerwick Harbour Extension (1881). He published a number of articles on harbour construction, including pioneering work on laying bagged concrete foundations. He was appointed a member of the Institute of Civil Engineers in 1872 and a fellow of the RSE in 1882 (Cay et al., *c*. 1947; DSA, 2014). The bridge at Glenlair still stands (CANMORE, event/578262).
50 He was with The Patent Shaft and Axletree Co. Ltd in Wednesbury, first as general manager then a director, by which time he was also a director of the Metropolitan Amalgamated Railway Carriage and Wagon Co. Ltd (Henderson & Grierson, 1914). His sister-in-law was Lady Florence Douglas, a daughter of the 8th Marquess of Queensberry, a woman who was much ahead of her time: a traveller, war correspondent, writer and feminist (Cay et al., *c*. 1947).
51 SSR: *Deaths*, 685/05 0005 Edinburgh. Facial neuralgia is an excruciatingly painful condition caused by pressure on the trigeminal cranial nerve, in this case as a result of a tumour.
52 DGA: GGD56/ (Unsorted Box), *Colin Mackenzie to James Clerk Maxwell*, 2/7/1866.
53 The last figure in the date is distorted, but 1835 seems to be the only reading that is consistent with the information in Dyce's congratulatory letter sent to Robert from Glenlair.

BRILLIANT LIVES

15. James Clerk Maxwell's Scottish Homes

1. William Wallace and his brother Alexander were listed at 'Back of Heriot Row' (1820–22), 44 India Street (1823–24) and 22 India Street (1825-29). Perhaps as they built up properties in the street they kept these particular properties as show-houses, if not actual residences. By 1830 their activities had moved on to St Vincent Street (nos 8 and 10). Wallace is also listed in the Canongate at 10 St John Street, which seems to have been their permanent base, if it is indeed the same William Wallace.
2. SOPR: *Marriages*, 685/01 0630 0360 Edinburgh, 4/10/1826.
3. When Mrs Molly Irving died in 1828, her unmarried daughters listed themselves in the directory under 'Irving, Mrs of Newton' ('Misses' was often mistranscribed as 'Mrs'), and continued to do so until they moved round the corner to Doune Terrace in 1830. In 1831 and 1832, the year of Lord Newton's death, a single 'Irving, Miss of Newton' was listed there, even though there were others then alive. We may therefore surmise they had lived together all along.
4. In 1967 number 14 India Street was listed by Historic Scotland as a Category 'A' building. Such buildings are legally protected from alteration without permission, with the most important examples being given Category 'A' status. See www.historic-scotland.gov.uk/index/ heritage/historicandlistedbuildings/listing.htm.
   Specific details for the house are given on: http://data.historic-scotland.gov.uk/pls/htmldb/f?p=2200:15:0::::BUILDING,HL: 29133,14%20 India%20Street.
   A photograph of India Street and the house as it was *c.* 1950 is to be found at: http://canmore.rcahms.gov.uk/en/site/135359/details/edinburgh+14+india+street/.
5. SOPR: *Marriages*, 685/01 0630 0360 Edinburgh, 4/10/1826 and *Blackwood's Edinburgh Magazine* (1826).
6. Nether Corsock was part of John Clerk Maxwell's estate in Kirkcudbrightshire, about twelve miles to the west of Dumfries. It was the name of the estate on the Gillon Plan of 1801 (DGA: GGD56/1, 1806). It did not then include Glenlair, which was to the south of the mutual boundary formed by the Glenlair burn. Glenlair was purchased by John Clerk Maxwell many years later, in 1839.

    On Ainslie's county map of 1821, the site of the present Glenlair House is marked as Nether Corsock, which is the name by which the Clerk Maxwells initially referred to it (although they also frequently wrote simply 'Corsock'). After the purchase of the Glenlair estate, however, John Clerk Maxwell decided to rename both his house and his estate as Glenlair (C&G, p. 25), and by 1850 it was so marked on the 6" OS map. The name Nether Corsock was then attached to a steading some way to the north of the original, while the original steading of Glenlair was renamed as Nether Glenlair to distinguish it from the new creation. The River Urr, to which James Clerk Maxwell frequently referred, forms the eastern border of the estate.
7. DGA: GGD56/ (Unsorted Box), 1/12/1830.
8. Time capsule dated 31/3/1831, see www.glenlair.org.uk/eureka-time-capsule-found-dated- 25th-march-1831.html
9. DGA: GGD56/ (Unsorted Box), *Letters*, 14–16/6/1898.
10. The tenants were successively:

NOTES TO CHAPTER 15

    1834–38   John McFarlan, FRCS, surgeon
    1839–42   Robert McIntosh, Esq., writer
    1843   James Scott, Esq.
    1844–63   Colonel Thomas Wardlaw (1787–1863), of the Bengal Native Infantry
    1864–69   Colonel Wardlaw's widow, Mrs Margaret Wardlaw (1812–1869)
    1868–69   During Mrs Wardlaw's later years, a Mrs Simson was also listed
    1869–71   Mrs Simson (d. 1871)
    1871–91   Miss Simson, presumably one of Mrs Simson's daughters

    John McFarlan was attached to the Royal Public Dispensary in West Richmond Street, not far from the University. Lt Col Thomas Wardlaw was listed in the *London Gazette* (6/2/1855, pp. 163–164) along with Lt Col Hugh Morrieson, Lewis Campbell's stepfather, when they were both posted as full colonels. Colonel and Mrs Wardlaw's details are given on their gravestone in Dean Cemetery.

11 RS: *Register of Sasines*, (886) PR 2057 82, 17/6/1856.
12 RS: *Register of Sasines*, (13,304) 431.188, 19/1/1874.
13 RS: *Register of Sasines*, (26,486) 1097.180, 31/12/1879.
14 RS: *Register of Sasines*, (6031) 1902.49, 15/2/1887 (under the terms of a trust set up on her behalf and of those in her own will).
15 This is just one example of his playing with words and codes in his letters home to Glenlair; in the same letter he writes Marine Villa as 'murrain vile' and signs off as 'JAS. ALEX. M^cMERKWELL').
16 DGA: GGD56/ (Unsorted Box), *Letter: Colin Mackenzie to JCM*, 1866.
17 James' will and testament (SPWT, 1879), as originally drawn up in 1866, left Katherine all his movable property and the lease of their house in London. The house at 14 India Street is not specifically mentioned. In July 1873 he added a codicil to his will revoking his share in the Prospect Hill farm, which he had been left by his mother, in favour of the Cay family, but mentioned nothing else. In 1874, however, he had a change of heart regarding 14 India Street and simply disponed it to Katherine outright. As this required no change to the terms of his will, there is no mention in it whatsoever of the house.

    In addition to Katherine and Colin Mackenzie, the other executors were Maxwell's friend and colleague Professor George Stokes and his physician Dr Paget, both of Cambridge. Colin Mackenzie's legal practice was very likely to have been Messrs Mackenzie and Black, WS, 28 Castle Street, which follows from the fact that Mackenzie and Black were involved in drawing up deeds for James Clerk Maxwell. It would have been natural to keep such business 'in the family', e.g. GGD/56/ (Unsorted Box), *Letter from Colin Mackenzie to James Clerk Maxwell*, 2/7/1866; and the disposition of 1874 on Katherine Dewer's behalf (RS: *Register of Sasines*, 13,304 431.188, 19/1/1874).
18 *Bosnia* was corrected to *Bothnia* in the second edition. The vessel was in fact a Cunard passenger liner that regularly plied from Liverpool to New York. Colin died of apoplexy on 15 July 1882 while the ship was within a day of Liverpool (Fairley, 1988, p. 77).
19 Cowper and McIver (1993) say Anne and Mary were their grand-daughters but the Simsons' only son died unmarried at the age of twenty-eight in 1865 (Writers to the Signet, 1890).
20 As summarised in RS: *Search Sheets*, 12991, 33506 (Series II) and 101878.
21 www.clerkmaxwellfoundation.org/index.html, registered charity SC015003.

16. Epilogue

1. Rautio (2006). Despite the proximity of the duck pond, which presumably was low in water that summer.
2. See www.glenlair.org.uk/gallery/glenlair-historical/Glenlair-House-1950.jpg
3. See www.glenlair.org.uk/gallery/glenlair-historical/1928-with-Col-Hannay.jpg
4. For example, Windsor Roofing (2014) informs us:
   > One of the loveliest historic homes that has corrugated roofing is in Scotland. It is called Glenlair, and was the home of Scottish scientist James Clerk Maxwell [...] Although Glenlair is still suffering from the effects of a huge fire in 1929, it and its corrugated roofing are still standing and still beautiful. The grand old house is slowly being restored by private donations.

   Many other such websites give the same information, which is no longer accurate since the restoration of the house by the Glenlair Trust.
5. CANMORE ID 212680. Shown in the Newall plans of 1868 as the 'business room'.
6. Captain Duncan Ferguson [current owner of Glenlair], 2014, private communication.

Appendix 3. James Clerk Maxwell's Legacy to Science and Mankind

1. It is often repeated that Maxwell did nothing to try to demonstrate the existence of electromagnetic waves, and while Heinrich Hertz did manage to do so he thought they had no practical application. However, according to Lewis Campbell, writing in 1901, 'wireless telegraphy is an application of ideas on which he [Maxwell] used to discourse to me' (Huxley, 1914, p. 386). It is true that Maxwell did not begin any quest for the waves and simply moved on, but it seems that did not stop him wondering about such things.
2. As any internet search will soon show, the quotations that follow are regularly attributed to Einstein. While there seems little doubt as to their veracity, the original sources are not easy to track down as they appear to have been reported, in one form or another, from various lectures and interviews that he gave over the years.

Appendix 4. The Birth Dates of George and Dorothea's Children

1. FamilySearch.org, ID=2:186BXG9.
2. NAS: GD18/5396, 24/9/1739.
3. FamilySearch.org, ID=2:1619Q71.
4. NAS: GD18/5396, 11 & 16/11/1741.
5. NAS: GD18/5396, 28/11/1742.
6. FamilySearch.org, ID=2:161B2HC.
7. FamilySearch.org, ID=2:161B78R.
8. FamilySearch.org, ID = 2:161BCT0.

Bibliography

Maxwell – Mainly Biographical

Domb, C., 1980. 'James Clerk Maxwell in London 1860–1865'. *Notes Rec. Roy. Soc. Lon.*, **35**, pp. 67–103

Everitt, C. W. F., 1975. *James Clerk Maxwell, Physicist and Natural Philosopher.* New York: Charles Scribner's Sons

James Clerk Maxwell Foundation, 1991. *James Clerk Maxwell 1831–1879. Scotland's Uncelebrated Genius.* Edinburgh: James Clerk Maxwell Foundation

James Clerk Maxwell Foundation, 1999. *James Clerk Maxwell Commemorative Booklet* [for the 4th International Congress on Industrial and Applied Mathematics, Edinburgh 1999] Edinburgh: James Clerk Maxwell Foundation. Available at: http://www.clerkmaxwellfoundation.org/html/media_library.html

Jones, R. V., 1980. 'The Complete Physicist – James Clerk Maxwell 1831–1879'. *Yearbook of the Roy. Soc. Ed.*, pp. 5–23

Smith-Rose, R. L., 1948. *James Clerk Maxwell, F.R.S. 1831–1879. A Physicist of the Nineteenth Century.* London: Longmans Green. Available at: https://books.google.co.uk/books?id=x51QAQAAIAAJ

Tait, P. G., 1910. 'James Clerk Maxwell'. *Encyclopaedia Britannica*, 11th edn, **17**, pp. 929–930. Cambridge: Cambridge University Press

Theerman, P., 1986. 'James Clerk Maxwell and Religion'. *Am. J. Phys.*, **34**, pp. 312–317

Tolstoy, Ivan, 1981. *James Clerk Maxwell: A Biography.* Edinburgh: Canongate. Available at: https://books.google.co.uk/books?id=Y5vvAAAAMAAJ

Maxwell – Mainly Scientific

Buchwald, J. Z., 1985. *From Maxwell to Microphysics.* Chicago, IL: University of Chicago Press

Cook, A. F. & Franklin, F. A., 1964. 'Rediscussion of Maxwell's Adams Prize Essay on the Stability of Saturn's Rings'. *Astron. J.*, **69**, pp. 173–200

Domb, C., ed., 1963. *Clerk Maxwell and Modern Science. Six Commemorative Lectures.* London: Athlone Press

Garber, E., 1969. 'James Clerk Maxwell and Thermodynamics'. *Am. J. Phys.*, **37**, pp. 146–155

Glazebrook, R. T., 1896. *James Clerk Maxwell and Modern Physics.* London: Cassell. Available at https://archive.org/details/jamesclerkmaxwel00glaziala

Harman, P. M., 1998. *The Natural Philosophy of James Clerk Maxwell.* Cambridge: Cambridge University Press

Hendry, J., 1986. *James Clerk Maxwell and the Theory of the Electromagnetic Field.* Bristol: Adam Hilger

Hopley, I. B., 1957; 1958; 1959. 'Maxwell's Work on Electrical Resistance' (3 parts), *Annals. of Science*, 13(4), pp. 265–272; 14(3), pp. 197–210; 15(2), pp. 91–108

Lodge, O., 1927. 'Clerk Maxwell and the Cavendish Laboratory'. *Nature*, 119(1), p. 46

Maxwell, J. C., 1936. 'Origins of Clerk Maxwell's Electric Ideas, as Described in Familiar Letters to William Thomson'. *Proc. Camb. Philos. Soc.*, 32(5), pp. 695–750

Simpson, T. K., 1997. *Maxwell on the Electromagnetic Field: A Guided Study.* New Brunswick, NJ: Rutgers University Press

Wright, W. D., 1961. 'The Maxwell Colour Centenary', *Nature*, 191, pp. 10–11

Maxwell and Other Electromagnetic Pioneers

Appleyard, R., 1930. *Pioneers of Electrical Communication.* London: Macmillan. Available at https://books.google.co.uk/books?id=15XkAAAAMAAJ

Forbes, N. & Mahon, B., 2014. *Faraday, Maxwell, and the Electromagnetic Field: How Two Men Revolutionized Physics.* New York, NY: Prometheus Books

Friedel, R. D., 1981. *Lines and Waves: Faraday, Maxwell, and 150 Years of Electromagnetism.* New York: IEEE Center for the History of Electrical Engineering

Hunt, Bruce J., 1994. *The Maxwellians.* Ithaca, NY: Cornell University Press

MacDonald, D., 1964. *Faraday, Maxwell, and Kelvin.* Garden City, NY: Anchor Books Science Study Series

Tricker, R., 1966. *The Contributions of Faraday and Maxwell to Electrical Science.* Oxford: Pergamon Press

Whittaker, Sir Edmund T., 1910. *A History of the Theories of Aether and Electricity from the Age of Descartes to the Close of the Nineteenth Century.* Dublin. Available at https://archive.org/details/historyoftheorie00whitrich. Republished 1951–53 in 3 volumes as *A History of the Theories of Aether.*

People and Places

Knott, C. G., 1911. *Life and Scientific Work of Peter Guthrie Tait.* Cambridge: Cambridge University Press. Available at: https://archive.org/details/cu31924064123148v

——, 1914. *Famous Edinburgh Students.* Edinburgh: T. N. Foulis. Available at https://archive.org/details/famousedinburghs00univ

Magnusson, M., 1974. *The Clacken and the Slate: The Story of the Edinburgh Academy from 1824 to 1974.* Edinburgh: Collins

Taylor, G. & Skinner, A., 1776. *Taylor and Skinner's Survey and Maps of the Roads of North Britain or Scotland.* London

Thompson, S. P., 1976. The *Life of Lord Kelvin*, 2nd edn, 2 vols. Providence, RI: AMS Chelsea Publishing

Wood, A., 1946. *The Cavendish Laboratory.* Cambridge: Cambridge University Press

BIBLIOGRAPHY

References

Aitken, J. C., 1887–88. 'Some Notes on the Abbey of Holywood and on the Welshes of Colliestoun and Craigenputtock'. *Trans. J. and Proc. Dumf. Gall. Nat. Hist. Antiqu. Soc.*, **6**, pp. 110–125

Allen, D. E., 2015. 'Alston, Charles (1685–1760), Physician and Botanist'. In: *ONDB*

Anderson, W., 1878. *The Scottish Nation; or, the Surnames, Families, Literature, Honours, Biographical History of the People of Scotland.* Edinburgh: A. Fullarton & Co.

Andrews, C. E., 2002. 'The Literary Club as Imagined National Community: Allan Ramsay and the Easy Club (1712–1715)'. *Eighteenth-Century Scotland*, (Newsletter of the Eighteenth Century Scottish Studies Society), **16**, pp. 8–12. Article available at: www.academia.edu/430358/The_Literary_Club_as_Imagined_National_Community_Allan_Ramsay_and_the_Easy_Club_1712-1715_

Armet, H., ed., 1951. *Register of the Burgesses of the Burgh of the Canongate 1622–1733.* Edinburgh: Scottish Record Society

Arnot, H., 1779. *The History of Edinburgh.* Edinburgh: William Creech

——, 1812. *A Collection and Abridgement of Celebrated Criminal Trials in Scotland from AD 1536 to 1784 [. . .].* Glasgow: Napier (Printer)

Arthur, J. W., ed., 2011. *Understanding Geometric Algebra for Electromagnetic Theory.* IEEE Press Series on Electromagnetic Theory. Piscataway, NJ: John Wiley: IEEE Press

Asiatic Journal and Monthly Register, 1841. 'Register – Calcutta: Military Appointments, Promotions &c.', **35** (June), p. 125

Baillie, J., 1801. *An Impartial History of the Town and County of Newcastle upon Tyne.* Newcastle: Anderson

Barbour, J., 1884–85. 'Notes on the Town's Common Mills and their History'. *The Transactions Journal and Proceedings of the Dumfriesshire and Galloway Natural History and Antiquarian Society*, **4**, pp. 58–72

Barringer, T., 2004. 'Dyce, William (1806–1864)'. In: *ODNB*

Bateson, E., 1895. *A History of Northumberland.* Newcastle Upon Tyne: Andrew Reid & Co. Ltd

Bell, B., 1881. *Lieut. John Irving RN of HMS Terror: A Memorial Sketch with Letters.* Edinburgh: David Douglas

Bertie, D., 2000. *Scottish Episcopal Clergy, 1689–2000.* Edinburgh: T. & T. Clark

Bertram, G., 2012a. *The Etchings of John Clerk of Eldin.* Taunton, Somerset: Enterprise Editions.

Bertram, G., 2012b. *The Etchings of John Clerk of Eldin, Lecture for the Old Edinburgh Club.* Available at: www.clerkofeldin.com/pdf/The-Etchings-of-John-Clerk-of-Eldin-OEC-Lecture.pdf

Blackett, J. B., 2010. 'Parish History'. Available at: http://kirkbeanheritagesociety.org.uk/history.html

Blackwood's Edinburgh Magazine, 1826. 'Register: Marriages', June–December

Boog Watson, C., 1929. *Register of Edinburgh Apprentices 1700–1755.* Edinburgh: Scottish Record Society

——, 1930. *Roll of Edinburgh Burgesses and Guild Brethren 1701–1760.* Edinburgh: Scottish Record Society

Bourne, J., 1852. *A Treatise on the Screw Propeller: With Various Suggestions of Improvement.* London: Longman, Brown, Green and Longmans

Boswell, James, 2013. *Boswell's Edinburgh Journals, 1767–1786*, ed. Hugh Milne. Edinburgh: Birlinn

Bowden, C. & Kelly, J., 2013. Search: Mary Hodshon. Available at: http://wwtn.history.qmul.ac.uk/counties/details.php?uid=MW085 [links to sisters and niece]
Boyd, L., 1908. *The Irvings and their Kin*. Chicago, IL: R. R. Donnelley & Sons Company
Brown, I. G., 1987. *The Clerks of Penicuik: Portraits of Taste and Talent*. Penicuik: Edinburgh Penicuik House Preservation Trust
Brown, J. & Anstruther, D., 1894. *The History of Sanquhar to which is added the Flora and Fauna of the District*. Dumfries: J. Anderson
Brown, J. L. & Lawson, I. C., 1990. *History of Peebles 1850–1990*. Edinburgh: Mainstream Publishing Company
Brown, N., 1910. 'Six North Country Diaries: The Diary of Nicholas Brown'. *The Publications of the Surtees Society*, 118, pp. 230–323
Brown, R., 2004. *A History of Accounting and Accountants*. New York, NY: Cosimo Classics
Brunton, G. & Haig, D., 1806. *An Historical Account of the Senators of the College of Justice, from its Institution in MDXXXII*. Edinburgh: The Edinburgh Printing Company
Brush, S. G. & Everitt, C. W. F., 1969. 'Maxwell, Osborne, Reynolds and the Radiometer'. *Hist. Stud. Phys. Sci.*, 1, pp. 105–125
Bryson, W. & Veitch, G., 1825. 'The Discipline of the Church of Scotland about Two Hundred Years Ago'. In: *The Antiquary's Portfolio, or Cabinet Selection Historical of Literary Curiosities* [...]. London: George Wightman
Buchan, J. W. & Paton, R. H., 1925. *A History of Peeblesshire*. Glasgow: Jackson Wylie and Co.
Bullen, A. H., 1888. 'Alexander Dyce'. In: *DNB00*
Bulloch, J. G. B., 1919. *A History and Genealogy of the Families of Bayard, Houstoun of Georgia, and the Descent of the Bolton Family from Assheton, Byron and Hulton of Hulton Park*. Washington, DC?: J. H. Dony (Printer)
Burke, B., 1894. *A Genealogical and Heraldic History of the Landed Gentry of Great Britain & Ireland*, vol. 2, 8th edn. London: Harrison and Sons
Burke, J., 1834–38. *Genealogical and Heraldic History of the Commoners of Great Britain and Ireland*, 4 vols. London: Henry Colburn
Burns Encyclopaedia, 2014. 'Douglas, Heron & Company'. Available at: www.robertburns.org/encyclopedia/DouglasHeronampCompany.294.shtml
Burns Encyclopaedia, 2014. '"Rosamond", Seizure of'. Available at: www.robertburns.org/encyclopedia/RosamondSeizureof.20.shtml [accessed February 2014]
Cahoon, B., 2014. *Isle of Man*. Available at: www.worldstatesmen.org/Isle_of_Man.htm
Cairns, J. W., 2009. 'Knight, Joseph (b. *c.* 1753), Slave and Litigant'. In: *ONDB*
Caledonian Mercury, 1833. 'Dreadful Accident at Edinburgh'. 18/3/1833.
Campbell, A., 1827. *The History of Leith, from the Earliest Accounts to the Present Period*. Leith: William Reid
Campbell, G., 2011. 'Two Bubbles and a Crisis: Britain in the 1840s'. *Proceedings of the ASSA Cliometrics Conference, January 2011*. Available at http://cliometrics.org/conferences/ASSA/Jan_11/Campbell.pdf
Campbell, L. & Garnett, W., 1882. *The Life of James Clerk Maxwell*. London: Macmillan and Co. Available at: https://archive.org/details/lifeofjamesclerk00camprich
Campbell, N. & Smellie, R., 1983, *The Royal Society of Edinburgh 1783–1983*. Edinburgh: Royal Society of Edinburgh
Carlyle, E. I., 1898. 'Stirling, James (1692–1770)'. In: *DNB00*

Cat, J., 2012. 'Glenlair: A Brief Architectural History'. Available at: www.glenlair.org.uk/glenlair-history

Cay Family, 1910. *Inventory of Cay Family Papers* [in private ownership]

Cay, J., 1696–1702. 'The Letters of Jabez Cay'. Available at: http://emlo.bodleian.ox.ac.uk/profile/person/b1c1fc67-8b30-4a94-aaf1-06e80533e2b6

Cay, R. D., 1857. *Application for Employment, 25 March*. Edinburgh: [in private ownership]

Cay, W. D., 1887. *Notes of Recollections of my Intercourse with James Clerk Maxwell my Cousin*. London: [By courtesy of Mr James Brown]

Cay, W. D., Dunn, C. W. & Dunn, E. M., c. 1947. *Cay Family Pedigree*. Edinburgh: James Clerk Maxwell Foundation [copy holder]

Chalmers, G., 1824. *Caledonia: Or an Account, Historical and Topographic, of North Britain: From the Most Ancient to the Present Times: With a Dictionary of Places, Chorographical and Philological*. Edinburgh: A. Constable

Chambers, R., 1825, 1980. *Traditions of Edinburgh*. Edinburgh: W. & C. Tait, W. & R. Chambers. [Apart from some front matter, the edition of 1980 is the same as 1912]

———, 1840. *A Biographical Dictionary of Eminent Scotsmen*. Glasgow: Blackie & Sons

———, 1861. *Domestic Annals of Scotland*. *Edinburgh*: W. & R. Chambers

Chase, M., 2004–13. 'Wedderburn, Robert (1762–1835/6?), Radical'. In: *ODNB*

Clerk Maxwell, G., 1756. *Two Letters to the Commissioners and Trustees for Improving Fisheries and Manufactures in Scotland*. Edinburgh: Sands, Donaldson, Murray & Cochran (Printers)

Clerk G. [George Clerk Maxwell], 1756. 'Drawings of Some Very Large Bones'. *Essays and Observations Physical and Literary*, II, p. 11 + plate

———, 1771. 'The Advantages of Shallow Ploughing'. *Essays and Observations Physical and Literary*, III, pp. 56–67

Clerk, J. [Lord Eldin], 1788. 'Account of Sir George Clerk-Maxwell Baronet'. *Transactions of the Royal Society of Edinburgh*, 1, pp. 51–56

Clerk, J, Bt [the Baron], 1737–38. 'An Account of the Observations of the late Solar Eclipse made at Edinburgh, on Feb. 18th, 1736/7'. *Philosophical Transactions*, 40, pp. 195–197

Clerk, J. & Gray, J. M., 1892. *Memoirs of the Life of Sir John Clerk of Penicuik, Baronet, Baron of the Exchequer . . . 1676–1755*. Edinburgh: T. A. Constable, at the University Press

Clerk of Eldin, J., 1827. *An Essay on Naval Tactics, Systematical and Historical, with Explanatory Plates*, 3rd edn. Edinburgh: Adam Black

Clerk of Eldin, J. & Laing, D., 1855. *A Series of Etchings Chiefly of Views of Scotland*. Edinburgh: Bannatyne Club

Cobbett, W., 1811. *Cobbett's Complete Collection of State Trials* [. . .]. London: Bagshaw

Cockburn, H., 1856. *Memorials of His Time*. New York, NY: Appleton and Co.

———, 1874. *Journal of Henry Cockburn, Being a Continuation of Memorials of his Time*. Edinburgh: Edmonston and Douglas

Coghill, H., 2010. *Lost Edinburgh*. Edinburgh: Birlinn

Colburn, H. and Bentley, R., 1832. *Letters of Eminent Men, Addressed to Ralph Thoresby*. London: Colburn and Bentley

Collins, A., 1806. *Baronetage of England*. London: John Stockdale

Committee on Scotch Entails, 1828. 'Report of the House of Commons Select Committee on Scotch Entails – Minutes of Evidence'. In: *House of Lords Sessional Papers 1801–1833*, (12/5/1828). Westminster: The House of Lords

Cooper, T., 1886. 'Buchanan, Dugald (1716–1768)'. In: *DNB00*
Corley, T. A. B., 2004. 'Anderson, Patrick (1579/80–c. 1660)'. In: *ODNB*
Cowper, A. S. & McIver, E. S., eds, 1993. *Pre-1855 Inscriptions in the Dean Cemetery*. Edinburgh: The Scottish Genealogy Society
Craig, G. Y., McIntyre, D. B. & Waterson, C. D., 1978. *James Hutton's Theory of the Earth: The Lost Drawings*. Edinburgh: Scottish Academic Press
Craik, A., 2008. *Mr Hopkin's Men: Cambridge Reform and British Mathematics in the 19th Century*. London: Springer Verlag
Cumberland Chronicle, 1777. [Advertisement]. 'Anderson's pills'. In: *The Cumberland Chronicle or Whitehaven Intelligencer*, 11/2/1777
Cumming, N. & Vereker, R., c. 2008. *An Illustrated History of Lindsay and Gilmour and Raimes and Clark & Co Ltd*. Available at: www.lindsayandgilmour.co.uk/wp-content/uploads/2011/03/An-Illustrated-History-of-Lindsay-and-Gilmour-and-Raimes-Clark-and-Co-Ltd.pdf
Cundall, F., 1904. *Biographical Annals of Jamaica*. Kingston: Institute of Jamaica
Curwen, J. F., 1932. 'Parishes (West Ward): St Michael, Lowther'. In: *The Later Records Relating to North Westmorland: Or the Barony of Appleby*. Kendal: T. Wilson & Son, pp. 325–334. Available at: www.british-history.ac.uk/n-westmorland-records/vol8/pp325-334
Cust, L. H., 1888. 'William Dyce', in *DNB00*
Daiches, D., 1986. *Edinburgh: A Travellers Companion*. London: Constable
Daiches, D., Jones, P. & Jones, J., eds, 1986. *The Scottish Enlightenment: A Hotbed of Genius 1790–1830*. Edinburgh: Edinburgh University Press / The Saltire Society (1996)
Dalton, C., 1896. *English Army Lists and Commission Registers, 1661–1714*. London: Eyre & Spottiswoode
Darlington, I., ed., 1955. 'Southwark Prisons'. In: *Survey of London: St George's Fields (The Parishes of St. George the Martyr Southwark and St. Mary Newington)*. London: City Council of London, vol. 25, pp. 9–21
Day, L. & McNeil, I., eds, 1996. *Biographical Dictionary of the History of Technology*. London: Routledge
DGFHS, 2008. *Old Parochial Registers: Deaths and Burials Index, Part I, 1617–79*. Dumfries: Dumfries and Galloway Family History Society
Dickson, J. D. H., 1912. Peter Guthrie Tait. In: DNB12
Dobson, D., 2005. *Huguenot and Scots Links, 1575–1775*. Baltimore, MD: Clearfield Company
Donnachie, I., 1971. *The Industrial Archaeology of Galloway*. Newton Abbott: David & Charles
Douglas, G., 1899. *A History of the Border Counties (Roxburgh, Selkirk, Peebles)*. Edinburgh, London: Blackwood & Sons
Douglas, R., 1798. *The Baronage of Scotland Containing an Historical and Genealogical Account of the Gentry of that Kingdom*. Edinburgh: Bell & Bradfute et al.
Douglas, S. E., 2013. *The Douglases of Bonjedward and Timpendean, Roxburghshire (Scottish Borders), Scotland*. Available at: www.douglashistory.co.uk/history/Documents/Douglas%20of%20Bonjedward%20and%20Timpendean-2013.pdf
DSA, 2014. *Dictionary of Scottish Architects*. Available at: www.scottisharchitects.org.uk
Dunlop, J. C., Dunlop, A. H. & Hole, W., 1886. *The Book of Old Edinburgh, and Handbook to the Old Edinburgh Street [...]*. Edinburgh: Constable
Dyce, W., 1837. *Letter to Lord Meadowbank*. Aberdeen: Dyce Archive, Aberdeen Art Gallery

Dyce, W. & Cay, W. D., 1888. *Dyce Papers I*. Birmingham Museums and Art Gallery

Dyce, W. & Dyce, J. S., aft. 1864. *Life, Correspondence and Writing of William Dyce RA 1806–64, Painter, Musician and Scholar by his Son James Stirling Dyce*. Aberdeen: Aberdeen Art Gallery and Museums

Earman, J. & Norton, J., 1998. 'Exorcist XIV: The Wrath of Maxwell's Demon. Part I. From Maxwell to Szilard'. *Studies in History and Philosophy of Science Part B: Modern Physics*, **29**(4), pp. 435–471

Edinburgh Calotype Club, 19th C. *Albums of the Edinburgh Calotype Club*. Edinburgh: NLS

Edinburgh Advertiser, 1798.[Advertisement] 'Anderson's Pills', 20/7/1798

Edinburgh Gazette, 1848. [Notice: Sequestrations, Albert Cay]. **5719**(2/1/1848), p. 54

Edinburgh Museums, 2013. *Slavery Panels*. Edinburgh. Available at: www.edinburghmuseums.org.uk/_MediaLibraries/Global/Slavery-panels.pdf

Edinburgh University Library, 2014. 'Walter Scott>Homes'. Available at: www.walter-scott.lib.ed.ac.uk/biography/homes.html

Einstein, A., 1905. 'On the Electrodynamics of Moving Bodies'. *Annalen der Physik*, **17**(June), pp. 891–921

——, 1954. *Ideas and Opinions*. New York: Crown Publishers Inc.

Ellis, A., 1956. *The Penny Universities: A History of the Coffee-houses*. London: Secker & Warburg

Emerson, R. L., 2004. 'Select Society (act. 1754–1764)'. In: *ODNB*

Encyclopaedia Britannica, 2013. 'Covenanter' Available at: www.britannica.com/EBchecked/topic/141118/Covenanter.

Enever, G., 2001. 'Newcastle Infirmary Time Line'. Available at: http://research.ncl.ac.uk/nsa/tl1.htm [accessed 1/11/2013]

Everitt, C. W. F., 1983. 'The Springs of Maxwell's Scientific Creativity'. In: *Springs of Scientific Creativity: Essays on Founders of Modern Science*. Minneapolis, MA: University of Minneapolis Press, pp. 71–141

Fairley, R., 1988. *Jemima: The Paintings and Memoirs of a Victorian Lady*. Edinburgh: Canongate Press

Falkirk Archives, 2012. 'Falkirk St Andrews Church Finding Aid'. Available at: www.falkirkcommunitytrust.org/heritage/archives/finding-aids/docs/churches/Falkirk_St_Andrew%27s_Church.pdf

Familysearch, 2015. familysearch.org. [NLS digital collections]. Available at: https://familysearch.org/

Faraday, M., 1846. 'Dr Faraday's Thoughts on Ray-vibrations'. *Philosophical Magazine*, **28**(May), pp. 345–50

Farmer's Magazine, 1811, 'Account of William Craik of Arbigland'. Available at: www.kirkcudbright.co/historyarticle.asp?ID=137&p=14&g=4

Farquhar, G., 1700. *The Constant Couple; or, a Trip to the Jubilee*. London: Longman, Hurst, Rees, and Orme [available as a Project Gutenberg e-book]

Ferguson, O., 2006. 'Introduction'. In: J. Melville (ed.), *William Dyce and the Pre-Raphaelite Vision*. Aberdeen: Aberdeen City Council, pp. 9–13

Ferguson, R. S., 1889. 'The Retreat of the Highlanders through Westmorland in 1745'. *Transactions of the Cumberland and Westmorland Antiquarian and Archaeological Society*, **10**, pp. 186–228

Fielding, H., 1987. 'The History of the Present Rebellion in Scotland'. In: W. B. Coley, ed., *The True Patriot and Related Writings*. Middletown, CT: Wesleyan University Press, p. 34ff.

Finlay, J., 2012. *The Community of the College of Justice: Edinburgh and the Court of Session 1687–1808*. Edinburgh: Edinburgh University Press

Fisher, D., ed., 2009. 'Clerk, Sir George, 6th Bt. (1787–1867) of Penicuik House, Edinburgh'. In: *The History of Parliament: The House of Commons 1820–1832*. Cambridge: Cambridge University Press. Available at: http://www.historyof parliamentonline.org/volume/1820-1832/member/clerk-sir-george-1787-1867#constituency

Fleming, A., 1938. *Scottish and Jacobite Glass*. Glasgow: Jackson, Son & Co.

Fletcher, H. R. & Brown, W. H., 1970. *The Royal Botanic Garden Edinburgh 1670–1970*. Edinburgh: HMSO

Flood, R., McCartney, M. & Whitaker, A., eds, 2014. *James Clerk Maxwell: Perspectives on his Life and Work*. Cambridge: Cambridge University Press

Forfar, D. O., 1992. 'The Origins of the Clerk (Maxwell) Genius'. *Bulletin of the Institute of Mathematics and its Applications*, 28(1&2), pp. 4-16. Also as 'Generations of Genius', see below

——, 1996. 'What became of the Senior Wranglers?' *Mathematical Spectrum*, 29(1), pp. 1–4

——, 1999. 'Generations of Genius'. In *James Clerk Maxwell Commemorative Booklet for the 4th International Congress on Industrial and Applied Mathematics, Edinburgh 1999*. Edinburgh: James Clerk Maxwell Foundation. Available at: http://www. clerkmaxwellfoundation.org/Maxwell_-_Origins_of_Genius.pdf

Forsyth, R., 1805. *The Beauties of Scotland*. Edinburgh: Thomson, Bonar and Brown

Foster, J., 1884. *The Royal Lineage of our Noble and Gentle Families, together with their Paternal Ancestry*. London: Hazell, Watson and Viney (printers – private distribution)

Fountainhall, Sir J. Lauder of, 1848. *Historical Notices of Scottish Affairs, Selected from the Manuscripts of John Lauder of Fountainhall*. Edinburgh: Constable

Fountainhall, Lord J. Lauder of, & Scott, Sir W., 1822. *Chronological Notes of Scottish Affairs, from 1680 till 1701: Being Chiefly Taken from the Diary of Lord Fountainhall*. Edinburgh: s.n.

Frandzen, T., 2014. 'Basil Cochrane, Commissioner of Excise'. Available at: www.jamesboswell.info/biography/basil-cochrane-commissioner-excise

Fraser, G. M., 1971. *The Steel Bonnets*. London: Harper Collins

Fraser, W., 1858. *The Stirlings of Keir, and their Family Papers*. Privately published. Available at: https://archive.org/details/stirlingsofkeirt00fras

——, 1869. *The Chiefs of Colquhoun and their Country*. Edinburgh: T. and A. Constable

Frey, A. R., 1888. *Soubriquets and Nicknames*. Boston, MA: Ticknor and Co.

Gazetteer for Scotland, 2013a. 'Ormiston'. Available at: www.scottish-places.info/towns/townfirst282.html

Gazetteer for Scotland, 2013b. 'Rossdhu'. Available at www.scottish-places.info/features/featurefirst5030.html

Gentleman's Magazine, 1823. [Obituary Notices] 93(July), p. 647

Gibson, A., 1927. *New Light on Allan Ramsay*. Edinburgh: W. Brown

Gibson, J., 1777. *The History of Glasgow from the Earliest Accounts to the Present Time*. Glasgow: R. Chapman and A. Duncan

Gifford, J., 1989. *William Adam, 1689–1748: A Life and Times of Scotland's Universal Architect*. Edinburgh: Mainstream Publishing

Gilpin, R., 1879. *Memoirs of Dr. Richard Gilpin, of Scaleby Castle in Cumberland*. London: B. Quaritch

Glasgow University Library, 2004. 'Househill (Hous'hill)'. Available at: www.the glasgowstory.com/image.php?inum=TGSB00285

Glendinning, M., MacInnes, R. & MacKechnie, A., 1996. *A History of Scottish Architecture: From the Renaissance to the Present Day*. Edinburgh: Edinburgh University Press

Glenlair Trust, 2012. 'Maxwell at Glenlair Trust'. Available at: www.glenlair.org.uk

Goodwin, G., 1887. 'Cay, John'. In: *DNB00*

——, 1890. 'Gordon, Alexander (?1692–1754)'. In: *DNB00*

Gordon, A., 1726. *Itinerarium Septentrionale; or, a Journey thro' most of the Counties of Scotland, and those in the North of England*. London: Printed for the author

Grant, A., 1884. *The Story of the University of Edinburgh during its First Three Hundred Years*, vol. 2. London: Longmans Green & Co.

Grant, F. J., 1898. *The Commissariot Record of Edinburgh, Register of Testaments, Part II*. Edinburgh: Scottish Record Society

——, 1908. *Index to the Register of Burials in the Churchyard of Restalrig, 1728–1854*. Edinburgh: Scottish Record Society

——, 1922. *Register of Marriages of the City of Edinburgh 1751–1800*. Edinburgh: Scottish Record Society

——, 1944. *The Faculty of Advocates in Scotland, 1532–1943 with Genealogical Notes*. Edinburgh: Scottish Record Society

Grant, J., 1850. *Memorials of the Castle of Edinburgh*. Edinburgh: William Blackwood and Sons

——, c. 1887. *Cassell's Old and New Edinburgh: Its History, its People, and its Places*. London: Cassell & Co. Ltd

Greenshields, J. B., 1864. *Annals of the Parish of Lesmahagow*. Edinburgh: Caledonian Press

Greig, J., 1911. *Sir Henry Raeburn, RA; his Life and Works, with a Catalogue of his Pictures*. London: The Connoisseur: Otto Ltd

Grierson, H., 1938. *Sir Walter Scott Bart*. New York: Haskell House

Griffenhagen, G. B. & Young, J. H., 1959. *Old English Patent Medicines in America*. Washington, DC: Smithsonian Institution. Available as e-book #30162, Gutenberg Project, 2/10/2009

Groome, F. H., 1885. *Ordnance Gazetteer of Scotland*. Edinburgh: Thomas Jack

Hallen, A. W. C., ed., 1891. 'Some Notes on the Attainted Jacobites, 1746'. In: *The Scottish Antiquary or Northern Notes & Queries*, vol. 5. Edinburgh: T. & A. Constable

Hamilton, H. B., 1901. *Historical Record of the 14th (King's) Hussars from AD 1715 to AD 1900*. London: Longmans Green & Co.

Hamilton, J. A., 1894. 'Mure, William (1718–1776)'. In: *DNB00*

Hamilton, T., 1896. 'Thomson, James (1822–1892)'. In: *DNB00*

Hanson, L., 2009, 2013. *Peebles through Time*. Stroud: Amberley Publishing

Harman, P. M. (ed.), 1990, 1995, 2002. *The Scientific Letters and Papers of James Clerk Maxwell*, 3 vols. Cambridge: Cambridge University Press

——, 1985. *Wranglers and Physicists: Studies on Cambridge Mathematical Physics in the Nineteenth Century*. Manchester: Manchester University Press

Harris, E., 2011. Index [with links to individual persons]. Available at: http://archive.stjohns-edinburgh.org.uk

Harvey, W., 2000. *Lead and Labour: The Miners of Leadhills*. [NLS: HP4.203.0481]

Harvey, W. & Downs-Rose, G., 1976. 'Lead & Leases; Production from the Wanlockhead Mines 1710–1780'. *British Mining*, **3**, pp. 21–28

Hayton, D. W. et. al., 2002a. 'CLERK, John (1676–1755), of Penicuik, Midlothian' in: *History of Parliament Online*, at: www.historyofparliamentonline.org/volume/ 1690-1715/member/clerk-john-1676-1755

——,. 2002b. 'Hutchinson, Jonathan (c. 1662–1711), of Newcastle-upon-Tyne, Northumb.' Available at: www.historyofparliamentonline.org/volume/ 1690–1715/member/hutchinson-jonathan-1662-1711

Hearnshaw, F. J. C., 1929. *The Centenary History of Kings College London 1828–1928.* London: G. Harrap & Son

Henderson, T. F., 1896. 'Pitcairne, Archibald'. In: *DNB00*

——, 1899. 'Tytler, James'. In: *DNB00*

Henderson, T. F. & Baker, A. P., 2004. 'Maxwell, Sir George Clerk, 4th Baronet (1715–1784)'. In: *ODNB*

Henderson, T. & Grierson, P. J. H., 1914. *Edinburgh Academy Register.* Edinburgh: Edinburgh Academical Club

Herbert, D., 1870. 'Life of Tobias Smollett'. In: *The Works of Tobias Smollett.* Edinburgh: Wm Nimmo

Hertz, H., 1893. *Electric Waves: Being Researches on the Propagation of Electric Action with Finite Velocity through Space.* London and New York: Macmillan and Co.

Hilts, V. L., 1975. 'A Guide to Francis Galton's English Men of Science'. *Transactions of the American Philosophical Society*, New Series, **65**(5), pp. 1–85

Historic Scotland, 1970a. '14 India Street, including Railings and Lamp'. Available at: http://data.historic-scotland.gov.uk/pls/htmldb/f?p=2200:15:0::::BUILDING,HL: 29133

——, 1970b. 'Lothian Road, St John's Church'. Available at: http://data.historic-scotland.gov.uk/pls/htmldb/f?p=2200:15:0::::BUILDING:27401

——, 1971. 'Mavisbank House, Midlothian Council, Lasswade Parish'. Available at: http://hsewsf.sedsh.gov.uk/hslive/portal.hsstart?P_HBNUM=7404

Hodgson, J., 1832. *A History of Northumberland.* Newcastle: T. & J. Pigg, for the author

Horn, B. L. H., 2004. 'Ramsay, John, of Ochtertyre (1736–1814), Writer on Scotland'. In: *ODNB*

Horsburgh, W., 1754. *Experiments and Observations upon the Hartfell Spaw.* Edinburgh: Hamilton and Balfour

Horsley, J., 1732. *Britannia Romana: Or, the Roman Antiquities of Britain: In Three Books.* London: John Osborn and Thomas Longman

House of Commons, 1738. *Journals of the House of Commons.* London: HMSO (reprinted 1803), pp. 170, 171, 197. Available at: http://books.google.co.uk/books?id =kBJDAAAAcAAJ

House of Lords, 1738. 'Journal of the House of Lords', vol. 25, 21–30/4/1738. Available at: www.british-history.ac.uk/lords-jrnl/vol25/pp222-233

——, 2007a. 'Act of Union 1707 – Background'. Available at: http://collections. europarchive.org/ukparliament/20090701100701/www.parliament.uk/actofunion/ 01_background.html

——, 2007b. 'Act of Union 1707 – Finalisation'. Available at: http://collections. europarchive.org/ukparliament/20090701100701/www.parliament.uk/actofunion/ 04_06_finalisation.html

Howat, J., 2013. 'Funeral Record, Methodist Episcopal Church, Rosario 1865–1893, Registers A & B'. In: *British Settlers in Argentina and Uruguay – Studies in 19th and 20th Century Emigration.* Available at: www.argbrit.org/Methodist_Rosario/MethRos_deaths1865-1893.htm

Hume, D., 1796. *The History of England: From the Invasion of Julius Cæsar to the Revolution in 1688*. London: T. Cadell

Hunter, J. J., 1864. *Historical Notices of Lady Yester's Church and Parish, Edinburgh*. Edinburgh: Johnstone Hunter & Co.

Hutton, J., 1788. 'Theory of the Earth'. *Trans, Roy. Soc. Ed.*, 1(2), pp. 209–304

Huxley, L., 1914. *Memorials in Verse and Prose of Lewis Campbell MA, LL.D, Hon. D. Litt.* London: Printed for private circulation by W. Clowes and Sons Ltd

Ingamells, J., 2004. 'Ramsay, Allan, of Kinkell (1713–1784)'. In: *ODNB*

Inglish, D., 1800. [Advertisement] 'Anderson's Pills'. London: *London Gazette*, 14/10/1800.

Inglish, I., 1689. [Advertisement] 'Anderson's Pills'. London: *London Gazette*, 8/7/1689, p. 2

Irving, A. S., 1841. *Original Songs*. Ed. Rev. W. Murray. Privately published

Irving, G. V. & Murray, A., 1864. *The Upper Ward of Lanarkshire Delineated*. Glasgow: Thomas Murray and Son

Irving, M., 1907. 'The Irvings of Newton'. In: J. B. Irving, *The Irvings, Irwins, Irvines, or Erinveines or Any Other Spelling of the Name* [Appendix]. Aberdeen: The Rosemount Press

Jackson, J., 1833. *Essays on Various Agricultural Subjects: And an Account of the Parish of Penicuik*. Edinburgh: William Blackwood

Jaffray Family, 1836–1926. 'MS1784, Papers of the Jaffray Family of Studley'. Available at: www.nationalarchives.gov.uk/a2a/records.aspx?cat=143-ms_1784&cid= 0&kw=john%20jaffray#0

James Clerk Maxwell Foundation, 1820. *John Clerk Maxwell, Charter for 14 India Street*. [Manuscript document]

Jeans, J. H., 1904. *The Dynamical Theory of Gases*, 2nd edn. London: C. J. Clay and Sons; Cambridge University Press

Jeffares, N., 2014. 'Cay, Mrs Robert Hodshon, née Elizabeth Liddell'. Available at: www.pastellists.com/Articles/Cay.pdf

Jefferson, S., 1838. *The History and Antiquities of Carlisle*. Carlisle: Samuel Jefferson

Jenkin, H. C. F. et. al., 1873. *Reports of the Committee on Electrical Standards' of the British Association for the Advancement of Science*. London & New York: E. & F. N. Spon

Johnson, S. & Boswell, J., 1984. *A Journey to the Western Islands of Scotland and the Journal of a Tour to the Hebrides*. London: Penguin Classics

Johnston, J. B., 1892. *Place Names of Scotland*, Edinburgh: David Douglas

Johnstone, C. L., 1889. *The Historical Families of Dumfriesshire and the Border Wars*, 2nd edn. Edinburgh and Glasgow: John Menzies & Co.

Jones, C., 1997. *The Edinburgh History of the Scots Language*. Edinburgh: Edinburgh University Press

Jones, R. V., 1973. 'James Clerk Maxwell at Aberdeen 1856–1860'. *Notes and Records of Roy. Soc. Lon.*, June, 28(1), pp. 57–81

Kay, J., 1838. *A Series of Original Portraits and Caricature Etchings*. Edinburgh: Hugh Paton

Kay, J. & Paton, H., 1877. *A Series of Original Portraits and Caricature Etchings by the Late John Kay with Biographical* [. . .]. Edinburgh: A. and C. Black

Kingsley, R. F. & Alexander, H. J., 2008. 'The Failure of Abercromby's Attack on Fort Carillon, July 1758, and the Scapegoating of Matthew Clerk'. *Journal of Military History*, 72(1), p. 43

Kirkwood, R., 1819. *Plan & Elevation of the New Town of Edinburgh*, 1819. Available at: http://maps.nls.uk/towns/detail.cfm?id=418

Knott, C. G., 1911. *Life and Scientific Work of Peter Guthrie Tait*. Cambridge: Cambridge University Press

Kohlrausch, R. & Weber, W., 1857. 'Elektrodynamische Maaßbestimmungen: insbesondere Zuruckfuhrung der Stromintensitats-Messungen auf mechanisches Maass'. *Abhandlungen der Mathematisch-Physischen Klasse der Koniglich-Sachsischen Gesellschaft der Wissenschaften*, vol. 3, Leipzig, pp. 221–90. Reprinted in: H. Weber, ed., 1893, *Wilhelm Weber's Werke*. Berlin: Springer, vol. 3, pp. 609–676

Ladies Magazine, 1781. 'From the London Gazette, Whitehall, Nov 6'. *Ladies Magazine*, 12

Laing, D., 1858. *A Catalogue of the Graduates [. . .] of University of Edinburgh [. . .]* Edinburgh: The Bannatyne Club

———, 1869. 'On the Supposed "Missing School of Design in the University of Edinburgh", 1784'. *Proceedings of the Society of Antiquities of Scotland*, **8**(January), pp. 36–40

Lamb, H., 1931. 'Clerk Maxwell as Lecturer'. In: *James Clerk Maxwell, A Commemoration Volume, 1831–1931*. Cambridge: Cambridge University Press, pp. 142–146

Lawson, J. P., 1849. *Historical Tales of the Wars of Scotland and of the Border Raids, Forays and Conflicts*. Edinburgh, London and Dublin: A. Fullarton & Co.

Leighton, H. R. ed., 1906–07. *Northern Notes and Queries*. Newcastle-Upon-Tyne: M. S. Dodds

Lim, P., 2011. *Forgotten Souls: A Social History of the Hong Kong Cemetery*. Hong Kong: Hong Kong University Press

Lim, P. & Atkins, C., 2011. 'Cemeteries in Hong Kong>Hong Kong Cemetery> Inscriptions for Cemetery Sections 17-47' In: *Gwulo: Old Hong Kong*. Available at: http://gwulo.com/node/8741

Lincoln, Henry, Earl of, 1715–16. *Declared Accounts: Army: Guards and Garrisons and Land Forces, 17 October 1715 to 24 December 1716*, London

Literary Gazette, 1825. 'Review of New Books: Memoirs of Mr William Veitch and George Bryson written by themselves [. . .]'. *The Literary Gazette and Journal of Belles Lettres, Arts, Sciences &c.*, **441**(2/7/1825), pp. 417–20, **442**(9/7/1825), pp. 438–440 and **443**(16/7/1825), p. 460

Lockhart, J. G., 1837. *Memoirs of the Life of Sir Walter Scott* (5 vols). Philadelphia, PA: Carey Lea and Blanchard

London Gazette, 1854. 'War Office Brevet, 20th June 1854'. Supplement to the *London Gazette* No. 21564.

———, 1855. 'Promotions to Colonel, 6/2/1855', p. 433

———, 1880. Supplement to the *London Gazette*, 25/2/1880, p. 1537

Lord Eldin, see Clerk, 1788

Lorimer, J., 1862. *Hand-book of the Law of Scotland*. Edinburgh: T. & T. Clark

Lowther, J., 1st Viscount, 1808. *Memoir of the Reign of James II*. York: T. Wilson & R. Spence, Printers

Lundy, D., 2015. 'Person Page – 17362 (Maria Anne Law)' Available at: www.thepeerage. com/p17362.htm#i173611

Macaulay, J., 2004. 'Adam, William (bap. 1689, d. 1748)'. In: *ODNB*

Macdonald G., 1920. 'The Romans in Dumfriesshire', *Trans. Dumf. Gall. Nat. Hist. Antiqu. Soc.*, **8**(3), pp. 67–100

MacIntire, D. B., 1999. 'James Hutton's Edinburgh: A Précis'. In: *The Lyell Collection*. London: Geological Society of London. Available at: http://sp.lyellcollection.org/content/150/1/1.full.pdf

MacIvor, I., 1978. *14 India Street: Ownership and Occupation*, Edinburgh: private report
Mackay, J. A., 1989. 'Two Hitherto Unrecorded Letters of Allan Ramsay'. *Studies in Scottish Literature*, 24(1), pp. 1–6
Mackenzie, E., 1827. 'Incorporated Companies: Companies Not of the Bye-Trades'. In: *Historical Account of Newcastle-upon-Tyne: Including the Borough of Gateshead*. Newcastle: Mackenzie & Dent, pp. 698–706
Mackenzie, W., 1841. *The History of Galloway from the Earliest Period to the Present Time*, 2 vols. Kirkcudbright: John Nicholson
Mahon, B., 2003. *The Man Who Changed Everything: The Life of James Clerk Maxwell*. Chichester: John Wiley
Martin, B., 1931. *Allan Ramsay, a Study of his Life and Works*. Cambridge, MA: Harvard University Press
Maxwell of Arkland, R., 1743. 'A List of the Members of the Society'. *Select Transactions of the Honourable Society of Improvers in the Knowledge of Agriculture in Scotland*
Maxwell, J. C., 1849. 'On the Theory of Rolling Curves'. *Trans. Roy. Soc. Ed.*, 16, pp. 519–540
——, 1851. 'On the Description of Oval Curves, and Those Having a Plurality of Foci. *Proc. Roy. Soc. Ed.*, 2, pp. 89–91 and plate II. [Read 1846, it was not reported in the *Transactions*. The *Proceedings* for 1844–51 were published together]
——, 1853. 'On the Equilibrium of Elastic Solids'. *Trans. Roy. Soc. Ed.*, 20, pp. 87–120
——, 1854. 'Solutions of Problems'. *Camb. & Dub. Math. J.*, 9, pp. 7–19
——, 1855. 'Experiments on Colour, as Perceived by the Eye, with Remarks on Colour-Blindness'. *Trans. Roy. Soc. Ed.*, 21(2), pp. 275–298 + plate
——, 1856a. 'Experiments on Colour Vision as Perceived by the Eye'. *Trans. Roy. Soc. Scot. Arts*, 4, pp. 394–400
——, 1856b. 'On the Transformation of Surfaces by Bending'. *Trans. Camb. Phil. Soc.*, 9, pp. 445–470
——, 1857–62. 'On Theories of the Constitution of Saturn's Rings'. *Proc. Roy. Soc. Ed.*, 4, pp. 99–101
——, 1858. 'On the General Laws of Optical Instruments'. *Quart. J. Pure Appl. Math.*, 2, pp. 233–246
——, 1859. 'On the Stability of the Motion of Saturn's Rings'. *Monthly Notices of the Roy. Astronom. Soc.*, 10, pp. 297–384
——, 1860a. 'Illustrations of the Dynamical Theory of Gases'. *Phil. Mag.*, 19 and 20, pp. 19–32 and 21–37
——, 1860b. 'On an Instrument for Exhibiting the Motions of Saturn's Rings'. *Report of the 29th Meeting of British Association for the Advancement of Science, September 1859, Aberdeen*, §62
——, 1860c. 'On the Dynamical Theory of Gases'. *Report of the 29th Meeting of British Association for the Advancement of Science, September 1859, Aberdeen*, §9
——, 1861. 'On the Theory of Three Primary Colours'. *Notices Proc. Roy. Inst. Gr. Brit.*, 11, pp. 370–374
——, 1861–62. 'On Physical Lines of Force'. *Trans. Camb. Philos. Soc.*, 21(139; 140; 141), pp. 161–175; 281–291; 338–348, and 23(151; 152), pp. 12–24; 85–95
——, 1864a. 'On Faraday's Lines of Force'. *Trans. Camb. Philos. Soc.*, 10, pp. 27–83 [Read December 1855 and February 1856: the earlier volumes were published belatedly]
——, 1864b. 'On Reciprocal Figures and Diagrams of Forces'. *Philos. Mag.*, 28, pp. 250–261

———, 1865. 'A Dynamical Theory of the Electromagnetic Field'. *Trans. Roy. Soc. Lon.*, **155**, pp. 459–512
———, 1866. 'On the Viscosity or Internal Friction of Air and Other Gases'. *Phil. Trans. Roy. Soc. Lon.*, **156**, pp. 249–268
———, 1867a. 'On the Theory of Diagrams of Forces as Applied to Roofs and Bridges'. *Reports of the British Association for the Advancement of Science, Dundee*, §156
———, 1867b. 'On the Dynamical Theory of Gases'. *Phil. Trans. Roy. Soc. Lon.*, **157**, pp. 49–88
———, 1868a. 'On a Method of Making a Direct Comparison of Electrostatic and Electromagnetic Force, with a Note on the Electromagnetic Theory of Light'. *Phil. Trans. Roy. Soc. Lon.*, **158**, pp. 643–658
———, 1868b. 'On Governors'. *Proc. Roy. Soc. Lon.*, **16**, pp. 270–283
———, 1869a. 'Experiments on the Value of v, the Ratio of the Electromagnetic to the Electrostatic Unit of Electricity'. *Reports of the British Association for the Advancement of Science*, pp. 436–438
———, 1869b. 'On Reciprocal Diagrams in Space and their Relation to Airy's Function of Stress'. *Proc. Lond. Math. Soc.*, **2**, pp. 58–60
———, 1870. *Theory of Heat* (2nd edn 1871). London: Longmans Green
———, 1871a. *Introductory Lecture on Experimental Physics*. London and Cambridge: Macmillan and Co.
———, 1871b. 'Remarks on the Mathematical Classification of Physical Quantities'. *Proc. Lond. Math. Soc.*, **3**, pp. 224–232
———, 1872. 'On Reciprocal Figures, Frames, and Diagrams of Forces'. *Trans. Roy. Soc. Ed.*, **26**, pp. 1–40
———, 1873a. *A Treatise on Electricity and Magnetism*, 2 vols. Oxford: Clarendon Press
———, 1873b. 'On the Condition that, on the Transformation of any Figure by Curvilinear Coordinates in Three Dimensions [. . .]'. *Proc. Lond. Math. Soc.*, **4**, pp. 117–119
———, 1876a. *Matter and Motion*. London: Society for Promoting Christian Knowledge
———, 1876b. 'On the Solution of Electrical Problems by the Transformation of Conjugate Functions'. *Proc. Camb. Philos. Soc.*, **2**(8), pp. 242–243
———, 1876c. 'On the Protection of Buildings from Lightning'. *Reports of the British Association for the Advancement of Science, Glasgow*, pp. 43–45
———, 1879. 'On Stresses in Rarefied Gases Arising from Inequalities of Temperature'. *Phil. Trans. Roy. Soc. Lon.*, **170**, pp. 231–256 & Appendix
———, 1880. 'On the Unpublished Electrical Papers of the Hon. Henry Cavendish'. *Proc. Camb. Phil. Soc.*, **3**(3), pp. 86–89
———, 1979. 'James Clerk Maxwell's Inaugural Lecture at King's College, London, 1860'. *Am. J. Phys.*, **47**, pp. 928–933
M'Crie, T., 1841. *Sketches of Scottish Church History: Embracing the Period from the Reformation to the Revolution*. Edinburgh: John Johnstone
McCulloch, I. M., 2008. '"A Blanket of Inconsistencies . . .": The Battle of Ticonderoga, 2008'. *The Journal of Military History*, **72**(3), pp. 889–900
McDowall, W., 1873. *History of the Burgh of Dumfries with Notices* [. . .], 2nd edn. Edinburgh: A. & C. Black
McKerlie, P. H., 1879. *History of the Lands and their Owners in Galloway*. Edinburgh: William Paterson
Melville, J., ed., 2006. *William Dyce and the Pre-Raphaelite Vision*. Aberdeen: Aberdeen City Council

Millar, A. H., 1909. *A Selection of Scottish Forfeited Estates Papers*. Edinburgh: T. & A. Constable

Mitchison, R., 2004. 'Clerk, Sir John, of Penicuik, Second Baronet (1676–1755)'. In: *ODNB*

Murdoch, A., 2004. 'Campbell, Archibald, third Duke of Argyll (1682–1761)'. In: *ODNB*

Murray, J., 2012. Private communication

Museum of Lead Mining, 2014. 'Home>Virtual Tour>Lochnell Mine'. Available at: www.leadminingmuseum.co.uk/mine.shtml

NAS, 2013. Scottish Government Records after 1707. Available at: www.nas.gov.uk/guides/scottishGovernmentAfter1707.asp

National Galleries of Scotland, 2014. 'Collection> . . .>Sir Henry Raeburn>Rear-Admiral Charles Inglis, c.1731–1791. Sailor'. www.nationalgalleries.org/collection/artists-a-z/r/artist/sir-henry-raeburn/object/ rear-admiral-charles-inglis-c-1731-1791-sailor-pg-1567

Nester, W. R., 2008. *The Epic Battles for Ticonderoga, 1758*. Albany, NY: SUNY Press

Newton, Sir Isaac, 1704. *Opticks*. Online version at The Newton Project (2009)

Nicolson, J. & Burn, R., 1777. *The History and Antiquities of the Counties of Westmorland and Cumberland*. London: W. Strahan and T. Cadell

Niven, W. D., 1890, *The Scientific Papers of James Clerk Maxwell* (2 vols). Cambridge: Cambridge University Press

Norton-Kyshe, J. W., 1898. *The History of the Laws and Courts of Hong Kong*. London: Fisher Unwin

O'Connor, J. J. & Robertson, E. F., 1998. 'James Stirling'. Available at: www-history.mcs.st-and.ac.uk/BiogIndex.html

——, 2000. Sir Isaac Newton, Available at: www-history.mcs.st-and.ac.uk/BiogIndex. html

——, 2003. 'Peter Guthrie Tait', Available at: www-history.mcs.st-and.ac.uk/BiogIndex. html

——, 2004a. 'James Hutton', Available at: www-history.mcs.st-and.ac.uk/BiogIndex. html

——, 2004b. 'Royal Society of Edinburgh'. Available at: www-history.mcs.st-and.ac.uk/Societies/RSE.html

Parson, W. & White, W., 1828. *History, Directory, and Gazetteer, of the Counties of Durham and the Towns and Counties of Newcastle-upon-Tyne and Berwick-on-Tweed*. Newcastle-upon-Tyne: William White & Co.

Paterson, J., 1847. *History of the County of Ayr*. Ayr: John Dick

Paton, H., ed., 1902. *Register of Interments in Greyfriars Burying-ground, Edinburgh, 1685–1700*. Edinburgh: Scottish Record Society

——, 1908. *Register of Marriages for the Parish of Edinburgh 1701–1750*. Edinburgh: Scottish Record Society

Paul, Sir J. B., 1909. *The Scots Peerage Founded on Wood's Edition of Sir Robert Douglas's Peerage of Scotland* [. . .] Edinburgh: David Douglas

Penicuik House Project, 2014. History of the family. Available at: www.penicuikhouse.co.uk/about-us

Pennecuik, A., 1815. *The Works of Alexander Pennecuik Esq. of New-Hall* [. . .] *Containing the Description of Tweedsdale*. Leith: Allardice, Constable et al.

Phillips, M., 1894. 'Notes on Some Forgotten Quaker Burial Grounds [Part 1]'. *Archaeologia Aeliana or, Miscellaneous Extracts Relating to Antiquities*, **16** (New Series), pp. 189–210

Pigot, J., 1837. *National and Commercial Directory of the Whole of Scotland and the Isle of Man*. London: James Pigot & Co.

Pitcairn, R., 1833. *Criminal Trials in Scotland from A.D. 1488 to A.D. 1624* (vols 1 & 3). Edinburgh: Wm Tait

Pittock, M., 2006. *Poetry and Jacobite Politics in Eighteenth-Century Britain and Ireland*. Cambridge: Cambridge University Press

Playfair, J., 1803. 'A Biographical Account of the Late James Hutton MD FRSEdin'. *Trans. Roy. Soc. Ed.*, 5(3), pp. 39–99

Pococke, R., 1887. *Tours in Scotland*. Edinburgh: Scottish History Society

Pointon, M. R., 1979. *William Dyce, 1806–1884: A Critical Biography*. Oxford: Oxford University Press

Potts, J. R., 2011. 'USS *Bonhomme Richard* 42-Gunner (1765)' Available at: www.militaryfactory.com/ships/detail.asp?ship_id=USS-Bonhomme-Richard-1765

Prebble, J., 1967. *Culloden*. London: Penguin

——, 1968. *Glencoe*. London: Penguin

——, 1971. *The Lion in the North*. London: Book Club Associates

——, 1982. *The Highland Clearances*. London: Penguin

Prevost, W. A. J., 1963. 'George Clerk and the Royal Hunters in 1745'. Kendal: *Trans. Cumb. Westm. Antiqu. Arch. Soc.*, 63, pp. 231–252

——, 1968. 'Dumcrieff and its Owners'. *Trans. Dumfr. Gall. Nat. Hist. Antiqu. Soc.*, 45 (3), pp. 200–210

——, 1969. 'Memorandum of a Journey to Dumcrieff'. *Trans. Dumfr. Gall. Nat. Hist. Antiqu. Soc.*, 46 (series 3), pp. 183–184

——, 1970. 'Mary Dacre, White Rose of Scotland'. *Trans. Cumb. Westm. Antiqu. Arch. Soc.*, 70, pp. 161–180

——, 1977. 'The White Cockade of Dame Mary Dacre or Clerk'. *Trans. Cumb. Westm. Antiqu. Arch. Soc.*, 7, pp. 182–183

Priestly, J., 1831. *Historical Account of the Navigable Rivers, Canals, and Railways, throughout Great Britain* [...] London: Longman, Rees, Orme, Brown and Green

Rae, J., 1895. *Life of Adam Smith*. London: Macmillan and Co.

Raeburn, Sir Henry., n.d. 'Robert Hodshon Cay'. Available at: https://commons.wikimedia.org/wiki/File:Sir_Henry_Raeburn_-_Robert_Hodshon_Cay_-_Google_Art_Project.jpg

Ramsay, A., *1719–21. Poems by Allan Ramsay*. Edinburgh: Allan Ramsay

——, 1808. *The Gentle Shepherd, A Pastoral Comedy*. Edinburgh: Martin, Creech, Constable et al.

Ramsay, A. et al., 1740. *The Caledonian Miscellany. Consisting of Select and Much Approved Pastorals, Choice Fables and Tales, with Other Occasional Poems*. Newcastle, London: William Cay

Ramsay, A. & Robertson, J. L., c. 1886. *Poems by Allan Ramsay. Selected and Arranged, with a Biographical Sketch of the Poet*. London: Walter Scott Publishing Co. Ltd

Ramsay, E. B., 1861. *Reminiscences of Scottish Life & Character*. Edinburgh: Edmonston and Douglas

Ramsay, J., 1888. *Scotland and Scotsmen in the Eighteenth Century*. Edinburgh: W. Blackwood

Randall, Sir J., 1963. 'Aspects of the Life and Work of James Clerk Maxwell'. In: C. Domb, ed. *Clerk Maxwell and Modern Science*. London: Athlone Press, pp. 1–25

Rautio, J., 2006. 'In Search of Maxwell'. *Microwave Journal*, July

Redington, J., ed., 1889. *Calendar of Treasury Papers, Volume 6: 1720–1728*. London: HMSO

Reid, J. S., 2012. 'Aberdeen, King's College: David Thomson – Local Educator and Reformer'. Available at: http://homepages.abdn.ac.uk/npmuseum/Scitour/KingDT.pdf

Rennie, J., 1837. *A New Supplement to the Pharmacopæias of London, Edinburgh, Dublin and Paris* [...]. London: Baldwin & Cradock

Riddell, J. O., 1930. *Clerk Maxwell of Glenlair. James Clerk Maxwell Centenary Booklet*. Galloway Series, No. 8. Castle Douglas: Adam Rae

Robertson, J., 2003. *Joseph Knight*. London: Fourth Estate (Harper-Collins)

Roberts, T. R., 1908. *Eminent Welshmen*. Cardiff and Merthyr Tydfill: The Educational Publishing Company

Robinson, G. G. et al., 1798. *A New and General Biographical Dictionary*. London: G. G. and J. Robinson et al.

Rock, J., 2013a. 'Newhailes New Research, Part One'. Available at: https://sites.google.com/site/researchpages2/home/newhailes-house-timeline

——, 2013b. 'Students and Associates of Cooper Senior'. Available at: https://sites.google.com/site/richardcooperengraver/students-and-associates-of-cooper-senior

Rodger, R., 2004. *The Transformation of Edinburgh: Land, Property and Trust in the Nineteenth Century*. Cambridge: Cambridge University Press

Rogers, C., 1856. *The Modern Scottish Minstrel*. Edinburgh: A. & C. Black

——, 1884. *Social Life in Scotland from Early to Recent Times*. Edinburgh: W. Paterson

Roscoe, S., Sullivan, G. & Hodgins, G., 2013. *A Brief History of Naval Communications*. Available at: http://jproc.ca/rrp/nro_his.html

Roxburghe Collection, n.d. 'EBBA ID 30565'. British Library. Available at: http://ebba.english.ucsb.edu/ballad/30565

Roy, W., 1747–55. *Military Survey of Scotland*. The Crown

Royal Academy of Arts, 1832. *The Exhibition of the Royal Academy MDCCCXXXII – The Sixty Fourth* [catalogue]. London

Royal Kalendar, 1822. *The Royal Kalendar, and Court and City Register for England, Scotland, Ireland, and the Colonies*. London: William Stockdale

Royal Scottish Geographical Society, 2014. *Images for All: Edinburgh 1773*. Available at: http://rsgs.org/exploring-geography/rsgs-collection/edinburgh-1773-2/

RSE, 2015. 'The Royal Society of Edinburgh' Available at: www.royalsoced.org.uk/

Runciman, A., *c.* 1772, *The Blind Ossian Singing and Accompanying himself on the Harp* (pen and ink wash on paper, Acc. No. D299), Edinburgh: National Galleries of Scotland

Scotia Mining, 2014. *History of Leadhills*. Available at: www.scotiamining.com/history-of-leadhills.html

Scotland.org, 2014. 'Disruption of 1843'. Online at: www.scotland.org.uk/history/disruption

Scotland, Privy Council, 1687. *Letters of Publication, in Favours of Thomas Weir Chyrurgeon in Edinburgh*, Edinburgh: Privy Council

——, 1694. *Letters of Publication, in Favours of Thomas Weir Chyrurgeon in Edinburgh*, Edinburgh: Privy Council

Scots Magazine, 1808. 'Deaths'. *The Scots Magazine and Edinburgh Literary Miscellany*, 70 (May)

Scott, H., 1915. *Fasti Ecclesiae Scotianicae : Synod Of Lothian and Tweeddale*, vol. 1. Edinburgh: Oliver & Boyd

——, 1925a. *Fasti Ecclesiae Scoticanae : Synods of Glasgow and Ayr*, vol. 3. Edinburgh: Oliver & Boyd

——, 1925b. *Fasti Ecclesiae Scoticanae, Synods of Fife and of Angus and Mearns*, vol. 5. Edinburgh: Oliver & Boyd

——, 1926. *Fasti Ecclesiae Scotianicae: Synods of Aberdeen and Moray*, vol. 6. Edinburgh: Oliver & Boyd

Scott, M. M. M., 1913. 'William Maxwell'. In: *Catholic Encyclopedia*. New York: The Encyclopedia Press

Scott, R., 1981. *The Politics and Administration of Scotland 1725–1748*. Edinburgh: Edinburgh University Press

Scott, Sir Walter, 1814. *Waverley*. Available as 'Waverley – Complete' at: https://archive. org/details/waverleycomplete04966gut

——, 1838. *The Poetical Works of Sir Walter Scott: First Series, Containing Minstrelsy of the Scottish Border; Sir Tristram; and Dramatic Pieces*. Paris: Baudry's European Library

Scottish Castle Association, 2013. 'Cramond Tower'. Available at: http://scottishcastlesassociation.com/rec-id-141-cat_id-1-highlight-2.htm

Scottish Covenanter Memorials Association, 2014. 'Who Were the Covenanters?' Available at: www.covenanter.org.uk/WhoWere/

Seitz, F., 2001. 'James Clerk Maxwell (1831–1879)'. *Proc. Am. Phys. Soc.*, **145**(1), pp. 1–44

Select Committee on Arts and Manufactures, 1836. 'Report from the Select Committee on Arts and Manufactures, September 1835'. In: *Selection of Reports and Papers of the House of Commons: Arts Connected with Trade*. London: The House of Commons, pp. 77–88

Seton, G., 1890. *The House of Moncrieff*. Edinburgh: Moncrieff Family Private Publication

——, 1896. *A History of the Family of Seton during Eight Centuries*. Edinburgh: T. & A. Constable

Shakhmatova, K., 2012. 'A History of Street Lighting in the Old and New Towns of Edinburgh World Heritage Site'. Available at: www.ewht.org.uk/uploads/downloads/Lighting%20project%20-%20publication%20ver%206%20Feb%202012.pdf

Shaw, W. A., ed., 1897. *Calendar of Treasury Books and Papers*. London

——, 1903. 'Warrants for Minor Appointments: 1744'. *Calendar of Treasury Books and Papers 1742–45*, **5**, pp. 643–655

Skempton, A. W., ed., 2002. *A Biographical Dictionary of Civil Engineers in Great Britain and Ireland*. London: Thos. Telford/ ICE

Sloan, K., 1997. 'Anne Forbes 1745–1802'. In: D. Gaze, ed., *Dictionary of Women Artists*. Chicago & London: Fitzroy Dearborn, pp. 537–539

Smeaton, J., 1754 (publ. 1812). *Report on the Drainage of Lochar Moss, near Dumfries; Drawn up for Charles Duke of Queensberry and others*

——, 1767. *The Practicability and Expence of Joining the Rivers Forth and Clyde by a Navigable Canal*. Edinburgh: Balfour, Auld and Smellie

Smeaton, O., 1896. *Allan Ramsay*. Famous Scots Series. Edinburgh: Oliphant, Anderson and Ferrier

Smellie, W., 1800. *Literary and Characteristical Lives of John Gregory MD; Henry Home, Lord Kames; David Hume Esq; and Adam Smith, LLD* [. . .] Edinburgh: Alexander Smellie

Smiles, S., 1904, *Lives of the Engineers: Smeaton and Rennie*, London: John Murray Available at: http://gerald-massey.org.uk/smiles/b_smeaton_and_rennie.htm

Smith, A., 1975. *The Forfeited Estates Papers, 1745* [thesis]. St Andrews: University of St Andrews

——, 1987. *The Correspondence of Adam Smith*. Glasgow Edition of the Works and Correspondence of Adam Smith. Indianapolis, IN: Liberty Fund

Smith, C., 2004. 'Thomson, James (1786–1849)'; 'Thomson, James (1822–1892)'; and 'Thomson, William, Baron Kelvin (1824–1907)'. In: *ODNB*

Somerville, T., 1798. *The History of Great Britain during the Reign of Queen Anne*. London: A. Strahan

SPWT, 1879. 'James Clerk Maxwell's Testament, first page', SC16/41/35, p. 116. Available at: www.scotlandspeople.gov.uk/content/images/famousscots/fstranscript115.htm

SRO, 1931. *Index to Particular Register of Sasines for Sheriffdom of Dumfries and Stewartries of Annandale and Kirkcudbright*. Edinburgh: HMSO

Statistical Accounts, 1791–99. 'Parish of Minnigaff, Kirkcudbright'. vol 7, pp. 52–61. Available at: http://stat-acc-scot.edina.ac.uk/link/1791-99/Kirkcudbright/Minnigaff/

——, 'Parish of Dalkeith, Edinburgh', vol. 12, pp. 18–27. Available at http://stat-acc-scot.edina.ac.uk/link/1791-99/Edinburgh/Dalkeith/

Stephen, L., 1890. 'Hamilton, William (1788–1856)'. In: *DNB00*

Stevenson, R. L., 1879. *Edinburgh: Picturesque Notes*. London: Seely, Jackson and Halliday

St Mungo's, 2013. 'St Mungo's Church'. Available at: www.stmungos.freeuk.com/history.htm [accessed 2013]

Storer, J. S. & Storer, H. S., 1820. *Views in Edinburgh and its Vicinity* [. . .]. Edinburgh: Constable & Co.

Surtees, R., 1820. 'Chapelry of Tanfield'. In: *The History and Antiquities of the County Palatine of Durham*. London: Nichols & Son, pp. 219–236. Available at: http://www.british-history.ac.uk/antiquities-durham/vol2/pp219-236

Swift, J., 1910. *The Poems of Jonathan Swift*. London: G. Bell & Sons. Available as a Project Gutenberg e-book

Sydney Morning Herald, 1881. 'Franklin's Arctic Expedition – Lieutenant John Irving'. Sydney 12/4/1881

Symson, J., 2002. *An Exact and Industrious Tradesman: The Letter Book of Joseph Symson of Kendal, 1711–1720*. Oxford: Oxford University Press

Tait, P. G., 1867. *Elementary Treatise on Quaternions*. London: Macmillan & Co.

——, 1880. 'James Clerk Maxwell'. *Proc. Roy. Soc. Ed.*, **10**, pp. 331–339

Tarrant, W., 1862. *The Early History of Hong Kong* [. . .]. Canton: The Friend of China

Taylor, E. F., 2006. *The Aeneid of Virgil Translated into English Verse*. EBook #18466, available from Project Gutenberg

Teevan, R., 2008–10. '1777 Andrew Blackburn'. Available at: http://bankingletters.co.uk/page123.html

Telford, T., 1838. *The Life of Thomas Telford by Himself*. London: Hansard and Sons (printers)

The Archive Hub, 2014. 'Douglas Heron and Company'. Available at: http://archiveshub.ac.uk/data/gb1830-ayr

The Argus, 1852. 'New Commission of the Peace: Territorial Magistrates'. 29/1/1852, p. 2. Available at: http://trove.nla.gov.au/ndp/del/article/4782913

—, 1888. 'Deaths'. Melbourne. 16/8/1888, p. 1. Available at: http://trove.nla.gov.au/ndp/del/article/6898114

The Falkirk Wheel, 2011. 'Lewis Hay Irvine [sic]' Available at: www.falkirk-wheel.com/features/people-of-the-falkirk-area/52-lewis-hay-irvine

The Glasgow Herald, 1832. 'Genuine Anderson's Pills'. 13/7/1832, p. 3

The Scotsman, 1823, 1831, 1834. 'Genuine Anderson's Pills'. 12/3/1823; 16/7/1831; 31/5/1834.

—, 1860. 'Edinburgh Bankruptcy Court: Examination of General Irving', 19/1/1860, p. 2

—, 1863. 'Grana Angelica', 21/2/1863, p. 1

The Society of Antiquaries of Scotland, 2014. 'The Founders of the Society'. Available at: www.socantscot.org/content.asp?Page=297

Thom, C., 2014, 'Robert Adam's first Marylebone House: The Story of General Robert Clerk, the Countess of Warwick and their Mansion in Mansfield Street', *The Georgian Group Journal*, 22, pp. 115–136

Thomson, C., 2005. 'Anniversary to Get Special Treatment', *Evening Chronicle*, 24/11/2005 [successor to the *Newcastle Courant*]

Thomson, J. J., Planck, M. et al., 1931. *James Clerk Maxwell: A Commemorative Volume, 1831–1931*. Cambridge: Cambridge University Press

Thomson, W. & Tait, P. G., 1867. *A Treatise on Natural Philosophy*. Oxford: Macmillan & Co.

Thoresby, R., 2009. 'Thoresby's Diary'. Available at: www.thoresby.org.uk/diary

Topham, E., 1899. *Edinburgh Life in the Eighteenth Century*. Edinburgh: William Brown

Turnbull, J., 2001. *The Scottish Glass Industry 1610–1750: 'To "Serve the Whole Nation with Glass'*. Edinburgh: Society of Antiquaries of Scotland

Turnbull, W. R., 1871. *History of Moffat: With Frequent Notices of Moffatdale and Annandale*. Edinburgh: W. P. Nimmo

UCL Department of History, 2014a. 'Profile & Legacies Summary: John Blackburn'. Available at: www.ucl.ac.uk/lbs/person/view/20601

—, 2014b. 'Jamaica St Thomas-in-the-Vale, Claim 24'. Available at: www.ucl.ac.uk/lbs/claim/view/20833

—, 2014c. 'Biography>Eliza Smart Bullock'. Available at: https://www.ucl.ac.uk/lbs/claim/view/15160

Venn, J., 1898. *Biographical History of Gonville and Caius College 1349–1897*. Cambridge: Cambridge University Press

Virgil, 2006. *The Aeneid of Virgil Translated into English Verse*. Project Gutenberg

Virtue & Co., 1868. 'Amesbury'. In: *The National Gazetteer: A Topographical Dictionary of the British Islands*. Available at: https://archive.org/details/nationalgazettee01londuoft

Wallace, J. M., 1987. *Historic Houses of Edinburgh*. Edinburgh: John Donald

Ward, T., 2012. *Clydesdale Bastle Project: Shielings and Buchts in Southern Scotland*. Biggar: Biggar Archaeological Group

Waterston, C. D. & Shearer, A. M., 2006. *Biographical Index of Former Fellows of the RSE 1783–2002*. Edinburgh: Royal Society of Edinburgh

Watt, J., 1771–73. 'Part 2: Muirhead I – Notebooks and Papers of James Watt and Family'. In *Industrial Revolution: A Documentary History. Series One: The Boulton & Watt Archive and the Matthew Boulton Papers*. Birmingham: Birmingham Central Library. Catalogue available at: www.ampltd.co.uk/digital_guides/industrial_revolution_series_one_parts_2_and_3/detailed-listing-part-2.aspx

Wedderburn, A., 1898. *The Wedderburn Book: A History of the Wedderburns 1296–1896*, vol. 1, part 3. Printed for private circulation

Wedderburn, J., *c.* 1841. A series of watercolours. [Art] (James Clerk Maxwell Foundation)

Wedderburn, R., 1991. *The Horrors of Slavery and Other Writings by Robert Wedderburn*. New York and Princeton, NJ: Markus Wiener Publishing

Wedderburn-Maxwell, J., 1985. 'Wedderburn-Maxwell, John' (IWM interview). Available at: www.iwm.org.uk/collections/item/object/80008937

Williamson, A., 1895. *Glimpses of Peebles or Forgotten Chapters in its History*. Selkirk: George Lewis

Wilson, D., 1886. *Memorials of Edinburgh in the Olden Time*. Edinburgh: Thomas C. Jack, Grange Publishing Works

Wilson, J. J., 1891. *The Annals of Penicuik* (1985 edn). Stevenage: Spa Books

Windsor Roofing, 2014. 'Corrugated Roofing'. Available at: http://roofing.thewindsor-companies.com/archives/347 [accessed 23/6/2015]

Wood, J. P., 1794. *The Antient and Modern State of the Parish of Cramond*. Edinburgh: John Patterson

Woodrow, R. & Burns, R., 1824. *The History of the Sufferings of the Church of Scotland* [. . .]. Glasgow: Blackie Fullerton and Co.

Wootton, A. C., 1910. *Chronicles of Pharmacy*. London: MacMillan and Co.

Worling, H. D. & P., 2012. 'A Box of Chymical Medicines: An Italian Medicine Chest Presented to Sir John Clerk of Penicuik in 1698'. *J. Roy. Coll. Phys. Ed.*, **42**, pp. 361–367

Writers to the Signet, 1890. *A History of the Society of Writers to Her Majesty's Signet with a List of the Members of the Society from 1594 to 1890 and an Abstract of the Minutes*. Edinburgh: Writers to her Majesty's Signet, Scotland

Youngson, A. J., 1966. *The Making of Classical Edinburgh*. Edinburgh: Edinburgh University Press

Index

Aberdeen (University of, *or* King's College) *See* Marischal College
Act for Preventing the Grouth of Popery, 1701 107
Adam, Robert 178
Adam, William 76, 205–206
Adams, Grylls 46
Aikman, William 62, 86
Ainslie Place, 25 173, 265, 267–268
Allanson, William 125–126
Alnwick 219
Alston, Dr Charles 204, 207–208
American War of Independence 133, 137
Amesbury 134
Amyatt, John 193
Anderson, Dr Patrick 209–210
Anderson, Katharine and Lilias 210
Anderson's Pills **209**
Appleby Dacre, Mary *or* Rosemary *See* Dacre, Mary
Appleby-Dacre, Joseph *See* Dacre of Kirklinton Hall, Joseph
Areskine of Alva 118
Arthur's O'on 179, 314 n20
Atholl, Duke of 135

BA committee on electrical standards 43
Balmaine, John 187, 191
Barbour, James 48
Baron Mure *See* Mure, William, Baron
Baron Sir John Clerk *See* Clerk of Penicuik, Sir John, 2nd Bt
Barons of the Exchequer 123, 129
Baxter, John jnr 179–180
Baxter, John snr 76, 178, 180
Benwell, Benwell stone 222, 224, 229
Bernoulli, Daniel 38

Bickerstaffe, Isaac 84
Birnie of Broomhill 207
Birnie, Bethia and Rev. Robert 207–208
Black, Dr Joseph 137
Blackburn of Househill, John 168
Blackburn of Killearn, Peter, MP 168, 173
Blackburn, Andrew 168
Blackburn, Isabella (Mrs James Wedderburn snr) 168
Blackburn, Jemima *See* Wedderburn, Jemima
Blackburn, John 169
Blackburn, Peter 168
Blackburn, Prof Hugh 43, 168
Blackcraig 142–144
Blackwood 183
Blackwood of Pitreavie, Robert 209
Blackwood, Bethia 209, 214
Bliss School, Peebles 248
Boerhaave, Herman 67, 69
Boltzmann, Ludwig 38
Bonhomme Richard 138
Bonnie Prince Charlie *See* Stuart, Prince Charles Edward
Border feuds 92
Borron, John 206
Boteler, Wyvil 125–126
Bothwell Bridge, Battle of 63, 188
Brewster, Sir David 32, 292 n49
Bristo Street 200
Britannia Romana 224
Brown, Robert 204
Buccleuch Place 257
Buccleuch Street 230
Buchanan, George 84
Buchanan, Susan 159
Buckstone 62

INDEX

Bullock, Emily 169, 233
Bullock, William 169–170
Burns, Robert 136

Caerlaverock 90
Camelon Irving Memorial Church 203
Cammo **74**
Campbell, Archibald, 3rd Duke of Argyll 130
Campbell, John, 2nd Duke of Argyll 74
Campbell, Lewis **58**
  biography of Maxwell 10, 59
  friendship with Maxwell 19, 54–55, 58
Campbell, Robert 25, 30, 59
Carlisle 78, 126–127, 155–157, 159
Carlisle, Bishop of *See* Fleming, Sir George, Bishop of Carlisle
Carmichael, Archibald 19
Carr, William Holwell 240
Carruthers, William 141
Cavendish, Sir Henry 49
Cavendish, William, 7th Duke of Devonshire 49
Cay of Charlton Hall **219**
  B. K. Ward 224
  baking & brewing business 221–222
  calotypes of Sheriff John Cay and sons 234
  children of Sheriff John Cay 233
  family portraits by E. Liddell 230
  origins 219
Cay and Black 238
Cay, Albert **237**
Cay, Albert, 2nd 245, 251
Cay, Barbara, *née* Carr 221
Cay, Charles Hope 59, 245, 250
Cay, Elizabeth (Mrs George Mackenzie) 236
Cay, Elizabeth (Mrs Thomas Dunn) 244, 251
Cay, Elizabeth (*née* Liddell) *See* Liddell, Elizabeth
Cay, Frances (Mrs John Clerk Maxwell) 8, 11, 14, 231–232, 242, 261
Cay, Frances (*née* Hodshon) *See* Hodshon, Frances
Cay, Grace, *née* Woolfe 223
Cay, Hannah 222
Cay, Isabel (*née* Wilkinson) 221
Cay, Isabella 245, 251
Cay, Isabella (*née* Dyce) *See* Dyce, Isabella
Cay, Jabez 78, 221–223
Cay, Jane 232, 242, **248**, **251**
Cay, John Frederick 245, 250
Cay, John, d. 1640 221
Cay, John, d. 1782 225–226, 229–230
Cay, John, d. c.1730 221, 223
Cay, John, of the Marshalsea 225–226, 229, 318 n11 & n12
Cay, John, Sheriff of Linlithgow 19, 169, 231–232, **233**
Cay, John, WS 235
Cay, Robert and Edward, sheep farmers 236
Cay, Robert d. 1754 224–225
Cay, Robert Dundas **238**, **244**
  death of his wife 247
  friendship with William Dyce 241
  Hong Kong 245–246
  later life 252–254
  marriage and children 244
  portrait of his sister and her son 242, 252
  Registrar of the Supreme Court, Hong Kong 247
  return to Edinburgh 247–249
  Shepperton Cottage 249
Cay, Robert Hodshon 219, 225, 230–232
Cay, Robert Hodshon, 2nd 244, 248–249
Cay, Robert, d. 1682 221
Cay, Robert, d. 1754 221
Cay, Thomas, d. 1623 221
Cay, Thomas, d. 1868 237
Cay, William 224
Cay, William Dyce 46–47, 244, **248**, 250, 321 n49
Chancellor of Shieldhill, Alexander 199
Chancellor, Jean or Jane 199–200, 212–213
Chancellor, Mary or Molly 199–201, 207, 212, 214, 257
Charlton Hall 219–221, 235
Charlton Lodge 236, 249
Chessel's Court 213
Circus Place 233, 235–236
Cleghorn, Elizabeth 154, 178, 182
Cleghorn, Rev John 178
Clerk jewels 61

Clerk Maxwell 99, **110**, 119
Clerk Maxwell, Dorothea 10, 110–111
   'Act to Enable Dorothea Clerk to Sell […]' 118
   birth 88, 96
   children 123, **281**
   heiress of Middlebie 7, 106, 109, 114, 118–120
   illness and death 151, 162–163
   marriage 96, 112
   orphaned 96, 106
   religion 109
   widowhood 151
   will 161
Clerk Maxwell, Elizabeth 11, 263
Clerk Maxwell, Frances (*née* Cay) *See* Cay, Frances
Clerk Maxwell, George 7, 72, 78, **110**, 154, 157
   alternative names 120, 148, 182
   birth and given name 110
   career and family life 120
   character 120–121, 147, 148
   children 123, **281**
   Commissioner for Customs **134**
   Commissioner for Forfeited Estates 123, **128**
   Commissioner for the Burgh of Sanquhar 147–148
   Dalkeith grammar school 111
   death 150
   Dumcrieff 117, 122, 149–150
   Dumfries 12, 121–122, 145
   early life 110–111
   Edinburgh 123
   family tragedies 149–150
   financial downfall 148–149
   Forth and Clyde Canal **132**
   Highlands 145
   illness 150
   improver 121, 146
   Jacobite rebellion of 1745 126
   Leiden 112, 116–117, 175
   linen factory 121, 123, 175
   Lord Treasurer's Remembrancer 125–126
   Lowther School 111–112, 303 n4
   marriage 96, 112–113
   Middlebie 119–120, 149–150
   mineral spas **138**
   mining 140, **141**, 206
   other activities and interests 122, 145
   Philosophical Society of Edinburgh 146–147
   publications 121, 146
   Remembrancer to the Exchequer **125**, 134
   RSE 146
   Society of Antiquaries of Scotland 147
   spinning school 121
   succession as 4th Bt of Penicuik 150, 182
   travels 145, 147
   Trustees Drawing Academy 146
   Trustees for Manufactures 145
   University of Edinburgh 112
Clerk Maxwell, James
   1841 census 172
   'A Dynamical Theory of the Electromagnetic Field' 45, 276
   Aberdeen 34, 37
   Adams Prize 37
   '[…] as Puck' 241
   BA (British Association) 38, 43
   Bakerian Lecture 45
   bequest from Aunt Jane 252
   birth 9, 11
   British Asses 38
   Cambridge 50
   career, choice of 29
   Cavendish laboratory 49–50
   childhood 11–22, 240
   'Classification of Physical Quantities, Remarks …' 50
   Colour, interest in and work on 25, 32–36, 39
   commemorations **273**
   comparison of electric and magnetic forces 48
   conformal mapping 51
   Connexions to Napier and Stirling **204**
   'Daftie' 18
   death 266
   death of father 37
   death of mother 14–15
   displacement current 276–277

INDEX

dynamical top 37, 39
early interests 18–24
Edinburgh Academy **16**
elastic equilibrium in solids 28
Electrical Researches of the Electrical Researches of the Hon. Henry Cavendish 52
electromagnetic machines 19, 26
electromagnetic theory 44–45, 275
experiments when young 24–27, 234
fish eye lens 33
frames and reciprocal figures 48
friendship with Campbell and Tait 19–21, 54–55, 57–58
Gallowegian brogue 14
geometry, interest in 20
Glenlair, 1866–71 46, 271
governors 48
grandparents 162, 232
hesitancy and obscurity 19, 28, 31–32, 289–290 n10
illness 24, 31–32, 43, 46, 52–54
India Street 9, 265
inquisitiveness 13–14
inverse square law of electric force 52
Italy 47
lack of fame 10, 54–55
lawsuit by Sir George Clerk 40
lecturer, abilities as 37, 41–43, 46, 57–58
legacy to science and mankind **275**
light & electromagnetic waves 277
lightning conductors 51
London & King's College, 43
magnetism, interest in 25
Marischal College 34, 37
marriage 39, 41
Matter and Motion 51
Maxwell's Demon 45
Maxwell's equations 279
Middlebie, laird of 37
molecular theory 38, 45
molecular vortices 44
natural philosopy of 33
Nether Corsock 13
Nicol, William, visit to 234
obscurity of his written explanations 28

odd mannerisms 18
'On Faraday's Lines of Force' 36, 276
'On Physical Lines of Force' 44, 276
'On Stresses in Rarified Gasses …' 52
'On the Dynamical Theory of Gases' 38
'On the General Laws of Optical Instruments' 33
'On the Properties of Matter' 33
'On the Solution of Electrical Problems …' 51
'On the Transformation of Surfaces by Bending' 32
'On the Viscosity or Internal Friction …' 45
oval curves 21–23, **272**
Peterhouse College 29–30
poetry 20, 38, 290 n14
polarised light 291 n28
portrait with his mother 242
prizes, medals and honours 20, 48, **273**
'props' 21, 23, 27
provision for his wife 266
quaternions 57
rebellion against his tutor 14
relatives involved in slavery 168–169
return to Edinburgh 18
rolling curves 27–28
Royal Society of London 45–46
RSE 36–37, 48
Saturn's rings 37, 39
sister, Elizabeth 11
speed of light 48
statue in Edinburgh 54
The Man who Changed Everything 55
Theory of Heat 49
Treatise on Electricity and Magnetism 28, 48, 50
Trinity College 29–30, 36, 41
tripos and Smith's prize 30–31
'tubbing' at Glenlair 14–15, 41
University of Cambridge **29**
University of Edinburgh **25**, 42
wife See Clerk Maxwell, Katherine
will 252, 266, 323 n17
Clerk Maxwell, John 7–8, **261**
31 Heriot Row; 14 India Street 166–167, 256

349

birth and early years 162–163
character 262–263
death 37
Glenlair House; Nether Corsock 262–264
illness 34, 36, 292 n44
legal career 262
marriage 261
portrait of 37
Clerk Maxwell, Katherine 39–41, 43, 50, 53–54, 252, 269
Clerk of Dumcrieff, George *See* Clerk Maxwell, George
Clerk of Eldin, John 72, 137, 145, 153, 161, 163, 198
Clerk of Eldin, William 158, 198, 201–202
Clerk of Killiehuntly, John 60–61
Clerk of Listonshiels and Rattray 61
Clerk of Penicuik **60**, 198
Clerk of Penicuik, John, the first of 61
Clerk of Penicuik, Sir George, 4th Bt *See* Clerk Maxwell, George
Clerk of Penicuik, Sir George, 6th Bt 8, 40, 124, 162, 164–166, 241
Clerk of Penicuik, Sir James, 3rd Bt 78, 127, 144, 146, 149, 153, **174**
architecture 178, 180–181
art and antiquities 179, 181
birth 174–175
coal mining 177, 181
death 182
debt 181–182
disposition 176, 181
Edinburgh residence 177–178
education 174
Europe, 1739–45 175–176
Falkirk Muir 176
grand tour 175
health 153, 174, 182
Jacobite rebellion of 1745 176
Leiden 175
Penicuik 178–180
marriage 178
Newbiggin 177
Ossian 179
Rossdhu 181
succession to Penicuik 177
Clerk of Penicuik, Sir John, 1st Bt 61, **62**

attack by robbers 63
death 64–65, 89
marriages and children 63–64
Clerk of Penicuik, Sir John, 2nd Bt 60, **65**, 117–120, 125, 140, 207, 222
antiquities 68, 75, 78, 179
astronomy 79
astuteness 70–71, 113, 119
Baron of the Exchequer 71
children 70, 72, 75, 89, 119, 125
coal mining 75–76, 177
death 80
Durham 79
early career 69–70
early life 65–66
Easy Club 84
Exchequer Court 79
grand tour 67–69
health 67, 79
interests and scholarly pursuits 65, 67, 73–74, 78, 80
Jacobite rebellions 73, 79
marriages 69, 72
memoir 60, 174
Moffat 73, 122
Philosophical Society of Edinburgh 79
Royal Society of London 78
speech and writing 66, 80
Trustees for Manufactures 78
union of the parliaments 70–71
University 66–67
wardship of niece 96
Whithorn, MP for 70
Clerk of Penicuik, Sir John, 5th Bt 123–124, **152**, 164, 281
Clerk of Pennycuick, Mrs *See* Irving, Janet
Clerk, Agnes (Mrs John Craigie of Glendoick) 123
Clerk, Barbara (Baron's daughter) 154
Clerk, Capt James, HEICS 123–124, 151–152, **159**
bagpipes 161
birth and naming 159
career 160–161
marriage and children 162, 311 n39
date of birth 282
death 163
HEICS 160–161

heir to Middlebie 162
HMS *Hector* 160
Over Lasswade 162
shipwreck of the *Major* 161
Clerk, Col Robert 178
Clerk, Dorothea *See* Clerk Maxwell, Dorothea
Clerk, Dr John (Baron's cousin) 146
Clerk, Fr Cosimo 68
Clerk, George, advocate 123–124, 148
Clerk, Henry (Baron's brother) 63, 81
Clerk, Henry (Baron's son) 72, 127, 174
Clerk, Henry (Jacobite 'rebel') 156
Clerk, Henry (William Clerk's son) 81
Clerk, Isabella *See* Wedderburn, Isabella (*née* Clerk)
Clerk, Isabella, Miss 60
Clerk, James RN 160
Clerk, Mrs Janet *See* Irving, Janet
Clerk, Johanna (dau. of George and Dorothea) 125
Clerk, John (Baron's eldest son) 70
Clerk, John, Lord Eldin 198, 201
Clerk, Lady Elizabeth *See* Cleghorn, Elizabeth
Clerk, Lady Mary *or* Rosemary *See* Dacre, Mary
Clerk, Mathew (Baron's son) 72, 296 n11
Clerk, Miss Isabella 305 n29
Clerk, Patrick (Baron's son) 127
Clerk, Robert (son of George and Dorothea) 124
Clerk, Robert (surgeon apothecary) 61
Clerk, William (Baron's brother) 63, **81**, 95–96
    Allan Ramsay, links with 82–83
    attempts to find a wife 87
    character 81
    debts 105
    Easy Club 84
    education 81
    funeral 108
    illness and death 89
    marriage and daughter 88
    poetry 82, 86, 88
    religion 108
    son 81
Clerk, William (of Montrose) 61

Clerk, William (son of George and Dorothea) 124
Clifton College 250
Clifton, skirmish 127
Clyde, river and valley 90, 183–184
Cochrane, Basil 137
Colquhoun of Luss, Sir James 181
Colquhoun, Isabella 162, 199
Colquhoun, James and Janet 199
Commission for Forfeited Estates 13–32, 123, 128–133, 145
Corsock Church 47
Cotton, George (?) 216
Covenanters 63, 186–189, 295 n5
Craig, James 196
Craigie, Barbara 123
Craigie, Capt George 123
Craigtown Mining Company 143–144, 148
Craik of Arbigland, William 135–136, 138
Cramond 74
Crawford (parish) 184
Crawford of Restalrig, Ronald 140–141
Creetown shot mill 144
Crooke's radiometer 52, 295 n86
Culloden, Battle of 128, 167, 177
Cumberland, Duke of 126, 128

Dacre Appleby, Mary *See* Dacre, Mary
Dacre of Kirklinton Hall, Joseph 155, 157
Dacres of Lannercost & Naworth Castle 155
Dacre, Mary 124, 152–153, **155**, 164, 222
Dalkeith grammar school 111, 174
Dalrymple, Sarah 103
Dalzell, Maister Ninian 100
Day, Alexander 240
Devonshire, Duke of *See* Cavendish, William
Dewar, Katherine *See* Clerk Maxwell, Katherine
Dewar, Rev Daniel 39, 40, 293 n58
Douglas, Charles, 3rd Duke of Queensberry 76–77, 132, 134, 138–141, 146, 161
Douglas, Gavin 84
Douglas, James, 2nd Duke of Queensberry 70
    imbecile son 71

Douglas, James, Earl of Morton 91
Drum 187
Drumlanrig 77–78
Drummond, George 197
Drummond, James, Earl of Perth 188–189
Dryfe Sands 12, 92–93, 100
Duchess of Queensberry, Catherine, wife of 3rd Duke 160
Duke of Argyll *See* Campbell, John, 2nd Duke of Argyll
Duke of Atholl, *See* Atholl, Duke of
Duke of Northumberland 225
Dumcrieff **74**, 76, 117, 122, 149–150
Dunbar of Machermore, Patrick 143
Dundas, Henry, 1st Viscount Melville 153, 181
Dyce, Isabella 240–42, **244**
Dyce, Margaret (*née* Chalmers) 239
Dyce, William, MD 239
Dyce, William, RA **239**
　death and legacy 243
　design schools 242–243
　development as an artist 239
　Edinburgh 240
　electromagnetic theory 240
　family 239, 244
　friendship with Robert Dundas Cay 239, 241
　Glenlaer 242
　marriage 243
　Mrs John Clerk Maxwell (neé Frances Cay) and her Son James 13, 242, 252–254
　music 243
　Nether Corsock 240
　polymath 242
　Puck 241
　RSE 240

East India Company *See* HEICS
Easy Club **83**
Economic crisis of 1847–48 237, 239
Edinburgh 3–5
　2nd New Town 4–5, 197
　Burgh Loch 197
　Calotype Club 234, 241
　Castle 104
　Charlotte Chapel 232
College of Art 146
Dean Cemetery 202–203, 267
Enlightenment **193**
First New Town 4–5, 196–197, 231
gas lighting 258
George Square 4, 197
Heriot Row 4, 6, 197
James Craig's plan 196
New Towns 3–5, **196**
Nor' Loch 4, 151, 158
North Bridge 180
Old Town 4–5, 195–196
Philosophical Society 79
Physick Garden 207
Post Office directories 288 n2 (Preface)
Second New Town 4–7, 231
St Andrew's Church 6, 22
St John's Church 6, 232–233
the Mound 158
University, Old College 26
Edinburgh Academy 16, 19–21, 202
Einstein, Albert 275, 279–280
Elswick Colliery 223
Elvan Water; Elvanfoot 184–185
Enlightenment, Edinburgh **193**
Enlightenment, Scottish 2
Enlightenment, notable scots of **283**
Entails 96–97
Erskine of Alva *See* Areskine of Alva
Erskine, Charles 78

Faraday, Michael 275, 277
Farquharson, Alexander 149, 150
Fergus, John 84
Ferguson, Capt Duncan 270
Fleming, Catherine 155
Fleming, Sir George, Bishop of Carlisle 124, 155, 157
Forbes of Culloden, Duncan 78, 145
Forbes of Newhall, John 82
Forbes of Newhall, Sir David 62, 82
Forbes, Ann 153
Forbes, Prof. James 23, 26, 29, 42
Forfeited estates 123, 128, 130; *see also* Commission for Forfeited Estates
Forth and Clyde Canal 132
Franklin expedition 202
Frenchland 76

INDEX

Gale, Roger 78
Garnett, William 10, 51, 53
Garrison Church 203
Geordie Boyd's mud brig 158
George IV 157
George Square 4, 164, 197, 200–201
George Street 6, 170, 231, 240
Gilpin of Scaleby Castle, William 78, 157, 222
Gilpin, Dorothy 78, 157–158, 222
Gilpin, John 222
Gilpin, Susannah 158, 222
Glasgow 132
Glendorch 140, 142
Glendouran 142
Glenlair 12, 16–18, 47, 171, **261**, 322 n6
Glenlair House 17, 47, 252, 269–270, 294 n72
Glenlair Trust 270
Gloag, James 21–22
Gloucester Lane 258
Gordon, Alexander 174, 180
Grana Angelica 210
Gravelines 229
Gray, Mary 74
Great Stuart Street, 6 238, 268
Grenville, George 135

Habbie's Howe 82
Haidinger's Brushes 24
Hamilton of Gilkerscleuch, Sir William 189
Hamilton, James, Earl of Arran 91
Hamilton, Sir William (Edinburgh) 26, 33
Hamilton, Sir William Rowan 57
Hart Fell Spa 138–139
Hawkshaw 87
Hay, Agnes Clerk 202, 204, 207
Hay, David Ramsay 22–23
Heaviside, Oliver 279
Hedley, Nicholas 228
HEICS 305 n42
Henderson of Elvingston, Henry 63
Henderson, Elizabeth 63
Hepburn of Keith, James 84
Heriot Row 58, 197, 201, 231–235
    no. 31 3, 6, 16, **165**, 171–172, 255, 264
Heron of Heron and Kirroughtrie, Patrick 142–143, 308 n87

Hertz, Heinrich 28, 279
Highland dress 157
Highland Society of Edinburgh 154
Highlands 129–132
Hodshon, Catherine 230
Hodshon, Elizabeth 228
Hodshon, Frances (Mrs John Cay) 225–226, 229–30
Hodshon, nuns in family 229
Hodshons of Lintz 219, **228**
Hodshon of Lintz, Ralph 225, 229–230
Hodshon of Lintz, Richard 228
Hollywood Abbey 99
Holyrood Abbey sanctuary 216–217, 225–226
Hopetoun, Earl of 132, 142
Hopkins, William 30
Horsley's great map of Northumberland 225
Houston of Houston 72, 103–104, 207
Houston of Wester Southbarr 103–104
Houston, Anne 72, 104, 207
Houston, Elizabeth 103–104, 107, 109
Huguenots 102
Hume, David 196
Hutchinson, Jonathan, MP 22–23
Hutton, Dr James 133, 137–138, 145
Hynd, William 141

ICMS *See* International Centre for Mathematical Sciences
Ilay, Lord *or* Earl of *See* Campbell, Archibald, 3rd Duke of Argyll
India Street, 14 3, 6, 9, 165–166, 171, **255**, 322 n4
    description of the house 258–261
    disposition to Katherine Dewar 265
    early amenities 258–261
    James Clerk Maxwell Foundation 268
    Louisa Helen and Jean Charlotte Mackenzie 265, 267
    occupancy 256–257, 264, 268, 322 n10
    ownership after John Clerk Maxwell 265
India Street, 18 6, 34, 173
Inglis of Cramond 72, 74–75, 174
Inglis, Janet 72, 104, 178, 206, *See also* Colquhoun, James & Janet

353

Inglis, Rear Admiral Charles 152
Inglish, Isabella; James 210–211
International Centre for Mathematical Sciences 268
Inveresk 167–168
Irving of Drum 187
Irving of Newton **183**
Irving of Newton, Alexander 58, 163, 199–201, 205–206, 214–215, 262, 316 n10
Irving of Newton, George Vere 187, 201
Irving of Newton, George, 2nd 183, 199–201, 212
Irving of Newton, George WS, 1st 187, **190**, 209, 212
Irving of Newton, Mrs *See* Chancellor, Mary or Irving, Bethinia
Irving of Newton, Robert 191, 199
Irving, Margaret *See* Weir, Margaret
Irving, Agnes (*née* Hay) *See* Hay, Agnes Clerk
Irving, Archibald Stirling 203–204
Irving, Bethinia 199, 201, 214, 218
Irving, David Williamson 204
Irving, Dr Thomas 191, 199, 212, 214
Irving, Dr Thomas fl. 1695 192
Irving, Gen Alexander, CB 203
Irving, Gen James 213, 215–216
Irving, George WS (John Irving's son) 203
Irving, James (macer) 187
Irving, Janet 123, **162**, 171, 183, 199, 255–258, 312 n56
  death 166
  Edinburgh homes 165
  marriage and children 124, 162
Irving, John, WS 123, 158, 199, 201–202, 204, 257, 262
Irving, Lt John 202
Irving, Mary and Jeannie 200
Irving, Mrs Dr. *See* Chancellor, Jean
Irving, Rev Lewis Hay 203
Irving, Sarah 191
Irving, Thomas 199
Isle of Man 135
Isle of May 138

Jackson, Philip 143
Jacobites 2, 83
Jacobite rebellion of 1715 **73**, 104
Jacobite rebellion of 1745 2, 79, **126**, 155, 160, 167
James Clerk Maxwell Foundation 268
James VI 91
James VII 189, 222
James' Court 124
Jeans, Sir James 38
Jeddart justice 92
Jenkin, Henry Fleeming 44
Johnstone of Dunskellie, Sir James 93
Jones, Capt John Paul 138

Kay; Key *See* Cay
Keir, Joanna 171
Kelland, Prof. Philip 23, 26
Kelvin, Lord *See* Thomson, William
Killearn 173
Killiehuntly 60
Killingbeck, Mary 229
Kilpatrick, Christian 64
Kirklinton 124
Knight, Joseph 169

Lady Clerk of Penicuik *See* Clerk Maxwell, Dorothea; Cleghorn, Elizabeth; Dacre, Mary
Lasagette, Daniel 103
Lasswade 75
Lauderdale, Earl of 62
Lauriston Lodge 249
Law, Maria Anne 165
Le Blanc, Major James 96, **102**
  charge of false muster 104
  connections with the Baron 103
  disposition of his property 106–107
  Huegenot origins 102
  marriages 103–104
  glass making 102–103
  military service 102, 104
  naturalisation 103
  repayment of William Clerk's debts 105
Leadhills 82, 138–142, 205–206
Leiden 66, 69, 112, 116, 175
Leith, Citadel of 73
Lesmahagow 183–184
Liddell of Dockwray Square 230
Liddell, Elizabeth (Mrs Robert Hodshon Cay) 219, 230, 232

Lintz 226–228, 230
Loanhead 76
Lockerbie 12, 92–93
Lockerbie Lick 93
Lodge, Sir Oliver 279
Lord Elliock *See* Veitch, James, Lord Elliock
Lord Newton *See* Irving of Newton, Alexander
Lorentz, Hendrik 279
Lorraine, Dr 53
Lowther Hills 140
Lowther School 111, 157

Maccus 90
Macdonald of Kinlochmoidart 156
MacIvor, Iain and Marion 268
Mackenzie, Charles 30, 291 n32
Mackenzie, Colin 53, 170, 252, 265–268, 323 n17
Mackenzie, Henry 197
Mackenzie, James Hay 170, 265, 268
Mackenzie, Jean Charlotte and Louisa Helen **265**
Mackenzies, the **265**
Main, James 214
March law and March wardens 92, 219
Marconi, Guglielmo 279
Marischal College 41, 239
Mary Queen of Scots 61
Mavisbank **74**, 177–179
Maxwell
　feud with Johnstones 92–93
　of Middlebie **94**; *see also* Maxwell of Middlebie
　origins and variants of the name 90
　Maxwell Lords 90–91, 100
　Roman Catholicism 94, 99–100, 301 n8
　See also Clerk Maxwell
Maxwell, Agnes (of Middlebie) 81, 87–88, 95–96
　complexities of her estate 106–107
　death 106
　marriage to James Le Blanc 102, 104
　marriage to William Clerk 88
　Middlebie 106
　Roman Catholicism 107–109
　unwelcome at Penicuik 88, 108

Maxwell, Baillie Homer 99
Maxwell, James Clerk *See* Clerk Maxwell, James
Maxwell, John Clerk *See* Clerk Maxwell, John
Maxwell, John, 8th Lord 91–93
Maxwell, John, 9th Lord 91, 93
Maxwell, Marion (of Middlebie) 94, 99
Maxwell, Robert (Dorothea's uncle) 117
Maxwell, Robert, 10th Lord 91, 93–94
Maxwell of Middlebie, John, 1st 94
Maxwell of Middlebie, John, the entailer 87, 95
Maxwell of Middlebie, John, the younger or last 95
Maxwell of Middlebie, Robert 95
Maxwell of Pollok 90
Maxwell of Speddoch **99**
Maxwell of Speddoch, Homer 94
Maxwell of Speddoch, Provost Homer 99–101
Maxwell Johnstone feud 92
Maxwell-Boltzmann Distribution 38
McLellan, Robert 95
Michelson and Morley experiment 280
Middlebie 7, 87–88, 94, 110, 149–150
　debts on estate 99, 109, 113–114
　disentailment 270, 293 n71
　entail 8, 95–99
　estate 114–116, 293 n71, 288 n3
　loss of 149–150
　sale and repurchase 119–120
　Twenty Merk Land 94, 114–116, 118
Milne's or Miln's Court 200, **211**, 213, 218
Moffat 76, 118, 139
Morton, Earl of *See* Douglas, James, Earl of Morton; *see also* Maxwell, John, 8th Lord
Mowbray, Giles 74
Mure, Baron William 134

Napier of Merchiston, John 204, 207
Nether Corsock 8, 10–12, 17, 47, 114, 240, **262**, 289 n6
Newall, Walter 47
Newcastle-upon-Tyne 219, 221
Newhall 82
Newton, house and estate 183–187, 200–201, 204

355

Newton, Lord *See* Irving of Newton, Alexander
Newton, Sir Isaac 204–205, 207
Nicol prism 27, 234, 291 n28
Nicol, William 19, 25
Nithsdale, Earl of *See* Maxwell, Robert, 10th Lord
North Charlton 219–222, 225, 232, 235–236
North Esk River 75–76
North, Christopher *See* Wilson, Prof. John
Nova Scotia, baronetcies of 62

Oersted, Hans Christian 275
Oglethorpe, Maj.-Gen. 126–127
'Old 31' *See* Heriot Row, no. 31
Oughton, Lieut.-Gen. Adolphus 130
Oyster Club 137

Paget, Dr George 53, 323 n17
Parton Church 47, 54
Patrick, Bishop of Kelso 183
Penicuik 180
    baronetcy of 8, 61
    estate 61
    Newbiggin 63, 178
    Penicuik House 178–180
    St Mungo's kirk 64, 180
    village 61
Pennecuik, Dr Alexander 82, 86
Philosophical Society of Edinburgh 79, 298 n32
Pitcairne, Dr Archibald 83, 85
Pixii, Hippolyte 276
Porteous of Hawkshaw, Henrietta 87
Preston Lieut.-Gen. George 104
Pretender, the Old 73, 74
Primrose of Dunipace, Sir Archibald 159
Princes Street 151, 154, 158–159, 162–164
Privy Council (Scotland) 187–188, 191

Queensberry House, murder of kitchen boy 71
Queensberry, 2nd Duke of *See* Douglas, James . . .

Radio *See* Wireless, development of
Raimes, Blanchards & Co. 218
Ramsay, Allan 62, **81**, 84, 121, 224
    Easy Club 83
    friendship with the Clerk family 86
    monument to 178
    origins 82
    poetry 85–88
    The Gentle Shepherd 82
Ramsay, Allan jnr 86
Readshaw, Cuthbert 141
Readshaw, Joshua and Caleb 142
Reformation 100
Reiving 92, 186
Relativity, theory of 275, 280
Restalrig burial ground 227–228, 237, 249
Rogers, Rev. Charles 84
Roman Catholicism, supression of 100
Rosamond (schooner) 136–137
Rose Castle 155, 156
Rose of Scotland 159
Roshven 172
Ross, Christian 84
Ross, James 83, 84
Ross, James, SSC 240, 249
Ross, Margaret (*née* Dyce) 240
Roy, General William 147
Royal Hunters 126–128
Royal Society of Edinburgh (RSE) 234, 298 n32
Ruddiman, Thomas 83
Rullion Green, battle of 63
Runciman, Alexander 146, 179

Sandford, Bishop Daniel 233
Schwatka, Lt Frederick 202
Scots Pills *See* Anderson's Pills
Scott Moncrieff, Mary (*née* Irving) 203
Scott, Sir Walter 156–159, 163, 197, 201–202, 313 n71
Scott, William 102
Scott, William, Baron Stowell 153
Scottish Civil War 94
Scottish History Society 60
Scottish Parliament, dissolution of 71
Sempill's House 178
Shepperton 220–221, 235
Shepperton Cottage 249
Sheriffmuir, battle of 74
Shortcleugh 142–143, 200, 206
Silverknowe (Marine Villa) 53, 266, 313 n72

INDEX

Simpson, William 174
Simson family 268
Slavery 168–169
Sloane, Sir Hans 78, 207
Smeaton, John 132–133
Smellie, William 193
Smith and Ramage 39
Smith, Adam 137
Smith, Samuel 142
Smith's prize 30
Smollet of Bonhill; Smollett Helen 87
Smuggling 135–136
Society of Antiquaries of Scotland 147
Society of Antiquaries, London 78
Sommervile's Close 200
South Shortcleugh *See* Shortcleugh
Speddoch 12, 99; *see also* Maxwell of
St Luke's drawing school 146, 175
St Margaret's Church, Restalrig 216–217, 227–228
Steill v. Weir 192
Steill, Thomas 211
Steinstone (Stevenson), Janet 81
Stirling of Garden; Stirling of Keir 206–207
Stirling, Archibald 205
Stirling, James 204–206
Stokes, Prof. George 30, 32, 291 n33, 323 n17
Stonebyres 183
Stuart, Alexander, 3rd Earl of Galloway 69
Stuart, Alexander, 5th Earl of Moray 187
Stuart, John, 3rd Earl of Bute 134
Stuart, Lady Margaret (Baron's 1st wife) 69
Stuart, Prince Charles Edward 79, 155, 157
Stukeley, Dr William 78
Sturgeon, William 276
Swan, Prof. William 32, 42

Tait, Peter Guthrie 20–21, 30, 42, **56**
Tartan ribbon, the 36
Thermodynamics 275
Thomson, James 250
Thomson, Prof. David 42
Thomson, William, Lord Kelvin 30, 43, 45, 49, 54, 57, 250, 275
Thoresby, Ralph 222–223

Tod of Craigieburn 77
Tripos, mathematical 29
Trotter, Esther 168
Trustees Drawing Academy 146
Trustees for Manufactures 78, 121, 132
Tubbing at Glenlair 41
Tynemouth 221
Tytler, Alexander(?) 82
Tytler, James 226

Union, Articles of 71
Urr, River or Water; stone bridge 12, 47, 250, 322 n6

Veitch, James, Lord Elliock 133, 146, 161

Wallace, William (builder) 255–253
Wanlockhead 140
Wedderburn Colvile, Andrew 169
Wedderburn Maxwell, Andrew 54, 163, 171, 266, 269
Wedderburn of Blackness, Sir John 128, 167
Wedderburn of Inveresk 167
Wedderburn, Andrew *See* Wedderburn Maxwell, Andrew
Wedderburn, Dr James 128, 167–169
Wedderburn, Dr John 168–169
Wedderburn, George, WS 170, 173, 267
Wedderburn, Isabella (*née* Clerk) 41, 128, 162–163, 166, **167**
Wedderburn, James (advocate) 128, **167**, 170, 172
Wedderburn, Janet Isabella (Mrs James Hay Mackenzie) 170, 265, 268, 313 n72
Wedderburn, Jean (Mrs Peter Blackburn) 168, 170
Wedderburn, Jemima (Mrs Hugh Blackburn) 16, 41, 166, 168, 171–172
Wedderburn, Maj.-Gen. James 172
Wedderburn, Maj.-Gen. John 170
Wedderburn, Robert 168–169
Wedderburn-Maxwell, Brig John 269
Wedderburn-Maxwell, Maj James Andrew Colvile 269
Weir (Were or Vere) of Newton, William 184
Weirs of Blackwood, Kirkton, Newton & Stonebyres **183**

Weir of Blackwood, William 188
Weir of Newton, John 183, **186**, 209
Weir, Dr Thomas 183, 190–192, **209**
Weir, Maj Thomas 183
Weir, Margaret 183, 187, 190–191
Weir, Rotaldus or Rothald 183
Weir, Sarah 190, 209, 212
Wheatfield 216
White cockade (of Mary Dacre) 156, 159
Whitworth, Robert 133
Wilde, Prof John 206

Williamson, John 139–140
Wilson, John W 60
Wilson, Prof. John 29
Wireless, development of 59, 279
Woolfe, Grace *See* Cay, Grace *née* Woolfe
Woolfe, Henry 223
Wranglers 30
Wright's Houses 61, 207

Yorkshire Hunters *See* Royal Hunters
Young of Killicanty, Dr Thomas 107, 109